Semiconductor Quantum Dots
Organometallic and Inorganic Synthesis

RSC Nanoscience & Nanotechnology

Editor-in-Chief:
Paul O'Brien FRS, *University of Manchester, UK*

Series Editors:
Ralph Nuzzo, *University of Illinois at Urbana-Champaign, USA*
Joao Rocha, *University of Aveiro, Portugal*
Xiaogang Liu, *National University of Singapore, Singapore*

Honorary Series Editor:
Sir Harry Kroto FRS, *University of Sussex, UK*

Titles in the Series:

How to obtain future titles on publication:
A standing order plan is available for this series. A standing order will bring delivery of each new volume immediately on publication.

For further information please contact:
Book Sales Department, Royal Society of Chemistry, Thomas Graham House, Science Park, Milton Road, Cambridge, CB4 0WF, UK
Telephone: +44 (0)1223 420066, Fax: +44 (0)1223 420247
Email: booksales@rsc.org
Visit our website at www.rsc.org/books

Semiconductor Quantum Dots:
Organometallic and Inorganic Synthesis

Mark Green
King's College London, London, UK.
Email: mark.a.green@kcl.ac.uk

THE QUEEN'S AWARDS
FOR ENTERPRISE:
INTERNATIONAL TRADE
2013

RSC Nanoscience & Nanotechnology No. 33

Print ISBN: 978-1-84973-985-6
PDF eISBN: 978-1-78262-835-4
ISSN: 1757-7136

A catalogue record for this book is available from the British Library

Published by The Royal Society of Chemistry,
Thomas Graham House, Science Park, Milton Road,
Cambridge CB4 0WF, UK

Registered Charity Number 207890

For further information see our web site at www.rsc.org

Printed in the United Kingdom by CPI Group (UK) Ltd, Croydon, CR0 4YY, UK

For Jo, Connor and Ruby.

Preface

The rationale for this book is twofold: to provide a useful guide to synthetic chemists and materials scientists who wish to prepare quantum dots and related materials by solution methods, and to highlight the evolution of the chemistry involved in quantum dot synthesis.

Usually, when one hears of the history behind quantum dots, the pioneers such as Brus and Efros are mentioned but little credit is given to the equally important development of the synthetic chemistry. Although the emergence of novel physical properties such as quantum charge carrier confinement is rightfully considered as an inaugural landmark, few acknowledge the independent evolution of inorganic precursor chemistry by, for example, Steigerwald, who developed many of the synthetic pathways we now routinely use in quantum dot synthesis and take for granted. The seminal paper in 1993 that opened up quantum dots to synthetic chemists, leading to the vastly improved structures and materials described in this book, described the key organometallic reactions that can be traced directly back to Steigerwald's work on phosphine chalcogenides, although little recognition has been given to this leading work in solid-state chemistry. The reasons behind this are unclear, although it may be because the most obvious attributes of quantum dots—their tuneable optical properties—are inherently physical rather than synthetic in nature, and the developments in preparative chemistry emerged after the optical and physical properties were initially uncovered, explained and explored. Whatever the reasons, the amalgamation of solution-based inorganic precursor chemistry, micelle chemistry and the optical studies of quantum confined systems (previously termed artificial atoms, colloidal semiconductors, *etc.*—it was not until 1988 that Mark Reed, then at Texas Instruments, coined the term 'quantum dot') resulted in the emergence of a robust discipline of interest to scientists across a broad range of subjects.

RSC Nanoscience & Nanotechnology No. 33
Semiconductor Quantum Dots: Organometallic and Inorganic Synthesis
By Mark Green
© Mark Green 2014
Published by the Royal Society of Chemistry, www.rsc.org

This book, in parts, reads like a selection of recipes, and this is intentional. Several excellent publications exist to describe the quantum size effects, the physics, physical chemistry and mathematics involved in quantum dots, but few directly tackle what inorganic and materials chemists might want: a tool kit to describe how to make quantum dots, precursors, conditions, and what properties (usually optical) might result. These descriptions also include interesting points that I felt might be useful to chemists, and although they are intentionally brief to allow the key papers to be described, I hope these comments will encourage the reader to follow up with the referenced publications. With thousands of papers now available, I hope to induce some order and ensure earlier key studies are not forgotten.

Mark Green
King's College London, UK

Abbreviations

DDA	Dodecylamine	PVP	Poly(Vinylpyrrolidone)
DDAB	Dodecyldimethyl Ammonium Bromide	QD	Quantum Dot
		SILAR	Successive Ion Layer Adsorption and Reaction
DHLA	Dihydrolipoic Acid		
DOA	Dioctylamine	TBP	Tributylphosphine
Et	Ethyl	TBPO	Tributylphosphine Oxide
FRET	Förster Resonance Energy Transfer	TBPS	Tributylphosphine Sulfide
		TBPSe	Tributylphosphine Selenide
FTIR	Fourier Transform Infrared	TBPTe	Tributylphosphine Telluride
FWHM	Full Width at Half Maximum	t-Bu	Tertiary Butyl
HDA	Hexadecylamine	TDPA	n-Tetradecylphosphonic Acid
HPA	n-Hexylphosphonic Acid		
i-Pr	Isopropyl	TEM	Transmission Electron Microscopy
Me	Methyl		
NMR	Nuclear Magnetic Resonance	TOA	Trioctylamine
		TOP	Tri-n-Octylphosphine
OA	Octylamine	TOPO	Tri-n-Octylphosphine Oxide
OAm	Oleylamine	TOPS	Tri-n-Octylphosphine Sulfide
ODA	Octadecylamine	TOPSe	Tri-n-Octylphosphine Selenide
ODE	Octadecene		
ODE/S	Sulfur in Octadecene	TOPTe	Tri-n-Octylphosphine Telluride
PAA	Poly(Acrylic Acid)		
PEG	Poly(Ethylene Glycol)	XPS	X-Ray Photoelectron Spectroscopy
PEI	Poly(Ethylenimine)		
Ph	Phenyl	XRD	X-Ray Diffraction

RSC Nanoscience & Nanotechnology No. 33
Semiconductor Quantum Dots: Organometallic and Inorganic Synthesis
By Mark Green
© Mark Green 2014
Published by the Royal Society of Chemistry, www.rsc.org

Contents

RSC Nanoscience & Nanotechnology No. 33
Semiconductor Quantum Dots: Organometallic and Inorganic Synthesis
By Mark Green
© Mark Green 2014
Published by the Royal Society of Chemistry, www.rsc.org

CHAPTER 1

The Preparation of II–VI Semiconductor Nanomaterials

1.1 Origins of Organometallic Precursors

Although nanoparticles (notably metals) can be traced to antiquity, the origins of modern semiconductor quantum dots (QDs) is a much later development. Specific advances in quantum-confined semiconductor particles can be traced to Grätzel,[1] who reported the synthesis of colloidal CdS to examine photo-corrosion, and Brus[2] who notably reported the band edge luminescence of the same material. At approximately the same time, Ekimov reported quantum confinement in CuCl prepared in a silica glass,[3,4] while Henglein, an early pioneer of colloidal semiconductors, reported the synthesis of CdS on colloidal SiO_2.[5] Fendler then reported the use of a reverse micelle to prepare CdS nanoparticles,[6] which was importantly improved on by Henglein, who used polyphosphates as a well-defined passivating agent, allowing nanoparticles to be processed and redispersed.[7]

The work described above utilised mainly inorganic salts as precursors in aqueous-based reactions. Consequently, the low temperatures used and the presence of air and water often resulted in polydispersed materials with relatively poor optical and crystalline properties. The small number of available suitable precursors also reduced the number of systems that could be explored. The need to improve these reactions by removing air and water dictated the use of organic-based starting materials; however, inorganic reagents such as Na_2Se could not easily be used in organic solvents so alternatives were required.

Independently, Steigerwald pioneered many of the early chemical routes to bulk solid-state materials,[8] inspired by the early work on the deposition of

RSC Nanoscience & Nanotechnology No. 33
Semiconductor Quantum Dots: Organometallic and Inorganic Synthesis
By Mark Green
© Mark Green 2014
Published by the Royal Society of Chemistry, www.rsc.org

semiconductors by metal organic chemical vapour deposition (MOCVD).[9] In a typical MOCVD reaction, metal alkyls and arsine gas were used; precursors that are hypothetically suitable in the preparation of nanomaterials in an organic solvent, but realistically improbable. Although effective in vapour processes, precursors designed for gaseous-phase reactions are not ideal for use in solution chemistry because of their toxic nature, air sensitivity, and the associated difficulties in handling. In addition, metal alkyls such as dimethyl- or diethyltelluride are also too stable to be effective precursors at temperatures as low as 200 °C, the region of interest for solution synthesis.

Initial reports on alternative precursors described the suitability of trimethylphosphine (Me$_3$P) as a tellurium transport agent[8,10] and as a potential replacement for dialkyltellurides.[11] The reaction between triethylphosphine telluride and mercury metal was investigated, giving mercury telluride (HgTe) in almost quantitative yields.[10] This reaction suggested the assignment of the phosphine telluride as a single coordinate complex of Te(0). Since elemental metals are not normally utilised as precursors, the reaction was repeated using the metal alkyl diethylmercury (Et$_2$Hg) and the less volatile diphenylmercury (Ph$_2$Hg). Refluxing an organic solution of the metal alkyl and a phosphine telluride complex lead to the preparation of HgTe *via* the formation of an intermediate, Hg(TePh)$_2$. It was also observed that Hg(TePh)$_2$ eliminated mercury metal during the reaction, which immediately reacted with excess phosphine telluride giving HgTe. This reaction is noteworthy due to the insertion of Te atoms into a Hg–C bond and highlights the important role phosphines can play in the low-temperature solution synthesis of semiconductors.

Steigerwald also prepared cadmium chalcogenides using silylated precursors and metal alkyls,[12] providing a solution-based organometallic route to CdSe. The reaction between metal alkyls and hydrogen chalcogenide gases had previously been reported,[13] but most silylated chalcogens exist as a liquid at room temperature and are therefore easier to handle than a vapour-phase precursor. The reaction proceeded in simple organic solvents providing an amorphous red/brown solid, which could be converted to bulk crystalline CdSe upon annealing at 400 °C. The work was extended to cover ZnSe and CdTe. The solvent was found to have a distinct effect on the reaction rate; dealkylsilylation in dichloromethane proceeded almost instantaneously, while reactions in toluene took days to complete and reactions in saturated hydrocarbons required weeks. Interestingly, there was no reaction between Me$_2$Cd and the silylated precursor in the absence of solvent. Although the materials resulting from these reactions were not necessarily nanometric in size, this molecular precursor chemistry can be thought as the origin of the solution-based organometallic routes to nanomaterials.

1.2   Inverse Micelles

Steigerwald then used the developed chemistry to prepare discrete CdSe nanoparticles in inverse micelles, which can be thought of as the first pseudo-organometallic route to nanomaterials.[14] A dioctyl sodium sulfosuccinate/water/heptane microemulsion was prepared and an aqueous solution of cadmium perchlorate added, followed by a heptane solution of $Se(SiMe_3)_2$. The room-temperature reaction yielded particles of CdSe, which had to be reduced to dryness to remove water and avoid flocculation. The powders were then redissolvable in hydrocarbons. Addition of Cd^{2+} stock solution followed by phenyl(trimethylsilyl)selenide resulted in a phenyl-passivated surface, following which, the particles could be isolated by centrifugation. Importantly, this is the first example of monomer passivation of a nanoparticle surface and the phenyl-passivated clusters were soluble in pyridine, but insoluble in petroleum ether. Absorption spectroscopy showed materials with a shift in bandgap with decreasing particle size and a slight excitonic shoulder. The paper also reported growth of the particles with further addition of precursor, demonstrating Ostwald ripening-type growth. Following isolation, annealing of these particles in 4-ethylpyridine improved the crystallinity giving crystalline zinc blende (cubic) type clusters.[15] Annealing of small (20 Å) particles in a mixture of tributylphosphine (TBP) and tributylphosphine oxide (TBPO) resulted in further particle growth to *ca.* 40 Å, giving particles with a wurtzite (hexagonal) crystalline core. This highlighted the suitability of phosphine oxides as capping agents and was used as the basis for further work.

Clusters of CdSe prepared by the inverse micelle route were then used in core/shell studies, where a ZnS shell was deposited on the surface of the particles, inorganically passivating the surface.[16] The use of an inorganic layer rather than a surfactant to protect the surface is an area of immense interest, as the inorganic layer is not restricted by factors such as the surfactant cone angle, is generally more stable and is a more complete protecting layer. By choosing the correct materials, specific heterojunctions with engineered band mismatches can be prepared, which will be discussed later. Once the layer of ZnS had been deposited on the CdSe core, the emission spectrum, previously broad with luminescence attributed to surface defect states, became more band edge, suggesting blocking of the defects. In this case, the sulfide shell was deposited using inorganic rather than organometallic reagents: however, once the surface trapping states had been removed, the band edge emission observed originated from a nanoparticle core prepared by organometallic-based precursors.

1.3 Organometallic Routes to CdE (E = S, Se, Te)

The seminal paper describing solution-based organometallic routes to nanomaterials was published in 1993,[17] and reported the first *totally* organometallic/organic-based synthesis of CdE (E = S, Se, Te) nanoparticles which resembled a solution analogue of the MOCVD process, using an inert atmosphere, appropriate precursors with origins in vapour deposition and coordinating solvents suitable for high-temperature reactions. Although organometallic precursors had previously been investigated as discussed, this route relied entirely on organometallic reagents thermolysed directly in a hot coordinating solvent (long-chain phosphine oxides, such as tri-*n*-octylphosphine oxide, TOPO) rather than a room-temperature reaction in an inverse micelle. The precursors employed, Me$_2$Cd and either a silylated chalcogenide or a chalcogen dissolved in trioctylphosphine (TOP), were chosen with reference to the work of Steigerwald, and the route elegantly incorporated both the precursor (phosphine chalcogenide) and capping agent (phosphine oxide) chemistries described above. It should be assumed that all reactions and preparations described hereafter are carried out under inert atmosphere conditions unless stated otherwise.

Cadmium selenide (CdSe) is generally considered the prototypical QD material as size quantisation effects result in tuneable emission across the entire visible range of the electromagnetic spectrum, making the material attractive for a wide range of optoelectronic applications. A notable report by Donegá *et al.* highlighted the importance of reaction temperature, reagents and the ratio of reagents. By carefully selecting optimum condition, CdSe particles with quantum yields as high as 85% were prepared.[18] In contrast, CdS, one of the earliest materials to be studied because of its ease of preparation, demands less attention as it displays only small changes in the optical properties when prepared on the nano scale. This can be attributed to the difference in Bohr radius of the exciton (CdSe, $a_B = 32$ Å; CdS $a_B = 19$ Å) which means CdS particles have to be significantly smaller than CdSe to exhibit size quantisation effects and therefore display a smaller shift in the band edge. (In comparison, TiO$_2$, a wide-bandgap semiconductor, has an excitonic radius of only 8 Å and would require crystals to be little more than clusters to display any optical effects from the confinement of charge carriers. TiO$_2$ displays little, if any, shift in the optical band edge. In this case, size quantisation effects are manifest as variations in the oscillator strength.[19])

In a typical reaction, Me$_2$Cd and a trioctylphosphine chalcogenide (such as trioctylphosphine selenide, TOPSe) were dissolved in TOP. The Lewis base phosphine, a liquid at room temperature, served as both a solvent for precursor delivery and a capping agent once the nanoparticles had formed. The precursor solution was then injected into TOPO (which had been rigorously dried and degassed), under an inert atmosphere at temperatures of between 100 and 350 °C, with lower temperatures producing smaller particles. The sudden introduction of reagents into a hot solvent and the subsequent immediate supersaturation resulted in the formation of nuclei. The drop in temperature after the injection of room-temperature reagents prevented further nucleation, and further heating resulted in growth of particles by Ostwald ripening. The sudden nucleation and slow growth steps, originally described by La Mer,[20] resulted in a monodispersed product. The surfactants, TOPO and TOP coordinated to the surface of the nanoparticles, providing physical and electronic passivation. The labile nature of the surfactants is a key requirement, desorbing from the particle surface to allow growth, yet coordinating strongly enough to allow particle isolation and provide the required protection for the nanoparticle.

After injection and during the growth stage, the reaction could be monitored by removing aliquots and recording the emergence of a band edge *via* absorption spectroscopy, and the growth could also hence be tuned by altering the reaction temperature. After the particles had grown to the required size, the reagents were then left to cool to *ca.* 60 °C, followed by addition of a polar solvent (termed a non-solvent) such as methanol, which induced precipitation. The precipitate, collected as a waxy powder by centrifugation, was dispersed in non-polar solvents producing an optically clear solution. Addition of small amounts of non-solvent increased the average polarity of the solvent and resulted in the precipitation of the larger particles. This size-selective precipitation resulted in a solution of nanoparticles with an extremely narrow size distribution.

The materials prepared were monodispersed (<5% standard deviation), 1.2–11.5 nm in diameter and slightly prolate with aspect ratios up to 1.3. The particles were also crystalline, capped with a monolayer of surfactant molecules (TOPO and TOP) and displayed excellent optical properties. Figure 1.1 shows typical absorption spectra from CdE (E = S, Se, Te) nanoparticles prepared by this method, with the first excitonic transitions clearly visible.

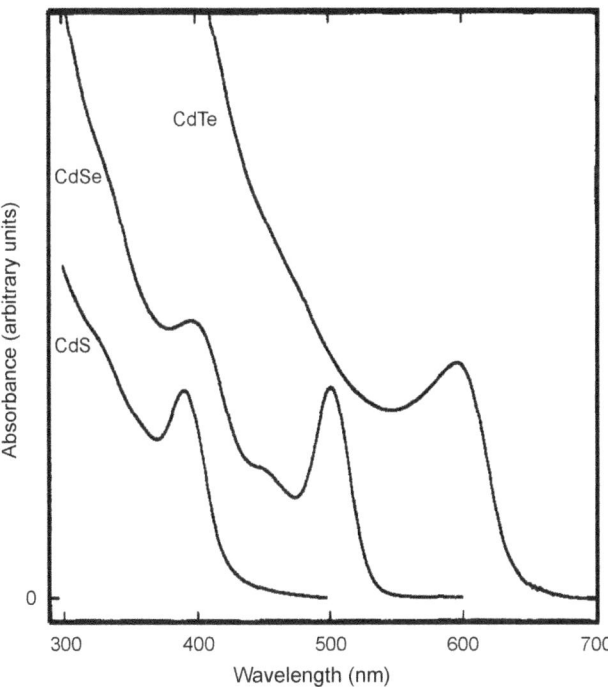

Figure 1.1 Absorption spectra for CdS, CdSe and CdTe nanoparticle prepared by organometallic chemistry. Reprinted with permission from C. B. Murray, D. J. Norris and M. G. Bawendi, *J. Am. Chem. Soc.*, 1993, **115**, 8706. Copyright 1993 American Chemical Society.

The emission from TOPO-capped particles was initially band edge without the need for a further inorganic shell, highlighting the high optical quality of materials prepared under an inert atmosphere.

In this case, the steric properties of TOPO/TOP play a key role; reducing the chain length to butyl groups or smaller resulted in uncontrolled growth. The actual coordination of the capping agent to the crystallite slowed growth kinetics, allowing for the controlled growth at elevated temperatures. The affinity of the phosphine oxide group for the surface metal (controlled by tuning the chain length of the alkyls chains)[21] also effected the growth rate; the increased Lewis base character, *i.e.*, the more electron donating, the stronger the binding and the slower the growth. The presence of the capping agent was also essential for the band edge emission and relatively high initial quantum yields of *ca.* 10% (for CdSe) were observed, which dropped by orders of magnitude upon removal of the surface ligand. Notably, it has since been discovered that the emission quantum yield of QDs can be increased up to a maximum of 75% *via* the surface treatment with sodium borohydride ($NaBH_4$). The addition of small amounts of the reagent reduced the ligand, removing the surfactant while allowing the surface cadmium to oxidise, yielding a cadmium oxide layer that enhanced the emission. Prolonged exposure or an excess of reducing agents resulted in the precipitation of the nanoparticles, although addition of the optimum amount of $NaBH_4$ was achievable and the treated nanoparticles were found to be stable for up to a year in ambient conditions.[22]

The surface ligand could be removed by refluxing the nanoparticles in pyridine, which resulted in ligand exchange, removing the TOPO and leaving the more labile pyridine coordinated to the particle surface. This was initially used as in intermediate for other capping agents and will be discussed later. Murray also reported that although Lewis base ligands are generally utilised for capping agents, Lewis acids, such as alkylboranes and alkylaluminium species could also be used to passivate the surface.[21]

The particles were crystalline as determined both by powder X-ray diffraction (XRD) and by transmission electron microscopy (TEM). The XRD patterns displayed reflections consistent with wurtzite (hexagonal) structured nanoparticles, with the distinction between hexagonal and cubic structure being lost in particles below *ca.* 2 nm in diameter (Figure 1.2). Modelling studies of the diffraction patterns suggested each nanoparticle contained one stacking fault.

This route quickly became the standard method to prepare high-quality II–VI nanoparticles, with CdSe becoming the most studied material. A notable amendment to the preparation of CdSe was reported by Bowen Katari,[23] where the selenium was dissolved in TBP rather than the longer TOP in an attempt to achieve a higher surface coverage, although comparisons between QDs prepared by both methods revealed little difference. In an attempt to prepare larger samples with absorption features beyond *ca.* 580 nm, the synthesis temperature was increased to 350 °C, which was followed by several minutes' continuous heating at *ca.* 320 °C after precursor

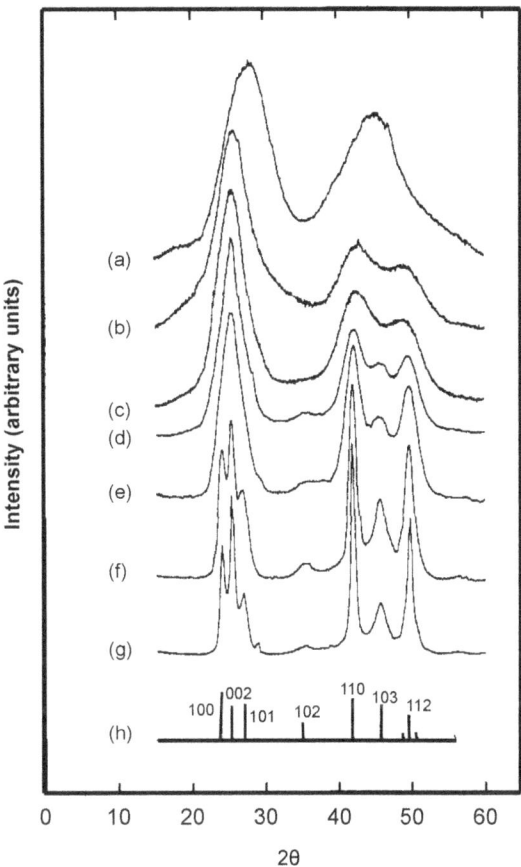

Figure 1.2 X-ray diffraction patterns for CdSe nanoparticles of varying diameters: (a) 1.2 nm; (b) 1.8 nm; (c) 2.0 nm; (d) 3.7 nm; (e) 4.2 nm; (f) 8.3 nm; (g) 11.5 nm; (h) bulk CdSe. Reprinted with permission from C. B. Murray, D. J. Norris and M. G. Bawendi, *J. Am. Chem. Soc.*, 1993, **115**, 8706. Copyright 1993 American Chemical Society.

injection. It is worth noting, however, that TOPO slowly decomposes above 330 °C. The nanoparticles prepared were larger, but possessed a larger size distribution. This was remedied by adding the precursor stock solution in 0.1 mL amounts. Particles prepared by this route were more spherical[24] and had fewer stacking faults, although small variations in temperature resulted in large differences in size distribution.

The actual size range of particles from a single reaction is a key parameter and the literature offers conflicting reports as to which route provides the lowest size distribution. Alivisatos addressed the issue, based on the Gibbs–Thompson equation and reported a detailed study on the kinetics and growth model of CdSe particles, suggesting that diffusion-limited growth could reduce the size distribution if the diffusion area and particle size were considered.[25] Alivisatos highlighted that the kinetics of crystal growth were also dependent on the variation of surface energy with size, and showed prolonged growth led to growth focusing, refocusing and defocusing (Ostwald ripening), with an intimate link to monomer concentration. During the focusing stage, the high concentration of the monomers exceeds the solubility of the nanoparticles: all particles grow, but the smaller particles grow faster than the larger ones and the size distribution can be focused to an almost monodispersed sample. Below a specific monomer concentration, the larger particles grow at the expense of the smaller ones, defocusing the size distribution. This highlighted that the size distribution could be controlled experimentally by maintaining a high monomer concentration. Further nucleation studies suggest the growth of CdSe nanoparticles is reaction limited,[26] which is supported by the earlier detailed study by Dushkin *et al.*[27] To complement experimental work, detailed theoretical studies of particle growth based on Monte Carlo simulations have also been carried out which make suggestions as to the favoured growth regime.[28] Various models have been utilised in examining the kinetic and thermodynamics of particle growth; the different stages of nucleation and growth have been discussed in the context of classical nucleation theory,[29] and a barrier diffusion model has also been developed that describes the growth kinetics in different solvents, highlighting Arrhenius behaviour dictated by solvent activation energies, which increase with molecular weight.[30] The kinetics of nanocrystal growth is complicated and covers many variables, yet if controlled it can be tuned to allow the growth of different-sized particles.[31,32] For example, controlling the growth of CdSe particles is possible by choosing ligands which form complexes of differing solubility, that act either as nucleating agents (resulting in higher particle yields and smaller particles) or as growth agents (resulting in early time ripening), allowing extremely small particles to be prepared and stabilised.[33] Other work has suggested that the presence of water and oxygen in the reaction mixture can lead to the etching of the small particles that aggregate into larger particles during the Ostwald ripening process.[34]

Emission from nanometre-sized CdSe can be tuned to *ca.* 600 nm while retaining a relatively high quantum yield, and emission beyond this region requires a different material such as cadmium telluride, CdTe, a material with a room-temperature bulk bandgap of 1.44 eV and an excitonic radius of 38 Å.[35]

Although the preparation of CdTe was described in the seminal organometallic synthesis paper, Mikulec reported the first serious study of CdTe nanoparticles prepared by a synthetic procedure based on the TOPO route.[36] The CdTe nanoparticles initially prepared by Murray *et al.* were not investigated in depth and had an estimated quantum yield of below 1%, attributable to the instability of the triphosphine telluride precursor. In the case of the original TOPO route, lower growth temperatures were used to maintain controlled growth of CdTe, which in turn led to materials with a lower quantum yield.

Mikulec reported essentially the same procedure as described above, but replaced TOP with hexapropylphosphorous triamide, $P(N(C_3H_7)_2)_3$, which exhibited an improved tellurium affinity, attributed to the lone pair donation from the phosphorus-bound nitrogen atoms.[26] The hexapropylphosphorous triamide telluride complex was then diluted in TOP, and injected into hot TOPO with Me_2Cd at 350 °C under an inert atmosphere, upon which the temperature dropped to 270 °C. Initial sampling after injection revealed a wide size distribution, which tightened after further particle growth at 290 °C. In this case, methanol was found to be an unsuitable non-solvent and acetonitrile was used instead, with the particle being soluble in *n*-butanol and tetrahydrofuran. Particles, once isolated in solution, had a drop of TOP to maintain the high quantum yields, presumably capping surface sites that were stripped of surfactant during isolation. Mikulec also reported the importance of precursor purity, with improvements in reaction reproducibility observed when using Me_2Cd which had been vacuum transferred through a 0.2 µm filter and phosphine which had been distilled. Increasing the cadmium : tellurium ratio to 2 : 1 appeared to provide optimum materials.

Materials prepared by the Mikulec method could be tuned to display band edge emission between 590 and 760 nm with consistently high quantum yields (average of 60%, a highest recorded value of 70%), and had a zinc blende (cubic) crystal structure. The emission quantum yield was found to increase with reaction time, attributed to annealing effects. The material, which could be grown between 4.4 and 13 nm in diameter, was found to be extremely air sensitive with quantum yields dropping to practically zero after 2.5 hours exposure to air.

Talapin *et al.* also reported an organometallic route to CdTe nanoparticles between 2.5 and 7 nm in diameter, based on the TOPO route.[37] In this route, two synthetic procedures were reported; the first involved mixing TOP and dodecylamine (DDA) under an inert atmosphere, followed by room-temperature addition of tellurium powder and Me_2Cd (molar ratio 1 : 1.47). The materials were slowly heated to 180 °C for 30 minutes, followed by further heating at 200 °C for a further 20 hours, allowing the tellurium to slowly dissolve in the reagents, promoting a slow growth of the particles. Nanoparticles prepared by this method were between 4 and 6 nm in diameter and exhibited quantum yields of up to 65%. This method is notable for the small size distribution range without employing a hot injection delivery of the precursors.

The second method involved heating tellurium powder with TOP at 220 °C to affect dissolution, followed by cooling and the addition of Me_2Cd and more TOP. Notably, in this case the molar ratio of Cd : Te was *ca.* 1 : 3.5. The reagents were then injected into DDA at 150 °C. Prior to injection, the flask was removed from the heat source which allowed nucleation with little growth. Heat was reapplied after *ca.* 15 minutes, allowing the particles to grow slowly. The temperature and growth time determined particle size, with particles averaging 3 nm in diameter when grown for a few hours at between

150 and 180 °C. Prolonged growth (up to 12 hours) at 200 °C resulted in larger particles up to 5 nm in diameter (7 nm with further addition of precursor). Particles prepared by this route exhibited band edge emission but with substantially lower quantum yields of up to 30%.

The higher quantum yield of the first route described by Talapin can be attributed to the prolonged heating, improving the crystallinity of the samples and the constant delivery of dilute tellurium. All samples displayed a cubic crystalline core as opposed to the hexagonal CdTe samples reported originally by Murray, and showed evidence, although only slight, of trap emission manifesting as a slight red-shifted shoulder in the luminescence spectra. The use of DDA as a capping agent was found to be a key parameter in the preparation of CdTe particles. Other long-chain amines were found to be inferior surfactants; octylamine (OA) was found to have too low a boiling point and dioctylamine (DOA) resulted in immediate precipitation. Hexadecylamine (HDA) allowed particle growth but the resulting QDs had a large size distribution relative to DDA-capped particles. A similar methodology was applied to the preparation of CdSe and InP particles.[38] An in-depth study of the reaction conditions and growth of the DDA-capped CdTe particles was reported,[39,40] highlighting that the small initial particles formed, emitted from a defect state, leading to exciton band edge emission after several minutes growth. It was also suggested that the precursors formed small, stable clusters at room temperatures, and that upon heating, particle growth was found to be fast over the initial 30 minutes, with the final size being reached in *ca.* 3 hours. In related work, stable, small (magic) clusters of CdTe (1.9 nm in diameter) have been prepared using *n*-hexylphosphonic acid (HPA) as a capping agent. The stable particles could be isolated due to the surface passivation and were found to exhibit broad emission with a quantum yield of 4% in methanol, although they aggregated due to the permanent dipole in the zinc blende cluster.[41]

It is worth noting at this stage that a room-temperature aqueous route can also be used to prepare CdTe materials of a similar optical quality to those prepared by organometallic precursors.[42] In a typical example, a cadmium salt and H_2Te gas are used as precursors, and thiols are utilised as capping agents. The routes are scalable, relatively cheap and easily reproduced, but the resulting materials lack the degree of crystallinity obtained using organometallic precursors. The key advantage of using aqueous methods to make CdTe is the increased stability in ambient conditions when compared to the extremely air-sensitive particles prepared by the organometallic route. Thiol-capped CdTe exhibited band edge emission in a similar spectral range as the organometallic analogues, with quantum yields up to 85% with stability in air for months.[43] The anomalously high quantum yield for this aqueous system is attributable in part to the chemistry and physics of the capping agent. One explanation involves the reaction of the capping agent with the particle; upon prolonged illumination, the thiol-capping reportedly decomposed giving S^{2-} ions in solution, which reacted with the cadmium surface of the nanoparticle.[44] This, in effect, resulted in a CdTe/CdS core/shell structure where the CdS electronically passivated the surface and protects the emitting core, and similar observations have been made with gold nanoparticles.[45] This, however, does not take into account the fact that thiol-capped CdSe particles are not as luminescent as thiol-capped CdTe, and in fact have relatively poor optically properties which one would not expect from CdSe/CdS core/shell particles. Another explanation is based on the redox energy levels of the thiol, which act as inhibitors for hole-trapping processes for the CdTe particles.[46] This highlights the important role of the capping agent, which should be viewed not as a simple inert surfactant, but as an integral constituent of any nanostructure with its own discrete chemistry and physics.

1.3.1 Alternatives to Metal Alkyls

Although the use of organometallic precursors resulted in high-quality materials, this route still utilised metal alkyls as precursors, which are notoriously toxic and difficult to handle. One method of circumventing the problem is the use of simple Lewis base adducts of the metal alkyls, related to the 'diphos' method of purifying metal alkyls for chemical vapour deposition.[47] Adducts are generally more stable, (and in some cases, crystalline solids whereas the parent metal alkyls are liquids), making the precursors easier to handle. Replacing Me_2Cd with a triethylamine adduct of Me_2Cd appeared to have no remedial impact on the quality of nanoparticles prepared.[48] Likewise, a study into the use of the solid adduct (2,2'-bipyridine) Me_2Cd, the adduct bis(3-diethylaminopropyl)-cadmium or the more stable metal alkyl dineopentylcadmium, $((CH_3)_3CH_2)_2Cd$, as substitutes for Me_2Cd revealed no detrimental effects on the quality of the final product.[49] Similarly, cadmium alkoxide complexes have been used as replacements for metals alkyls in CdE (E = S, Se, Te) nanoparticle synthesis.[50]

Lazell and O'Brien reported one of the simplest methods of preparing CdS nanoparticles using $CdCl_2$ and sulfur, and although not organometallic, this and related routes found their genesis in the original organometallic synthetic pathways. In this method, sulfur was dissolved in TOP giving trioctylphosphine sulfide (TOPS), which was mixed with the cadmium salt in excess TOP, then injected into TOPO at an elevated temperature. This can be seen as the first 'green' TOPO-related synthesis of nanoparticles, avoiding toxic metal alkyls or chalcogen species,[51] although Stuczynski *et al.* first used tributylphosphine sulfide (TBPS) in the preparation of cobalt sulfide *via* cluster precursors.[52] (It is worth noting a report that suggested TOPS was too unreactive towards zinc salts to be used in the synthesis of nanomaterials.[53])

This method has been developed, noticeably by Peng, who has reported several new green routes to passivated particles.[54] Peng made a detailed study of the decomposition of dimethylcadmium and observed metal precipitates that coordinated to impurities in technical-grade TOPO, giving complexes such as cadmium–hexadecylphosphonic acid.

The discrete cadmium complexes could be prepared from either Me₂Cd or CdO (but not CdCl₂, in contrast to the work reported by Lazell) and used in the synthesis of high-quality nanoparticles, suggesting the metal alkyl was not necessary. Therefore, a one-pot method of preparing particles was developed, based on the high-temperature dissolution of CdO in TOP with a known 'impurity' such as tetradecylphosphonic acid (TDPA). Injection of the chalcogen precursor resulted in the formation of cadmium chalcogenide particles that followed the growth kinetics suggested earlier. A notable feature of this method is the slow nucleation due to the stability of the cadmium–phosphonic acid complex, which has important advantages; the injection temperature need not be 350 °C to obtain a monodispersed sample and the reaction is very reproducible and controllable,[55,56] even allowing the growth of bimodal and trimodal size distributions through kinetic perturbation.[57] The slow nucleation has allowed studies into the crystallisation process, which have revealed that when using cadmium stearate as a precursor and a growth temperature of 200 °C, the initial particles formed are amorphous and required 10 minutes of annealing to undergo the required reconstruction and crystallisation.[58]

Work by Stucky has revealed that CdCl₂, CdI₂, and Cd(CO₂CH₃)₂ could also be used as precursors.[59] Peng further showed that cadmium acetate was the most useful precursor when used with weakly anionic fatty acids, and that a number of solvents such as phosphines, phosphine oxides and amines could be used, while the use of thiols was found to hinder nucleation due to the stability of cadmium–thiolate complexes. Particles between 4 and 25 nm in diameter could be prepared, although the larger particles showed a wide size distribution. Controlling the rate of growth was achieved by varying the chain length of the acid, with shorter chains resulting in faster growth. As grown, the highest quantum yields observed were 20–30%, a significant increase on emission reported from particles prepared by metal alkyl. This route is notable for producing particle sizes not normally accessible by the Me₂Cd route.[60] Kumar also reported the preparation of CdTe nanoparticles of varying morphology by the use of cadmium stearate, Cd(CO₂(CH₂)₁₆CH₃)₂, as a precursor.[61] The particles prepared exhibited a hexagonal crystalline core, as opposed to the cubic materials reported by Talapin.

Detailed work by Liu *et al.* has uncovered the reaction mechanism behind the formation of CdSe QDs using CdO as a precursor in a non-coordinating solvent (discussed later).[62] As suggested, the reaction between CdO and a phosphonic acid (present as an impurity in the capping agent or a specific ligand) resulted in the formation of phosphonic acid complex and the generation of water. The addition of TOPSe resulted in the formation of an TOPSe–Cd(X)₂ intermediate (where X was various groups), which weakened the P=Se bond. The P=Se bond was cleaved by the nucleophilic attack of either the carboxylic acid or phosphonic acid, leaving CdSe, a phosphine oxide and an acid anhydride. Complementary work by Steckel[63] confirmed this mechanism and also suggested another competing mechanism based on experimental observations, in which the surfactant (TOP) reduced the metal precursor from M^{2+} to M^{0}, which then reacted with a trialkylphosphine chalcogenide eliminating the phosphine, which reacted further to provide the particles, phosphine oxide and metal counterion anhydride. The use of trialkylphosphine chalcogenides is considered the standard group-VI precursor system, but the use of other chalcogen sources has also been described, such as NaHSe[64] and NaHTe.[65]

1.3.2 Alternatives to TOPO and Novel Solvents

The search for novel solvents is often driven by the need for a phosphine-free system, as TOP and TOPO are considered toxic,[66] and to a lesser degree, by cost. Peng demonstrated that non-coordinating solvents could also be used in nanoparticle synthesis, as opposed to the thermolysis in TOPO. In a typical example, CdO and oleic acid were dissolved in octadecene (ODE), a room-temperature liquid with a boiling point of *ca.* 320 °C. Elemental sulfur dissolved in ODE was then injected into the reagents at elevated temperatures, forming CdS nanoparticle capped with oleic acid. An interesting examination revealed that upon isolation of CdS particles prepared in ODE, a decrease in the emission intensity was observed. Redispersion of the particles in $CHCl_3$ resulted in an increase of trap-related emission associated with the removal of surface ligands, while redispersion in ODE resulted in the recovery of band edge emission, attributable to the solvent possibly blocking surface trapping states.[67]

The delicate balance between forming too many nuclei, which resulted in the defocusing of the size distribution by Ostwald ripening, or too few nuclei, which resulted in growth that was too fast to reach the required size and distribution, was easily achieved by simply altering the concentration of the capping agent and thus the amount of metal complex available for the reaction (the monomer concentration). Another advantage is the extreme control over particle size; the synthesis of CdSe particles less than 4 nm in diameter is difficult to achieve using methods discussed previously, but using ODE as a solvent, particles in the size range 1.5–20 nm were routinely prepared.[68] The role of ODE was envisaged as a non-coordinating solvent and a simple chalcogen delivery agent. Further work has shown that in the case of selenium in ODE, the viability of ODE/Se is dependent on the length of time the reagents are heated and the temperature. Selenium reportedly formed rings and chain species that formed bridged structures between alkene/alkane chains consisting of multiple selenium atoms. This form of precursor was found to be highly reactive and useful, unlike the monoselenide species obtained upon prolonged heating.[69] Using this route, zinc blende-structured CdSe could be prepared by adding a phosphonic acid in the TOPSe solution.[70] The generation of zinc blende-structured CdSe by the organometallic route is unusual, and few reports describe CdSe with such a crystalline core.[71] The absence of phosphine or phosphine oxide capping agents, or the use of a non-coordinating solvent, appears to be a constant factor with such a synthesis; for example, the use of H_2Se gas as a chalcogen precursor and the absence of a phosphine capping agent in a similar reaction reportedly yielded CdSe with a cubic crystalline phase.[72] Likewise, the use of cadmium carboxylates and long-chain amines in ODE also resulted in small zinc blende CdSe (although TOPSe was used as a precursor).[73] Other reports describe the synthesis of high-quality tetrahedral zinc blende-structured CdSe[74] onto which have been grown zinc blende-structured CdS shells.[75,76] The growth of thick (5 nm) shells on a zinc blende CdSe core has also reportedly resulted in wurtzite structures, with a remarkable suppression of fluorescence intermittency,[77] a phenomena also observed in other giant shelled QDs.[78] The optics of zinc blende-structured CdSe has been investigated: although the majority of parameters were comparable to wurtzite-structured CdSe, at larger particle sizes (>*ca.* 4 nm), a weaker confinement effect was observed for the zinc blende materials, resulting in a slightly different sizing curve.[79] Interestingly, shape control of zinc blende CdSe has been achieved using ODE and oleic acid as a solvent system, with a variation in reaction temperature being responsible for the impressive variety of shapes obtained, such as monodispersed cubes and tetrahedrons.[80]

Peng also reported an in-depth study into the preparation of CdTe particles in ODE, using fatty acids and phosphines as capping agents, with cadmium oxide and tellurium powder as precursors.[81] Initial studies showed that by altering the capping ligand concentration and chain length, the monomer concentration could be fine-tuned which allowed for controlled nucleation and growth of particles. In the study on CdTe particles, ligand effects were found to have a significant role in the development of particle size, size distribution, morphology and crystal phase (which was later utilised in the synthesis of anisotropic structures[82]). The term 'monomer concentration' was replaced with 'effective monomer concentration', which more accurately described monomer activity with respect to ligand stability and ultimately how this affected the formation of the particles. The term 'activity coefficient' was used to describe both the stability and steric effects on the monomers. The preparation of samples with various shapes and crystallinities could therefore be achieved by selecting specific ligand systems and concentrations that drove the reaction to the desired product. The competition between ligand–monomer stability in fatty acids and chain length in phosphines was discussed, and a complicated relationship was uncovered.

The CdTe particles prepared in the non-coordinating solvent had quantum yields as high as 70%, and emission line widths significantly narrower than those previously reported. The particles were again air sensitive, although the oxidation was partially suppressed by using unsaturated fatty acids as capping agents. Saturated fatty acids were found to produce less stable nanoparticles due to their superior crystal packing, which provided gaps that acted as oxygen channels.

The ODE route to CdTe has been extended, using *in situ* precipitated Cd^0 particles (100–150 nm in diameter) as precursors, generated by simply prolonging the heating step of CdO until a black precipitate was observed.[83,84] Injection of trioctylphosphine telluride (TOPTe) in the solution shortly after the observation of the precipitate resulted in QD formation, approximately 4 nm in diameter and with emission quantum yields of up to 80%. Injection of the chalcogen precursor before observation of the precipitate resulted in CdTe tetrapod formation. The formation of the Cd^0 particles results in a regulation of monomer supply, avoiding fast growth into the tetrapods.

The use of green chemistry has attracted a great deal of attention, and numerous adaptation of the routes have been reported, including routes to PbS, MnS[85] and CdS,[86] the use of two phases[87–90] and the use of initiators to control the size distribution.[91] Although some systems avoid the use of TOPO, several still use TOPSe as a convenient chalcogen precursor. Probably the most notable replacement for TOPO as the reaction solvent is ODE, although the seminal reports of the related synthesis of cobalt nanoparticles utilised long-chain ethers as non-coordinating solvents.[92] A notable report is the use of heat transfer fluids, not normally used as reaction solvents, in the preparation of CdSe and CdSe/CdS QD structures.[93] In this report, TOPSe is still used as the selenium precursor and oleic acid is used as a capping agent, and a range of reaction solvents compared. Two heat transfer solvents, DTA (a mixture of biphenyl and diphenylether) and T66 (a mixture of terphenyls and polyphenyls), were found to be useful in the controllable preparation of small QDs relative to the TOPO-based synthesis, and a detailed kinetic study highlighted differences in viscosity, surface free energy and CdSe solubility as the reasons behind the differences when compared to the standard preparations. Similarly, the ionic liquid trihexyl(tetradecyl)phosphonium bis(2,4,4-trimethylpentylphosphinate) has been used as a replacement for TOPO in the preparation of CdSe, with both the anionic and cationic constituents coordinating to the particle surface sites.[94,95] A gradual emergence of strong band edge luminescence was observed, although only when TOPSe was used as a precursor, consistent with previous reports of TOPSe being a cause of enhanced emission.[96] Other methylimidazolium-based ionic liquids have also been used as a capping agent and solvent for single-source precursors.[97] Paraffin has also been used as a phosphine-free solvent for the preparation of oleylamine (OAm)-capped CdSe/ZnS QDs, with quantum yields of up to 36%.[98,99] Similarly, olive oil has been used as a reaction solvent, giving the cubic phase of oleic acid-capped CdSe, the diameter tuneable from 2.3 to 6.0 nm and exhibiting band edge emission with quantum yields up to 15%.[100] Also, the use of microwaves in synthesis,[101] and specifically the use of microwave-active trialkylphosphine-based precursors, have opened up the use of low-boiling-point, non-microwave-active solvents, such as pentane, heptane, octane and decane.[102–104]

1.4 ZnE (E = S, Se, Te)

Bulk zinc chalcogenides exist as wide-bandgap semiconductors (ZnS 3.54 eV, ZnSe 2.58 eV, ZnTe 2.26 eV) and are therefore ideal materials for emission in the near UV/blue end of the visible spectrum. The preparation of ZnE (E = S, Se, Te) nanoparticles was initially briefly reported by Murray, who described the thermolysis of Et$_2$Zn and phosphine chalcogenides in TOP.[17] It was observed that the use of TOPO as a capping agent was unsuccessful, with the dialkylzinc forming stable TOPO adducts. Hines and Guyot-Sionnest reported zinc selenide (ZnSe) particles with a cubic crystalline core, prepared by the thermolysis of Et$_2$Zn and TOPSe in HDA.[105] The use of TOPO was again found to be unsuccessful, attributed to TOPO binding too strongly to zinc and TOP binding too weakly. The use of long-chain amines was found to be ideal; they are slightly weaker Lewis bases than phosphines, so the coordination to the particle surface through the electron-rich nitrogen atom provided ideal binding characteristics that allowed growth while passivating the particles, and the long alkyl chain imparted solubility in organic solvents while allowing solvent–non-solvent interactions. The emission from ZnSe capped with HDA was found to be band edge with quantum yields of 20–50%, attributable to the high growth temperatures and efficient capping of the amine and the phosphine. The use of amines as capping agents has been investigated by Woo,[106] who suggested that the linear steric features of primary amines (TOP has a cone angle of *ca.* 130°)[107] allowed for more efficient packing on the particle surface resulting in a more efficient surface passivation, and that amines etched the particle surface removing defects and trapping sites.

The Hines method of preparing ZnSe particles was advanced by Cozzoli *et al.*, who amended the method to include the use of amines with longer chain length/higher boiling point.[108] The amended method also reported the additional slow introduction of further precursor to refocus the size distribution. The route is also interesting due to the growth conditions; after the rapid injection of precursor at *ca.* 300 °C, the reaction cooled to *ca.* 265 °C and was allowed to proceed at this temperature. Increasing the concentration of the precursor stock solution resulted in larger particles (up to 8 nm). The method also covered altering growth conditions to prepare particles of various morphologies, which will be described later.

Li *et al.* prepared cubic ZnS and ZnSe particles in ODE and tetracosane (a non-coordinating solvent with a boiling point of 391 °C) using zinc carboxylates and tributylphosphine selenide (TBPSe) or an ODE solution of sulfur (ODE/S) as precursors, and a variety of capping agents such as amines and fatty acids.[109] The growth of zinc chalcogenides demands a more stringent set of conditions; again using capping ligands with a long chain length slowed the growth by increasing steric effects of the zinc monomer. The growth of particles using just a fatty acid as a capping agent was possible, but gave substandard material. The use of a high concentration of fatty acids resulted in no particle nucleation, attributed to the activity coefficient being too low to form the required amount of nuclei. Activation of the zinc precursor by inclusion of an amine during the reaction increased the reaction rate and gave balanced growth and nucleation, observable by the sharp features in the optical spectra. Interestingly, ZnO nanoparticles were prepared if no selenium was added, suggesting that ZnSe particles may be prepared by a ZnO intermediate or small cluster. The amine was found to be unstable in the reaction mixture, so was added with the selenium reagents. This resulted in more reproducible reactions; however, the sharp profile of the absorption spectra was lost to some degree. ZnSe particles grown with just fatty acids had quantum yields of below 10%, whereas amine passivated particles had quantum yields of up to 50% with emission as narrow as 14 nm (full width at half the maximum, FWHM). Emission could be tuned to between *ca.* 370 nm to *ca.* 440 nm, making the particles ideal blue emitters. ZnS nanoparticles displayed a different chemistry, with fatty acid passivated particles being approximately 10 times brighter than amine/acid passivated ones, although both types of ZnS displayed significant deep trap emission. This route also utilised similar injection/growth temperatures, when convention normally dictates a high injection temperature, followed by a low-temperature growth step to separate nucleation and growth. This step was necessary due to the delicate balance between nucleation and growth in zinc chalcogenide particles.

Reiss *et al.* have reported an almost identical method,[110] which produced capped ZnSe nanoparticles with a hexagonal crystalline core as opposed to the cubic analogues reported by Li. The similarity between hexagonal and cubic diffraction patterns coupled with the broadened reflections make assignment difficult, although it is possible that particles reported by Li may exhibit a slight reflection from the 103 plane, making them more hexagonal in nature. Reports on similar experiments described by Chen *et al.* reported ZnSe prepared by the reaction between zinc oxide and lauric acid (giving the carboxylate salt), HDA and TOPSe resulting in hexagonal crystalline particles approximately 4 nm in diameter.[111] Surprisingly, a method of preparing ZnSe particles using TOPO has been reported, where $ZnCl_2$ in glycerol was mixed with HDA and TOPO before a selenium solution of hydrazine was added at 200 °C, followed by growth at 160 °C. The particles, which were clearly cubic in nature, were up to 17 nm in diameter and displayed broad band edge emission at *ca.* 400 nm.[112] Selenourea has also been used as a precursor for ZnSe QDs, using $Zn(O_2CCH_3)_2$ as a precursor, with TOPO and octadecyl-amine (ODA) as capping agents.[113] In this case, by varying the ratio between ODA and TOPO, the crystallinity could be tuned between zinc blende and wurtzite-structured materials and differing particle geometries.

Quasi-spherical and rod-shaped ZnS QDs have also been prepared by the injection of Et_2Zn into sulfur/HDA-containing ODE, followed by growth at 300 °C.[114] Low injection temperatures resulted in quasi-spherical particles of intermediate cubic/hexagonal crystallinity, but above 150 °C the formation of rods predominated. The use of a large excess of sulfur also drove the formation of rods. To increase the yield of the rods, the isolated particles were annealed in OAm at 60 °C for 1 day, resulting in the oriented attachment of the particles yielding anisotropic particles with a cubic crystalline core. Again, the use of TOPO did not result in rod formation, producing a random distribution of particle shapes and sizes. The optical properties of the rods and spherical particles showed excitonic absorption features at *ca.* 315 nm, with near band edge emission, not trap emission as the group previously reported.[85] The better optical properties were attributed to the use of the metal alkyl precursors.

In most cases, particles of ZnE (E = Se, S) appeared air sensitive and the photoluminescence quenched upon oxidation. The use of olive oil, described earlier as a suitable solvent for the preparation of passivated CdSe particles, has been applied to the synthesis of ZnSe particles.[115] Here, selenium was dissolved in hot olive oil (200 °C), cooled to room temperature, then injected into a solution of ZnO in olive oil at 330 °C followed by growth and isolation by quenching in cold acetone. Notably, this reaction was not conducted under an inert atmosphere. Band edge emission was observed at *ca.* 400 nm with a clear excitonic peak observed in the electronic spectra. The particles appeared monodispersed, and could be prepared in a flower-like morphology by varying the reaction conditions. For such a simple reaction, the resulting particles were of a notably high quality, and represent a significant method of preparing ZnSe, which requires relatively high synthesis temperatures when compared to other QDs. In this case, the olive oil presents a solvent system that remains in the liquid state for the entire range of temperatures, unlike the earlier work, which required the use of tetracosane.[109]

Zinc telluride (ZnTe) has also been prepared, by a precursor reduction method,[116] where either $Zn(CO_2CH_3)_2$ or $ZnCl_2$ was dissolved in benzyl ether with either oleic acid or OAm followed by the addition at relatively low temperatures (250 °C) of TOPTe and, if the oleic acid is employed, a reducing agent such as superhydride. The use of the reducing agent was needed as the use of TOPTe alone was an insufficient source of Te^{2-}-type intermediate. If the amine was used, the additional reducing agent was not required, as the amine was itself a weak reducing agent. The difference in reducing agents resulted in differing availabilities of tellurium monomer, and hence resulted in differing particle morphologies. The use of the acid resulted in quasi-spherical 5 nm particles, while the use of the amine resulted in tetrahedrons 15–18 nm in diameter, or rods, depending on reaction conditions.

1.5 HgE (E = S, Se, Te)

Mercury chalcogenides nanomaterials are of interest because of their unusual electronic structure[117] and the associated emission in the near infrared region of the spectrum, an area of immense interest.[118] Mercury sulfide (HgS) is a semiconducting material existing in two polymorphs; the stable α-HgS (trigonal, cinnabar) and metastable β-HgS (zinc blende, meta-cinnabar). Estimations of the band gap vary between −0.1 eV and 0.05 eV, making the material either a semimetal or a narrow-gap semiconductor.[119,120] Mercury selenide (HgSe) favours the zinc blende structure and has a bandgap with similarly undetermined values, of between 420 meV and −274 meV.[121,122] Mercury telluride (HgTe) has a reported negative bandgap of −0.32 eV [123] and favours a cubic crystalline form. Strong size quantisation effects are expected in these materials due to the large excitonic radius (*e.g.*, a_B for HgTe = 40 nm). Again, differing values for the bandgap of HgTe can be found ranging from 0.15 eV [25] to −0.32 eV.[56,124] The narrow bandgap and large exciton make the materials ideal for applications in telecommunications, as the emission can easily be tuned to the infrared region.[125]

Aqueous-based methods of preparing HgTe QDs at room temperature (using H_2Te and $Hg(ClO_4)_2$ as precursors and thiols as capping agents, resulting in particles 3–6 nm in diameter) have proved extremely successful, with particles exhibiting quantum yields of up to 50% and emission profiles in the 1050 nm region.[126] These particles were found to be extremely sensitive to thermal effects and the environment, with emission shifting to *ca.* 1200 nm upon standing, and up to 1500 nm upon refluxing, with quantum yields dropping below 1%.[127] Numerous studies have been carried out using this system, including the synthesis of stable core/shell materials,[76] alloys,[128] investigations into growth conditions,[129] incorporation into emitting devices[130] and photocurrent mechanisms associated with the capping ligand.[131]

The seminal investigations into solution routes to HgTe nanomaterials using organometallic precursors were reported by Steigerwald, who used complexes such as $Hg(TeR)_2$.[132,133] Thermolysis of the majority of precursor resulted in rapid uncontrolled growth of HgTe; however, the use of a specific precursor, $Hg(Te\text{-}n\text{-}Bu)_2$ stopped the reaction proceeding to the bulk material, giving nanoscopic materials. Few details of the optical properties were recorded. The rapid growth of materials appears to be common to most mercury chalcogenide species. Murray briefly investigated organometallic routes towards mercury chalcogenide nanoparticles capped with surfactants.[17,21] The QDs were prepared using the precursors described by Steigerwald (Ph_2Hg and trioctylphosphine chalcogenides) using TOPO as a surfactant at synthesis temperatures of *ca.* 100 °C, although the particles produced were not described in any depth.

The use of metal ions stabilised by surfactants as described earlier by Peng slows the effective mass transfer of reagent ions to the nanoparticle by either hindering the reaction with a strongly coordinating ligand or by altering the diffusion rate by the use of sterically demanding ligands.[134] Higginson reported that in the case of HgE nanoparticles, the ligand coordination to the metal ion is not sufficient to control particle growth and the stability constant, K, of the ligand must be considered when preparing the materials.[134] Weakly binding ligands such as fatty acids and amines ($\log K \approx 6$–18) do not hinder particle growth and allow bulk material formation, while strongly binding ligands such as polyamines and phosphines ($\log K \approx 17$–30), phosphine oxide and thiols ($\log K \approx 17$–30) result in the reductive elimination of mercury metal at temperatures associated with nanoparticle synthesis. Although this may not be a problem during nanoparticle preparation (previous results from Steigerwald have shown that eliminated mercury often reacts with excess chalcogen), it may need to be considered during any prolonged annealing step. From these results, a reverse micelle route was used to prepare HgS nanoparticles with a β-HgS crystalline core, in a similar manner to the early work on CdSe. An aqueous solution of mercury acetate and a strongly binding ligand (thioglycerol) were added to a solution of bis(2-ethylhexyl)sulfosuccinate in hexane. It is interesting to note that the mixture of thioglycerol and $Hg(CO_2CH_3)_2$ resulted in the almost instantaneous precipitation of HgS, therefore the order of addition was specified to minimise this side reaction, with thioglycerol added last. Gradual addition of a hexane solution of $S(SiMe_3)_2$ resulted in the slow growth over several minutes of the HgS particles, clearly observable by the colour change, until the particles precipitated when they reached 5 nm in diameter. In this case, the actual size of the nanoparticles produced did not rely on the size of the micellar water pool, unlike other micelle-based routes to nanomaterials. To further protect the particles, a layer of either zinc or cadmium was grown on the surface by replacing the sulfur precursor with the relevant metal alkyl or acetate at the required stage. After surface treatment, a surfactant was then added to passivate the particle; thiophenol for cadmium-capped particles, long-chain thiols for zinc-capped particles. The absorption and emission characteristics are shown in Figure 1.3. Emission quantum yields of uncapped particles were below 1%, whereas the metal-capped particles had

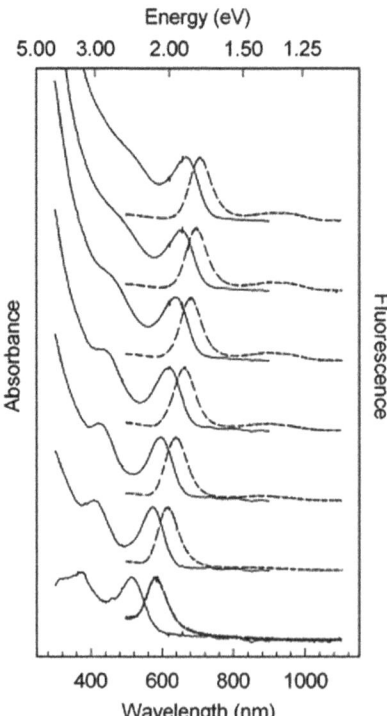

Figure 1.3   Absorption (solid line) and corresponding emission (dashed line) for HgS nanoparticles. Reprinted with permission from K. A. Higginson, M. Kuno, J. Bonevich, S. B. Qadri, M. Yousuf and H. Mattoussi, *J. Phys. Chem. B*, 2002, **106**, 9982. Copyright 2002 American Chemical Society.

quantum yields in the 5–6% region. The cadmium-capped particles were stable for months, while the zinc-capped particles displayed evidence of etching over a period of days, as determined by a blue shift in the absorption spectrum.

This route was amended by replacing the surfactant solution with TOP and polyoxyethylene 4-lauryl ether.[135] This resulted in clusters up to 3 nm in diameter, capped with TOP (which may have oxidised to TOPO), the poly-ether surfactant and thiogycerol. The use of $Se(SiMe_3)_2$ instead of the anal-ogous sulfide resulted in HgSe nanoparticles being formed, while a mixture of the two chalcogen precursors resulted in an alloy of $HgSe_{1-x}S_x$. Emission from the alloy showed evidence of surface trapping states. An unusually stable cluster of HgSe was observed with a band edge at 2.08 eV; even with the addition of further precursor, further particle growth was not observed, resulting in an increased concentration of clusters in solution. An in-depth analysis on the electronic structure of the cluster was reported, with five excited states being identified. Investigations into the high-temperature stability of β-HgS and HgSe have also been reported, confirming the crystal core of both materials remains stable up to 600 K, despite an expansion of the lattice.[136]

The first organometallic-type route to passivated HgTe utilised the ther-molysis of mercury halides and TOPTe in TOPO and amines at low temper-atures (*ca.* 100 °C) resulting in nanoparticles of crystalline HgTe, 3–6 nm in diameter with a near infrared band edge, although the growth was found to be extremely rapid and the use of temperatures normally associated with nanoparticle synthesis resulted in bulk material.[137] This route has been used to prepare materials incorporated in effective photodetectors that span the atmospheric mid-wavelength infrared transparency window.[138] The synthesis was then improved by reducing the reaction temperature significantly, and injecting TOPTe while the reaction flask was being cooled with dry ice after the initial heating required for the formation of the Hg precursor.[139] The particles of HgTe, *ca.* 3.5 nm in diameter, exhibited an excitonic peak in the absorption spectrum at *ca.* 1150 nm, with near band emission observed at *ca.* 1200 nm with quantum yields of 55–60%. The sample, upon ageing in toluene solution for 2 weeks, shifted further into the red but still maintained a quantum yield of 26%. A similar reaction was found to proceed at room temperature, using mercury oleate and an ODE solution of TOPTe as precursors, and dodecanethiol as a capping agent, giving HgTe particles *ca.* 2.3 nm in diameter with emission at *ca.* 830 nm and quantum yields of 20–30%.[140] This route was used to prepare HgTe suitable for incorporation into photodetectors.[141] Keuleyan developed the route in more depth, using $HgCl_2$, TOPTe and a long-chain amine at a range of temperatures to yield a range of triangular HgTe particles that emitted up at to 5 μm.[142] The method was again extended to the preparation of HgTe particles that were successfully used in the preparation of a mid-infrared photodetector.[143,144]

The room-temperature synthesis was utilised in the preparation of HgSe QDs, which were prepared by injecting TOPSe into an ethanol solution of

TOPO and Hg(O$_2$CCH$_3$)$_2$.[145] The shape and size of the particles could be controlled by varying the precursor : surfactant ratio. This report also suggested why mercury chalcogenides could be prepared at room temperature as opposed to the higher temperatures required for the analogous Pb, Cd and Zn chalcogenides. It was suggested that the positive redox potential for Hg$_2^{2+}$ made reduction of the acetate precursor favourable, allowing the reaction to proceed without heating. This also explained why Pb chalcogenides require low reaction temperatures and why Zn chalcogenides required high temperatures. Alloyed particles of HgSe$_x$S$_{1-x}$ with a cubic crystalline core have also been prepared by a simple one-pot reaction: CH$_3$Si–SeS–SiCH$_3$ was added to a frozen solution of TBP/Hg(O$_2$CCH$_3$)$_2$, whereupon a precipitate formed. The precipitate decomposed upon warming to room temperature, resulting in the alloyed particle formation, although few further details were supplied.[146]

Not all mercury chalcogenides can be prepared at room temperature, highlighting that the reaction is not necessarily driven by the reactivity of the cation precursor alone.[147] The reaction of Hg(O$_2$CCH$_3$)$_2$ and TOPS did not proceed at room temperature, due to the inert character of the chalcogen precursor. The TOPS was preheated to 150 °C then injected, while hot, into the TOPO/Hg^{2+} at room temperature, which was then heated at 120 °C for 15 minutes, forming HgS particles. The preheating was found to be essential, while the typical hot injection into a metal precursor solution resulted in reaction failure. The resulting cubic β-HgS particles were *ca.* 3.9 nm in diameter, exhibited an excitonic absorption feature at *ca.* 910 nm and broad, weak photoluminescence with a maximum at 1050 nm. The synthesis follows a similar route to that of Xu *et al.*, who prepared HgS nanomaterials using ODA as a capping agent, although no emission data was reported.[148]

1.6 Anisotropic Quantum Dots

The synthesis of QDs of differing shapes[149,150] is a key element in the control of the electronic structure of the materials, and numerous applications utilise non-spherical particles.[151] The focusing of a particle size distribution has been discussed, and the concentration of monomers has been identified as a key factor; a low monomer concentration results in Ostwald ripening and a defocusing of the size distribution, whereas an increased concentration results in the smaller particles growing faster, producing a narrower size range. CdSe of the wurtzite polytype is an inherently anisotropic material and was the focus of the seminal work into the synthesis of anisotropic QD structures, using dimethylcadmium and trialkylphosphine selenide as precursors. Slow growth in TOPO resulted in spherical particles, whereas a fast growth rate resulted in increased growth along the unique *c*-axis. A massive increase in the monomer concentration (kinetic overdrive) also resulted in the formation of elongated nanorods along the *c*-axis, due to the differing growth rates of differing crystal faces.[152]

Alivisatos reported that pure TOPO was unsuitable as a surfactant for the reproducible synthesis of anisotropic semiconductor nanoparticles, with growth proceeding too quickly at high monomer concentrations yielding insoluble materials. Technical-grade TOPO was found to be more suitable, because of the large amount of impurities that strongly coordinate to the monomer ions and retard the growth. Impurities present in technical-grade TOPO include HPA, which was therefore used as an additive to pure TOPO to controllably regulate the growth of nanorods. Rapid injection of high concentrations of precursor (Me_2Cd and TBPSe) at 360 °C into small amounts of TOPO with a small proportion (up to 3 mol%) of HPA resulted in spherical particles. Increasing the HPA content to between 5% and 20% reproducibly resulted in the preparation of rods; particles with the highest aspect ratio were obtained from a TOPO solution with 20% HPA. By controlling the HPA : TOPO ratio, the growth rates of differing crystal faces were controlled. Electron microscopy investigations revealed that the particles grew quickly along the long axis after precursor injection, consistent with the high monomer concentration. As the monomer concentration decreased, the short axis grew. However, by reinjecting further precursor, the monomer concentration could be kept high enough, resulting in larger quantum rods. The rods exhibited emission quantum yields of *ca.* 1%, which was increased up to 5% by the addition of a wide-bandgap shell, to be discussed later. Various shapes were obtained by controlling the of HPA : TOPO ratio, injection rates and number of injections, yielding rods with aspects ratio of up to 30, and particles exhibiting various morphologies such as arrows, teardrops and tetrapods.[153,154] By using a single injection with fixed conditions (injection volumes 1–2 mL, injection time <0.2 seconds) the HPA : TOPO ratio was varied to control rod morphology and aspect ratios. Rods of between 5 and 21 nm in length could be obtained, with aspect ratios of up to 5 when using the optimum amount of HPA as describe above. When using 60% HPA, particles with an arrowhead morphology were obtained.

By using the optimum growth solution, injection time of 0.2 seconds and an injection volume of 2 mL followed by slow addition of precursor to maintain growth, rods with an almost perfect wurtzite structure of more than 100 nm in length have been produced, with aspect ratios of up to 30. The growth of the crystals was monitored by TEM and XRD, where the crystal face along the long axis was distinguished by a sharp reflection in the diffraction pattern. The small amount of zinc blende structure could be observed in XRD analysis and, notably, by high-resolution transmission electron microscopy (HRTEM) as 'kinks' in the long rods where the zinc blende structure dominates.

From these results, three basic effects were noted: slow growth rates favour spherical particles; rapid growth rates resulted in anisotropic materials with growth along a specific face; and HPA emphasised growth of particle crystal facets. In CdSe, HPA was observed to increase the growth rate of the $(00\bar{1})$ face relative to the others. The range of shapes observed using high concentrations of HPA was attributed to the isolation of particles in different stages of growth. The particle morphology was tracked from rods, to pencils, to 'pine trees'. The teardrop-shaped particle was noted as a particularly good example of shape control; the shape developed as a particle grew at low monomer concentrations and slow injection volumes, followed by rapid injection of monomer; the particles favoured a spherical morphology, developing a tail when the monomer concentration was increased.

The growth of anisotropic particles has also been reported by gradual addition of the chalcogen precursor, not by increasing the concentration of precursors or by altering the reaction chemistry.[155] Another notable low-temperature route to anisotropic CdSe structures involved the thermolysis of a $CdCl_2$–octylamine complex and *in situ* generated $[CH_3(CH_2)_7NH_3]^+[CH_3(CH_2)_7NHC(=O)Se]^-$ at 70 °C.[156] The resulting wurtzite-structured CdSe nanoribbons were isolated by precipitation using ethanol (containing TOP), and were found to be 1.4 nm thick, 10–20 nm wide and micrometres in length. The nanoribbons had extremely clear excitonic transitions and band edge emission and a surprisingly high quantum yield of 1–2%.

Other studies have provided low-temperature (160 °C) routes to CdSe nanorods using cadmium naphthenate, TOPSe and HDA.[157] Differing structures have also been observed; intermittent addition of high concentrations of precursor ($CdCl_2$ and TBPSe) to a hot TOPO solution resulted in the formation of pyramid-type structures estimated at *ca.* 60 nm high, each side approximately 70 nm long.[158] It is also worth noting that almost all anisotropic CdSe structures have emission quantum yields significantly lower than spherical CdSe particles. This is attributed to the increased surface area resulting in more surface states which act as charge trapping sites, and the increased delocalisation (and hence reduced overlap) of charge carrier wave functions, resulting in a lower probability of recombination.

Other anisotropic cadmium chalcogenide particles have also been prepared. For example, maze-like structures have been synthesised using similar precursors, taking advantage of the inherent polytypism and long-range order in CdSe.[159] Hyperbranched particles of CdSe and CdTe have also been prepared by varying precursor concentration and type.[160] Other chemical routes have also been explored; cadmium diethyldithiocarbamate was used as a single-source precursor to CdS rods and bi-, tri- and tetrapods, when thermolysed in HDA, giving particles with zinc blende cores and wurtzite arms.[161] Examination of the facets suggested a similar growth mechanism to that of CdSe rods and tetrapods. Similarly, cadmium thiosemicarbazide gave nanorods when thermolysed in TOPO[162] (single-source precursors are discussed in a later chapter).

CdS rods, bipods and tripods were prepared by increasing the amount of sulfur precursor in the green chemical routes to sulfides described previously, utilising $CdCl_2$ and sulfur as precursors in OAm. Similarly, sulfur and $Cd(CO_2CH_3)_2$, when dissolved in OA and added sequentially to HDA at 140 °C, followed by further additions over approximately 3 hours, resulted in particles of varying morphologies.[163,164] Notably, the morphology and crystal phase of the CdS particles produced could be controlled between cubic and hexagonal, by tuning reagent concentrations, and also temperature. The anisotropy observed appeared to be driven by a small amount of hexagonal structure in the cubic particles. Wurtzite-structured CdSe rods have also been grown by redox-mediated Ostwald ripening.[165] In this case, spherical CdSe particles with a water-soluble surface amine ligand, when annealed in water at low temperature, grew asymmetrically at the expense of the smaller particles. The surface selenium oxidised and was then reduced in solution by the amine, thus generating Se^0. When this occurred in the presence of extra cadmium precursor, growth along the c-axis occurred. A similar phenomenon has been observed by the inclusion of oxygen in the growth of CdSe due to oxygen passivation of specific crystal facets.[166]

Cadmium telluride rods with a wurtzite crystal core were prepared by the thermolysis of $Cd(CO_2(CH_2)_{16}CH_3)_2$ and TOPTe in TOPO, although the concentration used was not particularly high (tens to hundreds of milligrams of precursor in 4 g TOPO).[167] Zinc selenide rods and polypods could also be prepared by varying the concentration of precursor solution (Et_2Zn and TOPSe), precursor delivery rate and temperature of a method described earlier.[108] This route is unusual when compared to the analogous CdSe system; in the case of ZnSe, a smaller amount of precursor was added to a larger amount of solvent, resulting in a dilute solution. A slower injection rate also promoted the growth of rods with a high aspect ratio of up to 8. Injection of precursor at a high injection rate and a lower temperature resulted in polypods, as did using the same conditions for rod growth except for a larger volume of precursor. Anisotropic ZnSe particles displayed diffraction patterns consistent with wurtzite crystals (as opposed to the zinc blende spherical particles), with growth along the c-axis, although HRTEM studies confirmed the existence of both zinc blende and wurtzite phases in all anisotropic ZnSe structures. Zinc selenide rods could also be grown using zinc oxide or carboxylates as previously described.[109]

1.6.1 Diffusion Model for Anisotropic Growth

Further work revealed a total of four growth stages, which could not be explained entirely by the directed binding of phosphonic acids described above, and a diffusion mechanism was suggested.[168] The first growth phase was described as the 1D growth stage, where the monomer concentration was 1.4–2% and resulted in rods growing along the long axis. The second phase, previously unobserved in the initial studies, occurred when the monomer concentration had dropped to 0.5–1.4% and was termed the 3D growth stage. In this region, the particles grew in all three dimensions, maintaining the aspect ratio from the first phase but increasing the overall volume. The third stage (1D/2D ripening stage) occurred when the monomer concentration stabilised at 0.5%, resulting in the nanoparticles increasing along the short axis and decreasing along the long axis. The volume and number of nanoparticles appeared constant, suggesting that the monomer actually shifted position (diffused) on the particle from the long axis to the short. Normal Ostwald ripening was also observed at monomer concentrations of 0.1–0.2%. Peng has also reported an in-depth analysis of the kinetic (diffusion) *versus* the thermodynamic models in CdSe rod growth and proposed a model that explained the growth of faceted crystals in terms of potential.[152,169]

In the CdSe model, at high monomer concentrations, the crystal grows in the wurtzite structure on the $(00\bar{1})$ facet due to the presence of negatively charged Se ions, uncoated by ligands because of the Lewis base character of the surfactants. When cadmium atoms are added to this layer, they posses three dangling bonds. This, combined with the dipole moment along the *c*-axis,[170] gives the (001), and especially the $(00\bar{1})$ facet a large chemical potential, making them the most active. The facets perpendicular to the *c*-axis (100) have only one dangling bond.[171] During the 1D growth stage, a limited amount of monomers diffuse into the diffusion sphere that surrounds the particle, and growth along both the $(00\bar{1})$ and (001) facets is rapid due to the increased chemical potential, making the *c*-axis the long axis. At this point, the diffusion flux is only enough for growth along the long axis of the particle. As monomer concentration drops, the overall chemical potential of the monomers drops, which affects the growth of the $(00\bar{1})$ and (001) facets. As a result, growth is consistent in all dimensions (3D growth stage). In the third 1D/2D ripening stage, the chemical potential in solution equals that of the surface atoms, which leads to an equilibrium. With the low monomer concentration and hence no more reagents for growth, the monomers start to diffuse from the $(00\bar{1})$ and (001) facets to the others, to minimise the overall surface energy of the particle.

The role of the 'strong ligand'; either HPA or *n*-tetradecylphosphonic acid (TDPA) was found to control the growth rate, not the particle shape directly, with growth in TOPO/HPA being the fastest. The presence of the strong ligand was essential, however, to keep the monomer concentration high enough to invoke rod growth.

Peng further investigated the growth of anisotropic CdSe by using cadmium oxide as a precursor instead of dimethylcadmium.[172] It was noted that structures with a slightly higher than normal aspect ratio (*ca.* 6.5 compared to 3) could be obtained if an aged solution of cadmium oxide–phosphonic acid complex was used. It was found that the cadmium–tetradecylphosphonic acid complex was the most stable (Cd–TDPA) and slowed the nucleation rate down relative to similar experiments with HPA complex of cadmium. During synthesis using the stable Cd–TDPA complex as a monomer, two sharp features in the absorption spectrum were observed. The first, at *ca.* 285 nm, was assigned as a molecular species. The second feature, at 349 nm, was assigned as a CdSe cluster, a thermodynamically stable, closed shell molecular clusters up to 2 nm in diameter[173,174] consisting of 17 Cd atoms, undetectable by TEM. These features in the absorption spectrum were used as a measure of monomer concentration. By monitoring the absorption spectra, the coexistence of clusters and small dots was observed, with an excitonic peak becoming evident at *ca.* 540 nm, consistent with particles with a 3 nm diameter. This explained the minimum small axis of the rods at 3 nm. The cadmium : selenium ratio was found to be a critical variable for the preparation of rods; the higher the Cd : Se ratio, the larger the aspect ratio of rods produced. The rate of the nucleation with differing ratios of monomer could be monitored by the evolution of the 349 nm feature, with the higher ratios of precursors taking longer to form the stable cluster, indicating that the Se–phosphine complex is more reactive that the cadmium monomer. These clusters have also been shown to be important in the synthesis of anisotropic CdSe particles.[175]

A further report confirmed the growth of magic clusters with similar optical properties,[176] and larger blue-emitting clusters of other materials have also been reported.[177,178] CdSe magic clusters have also been prepared with stable absorption profiles at 395, 463 and 513 nm, all of which exhibited band edge emission.[179,180] The organometallic-type growth of nanoparticles through various stable configurations, termed 'discontinuous growth', has been observed for a range of materials, including CdS, CdHgS, ZnSe and ZnO.[181]

The growth of teardrop-terminated rods has also been observed, similar to those described by Manna.[153] This structure is unusual as nanoparticles of this shape were suggested to be grown by the sudden increase in monomer concentration in a system already forming a spherical dot, as opposed to the consistently high monomer concentrations reported. The growth of 'close to perfect' rods was achieved by the use of high temperatures and the octade-cylphosphonic acid complex of cadmium, due to the fast diffusion and low reactivity of long-chain monomers. It is harder for long rods to undergo diffusion to the growing facets, hence the requirement for the higher growth temperature. The growth of rice-shaped rods was also observed, *via* the 3D growth phase at temperatures between 300 and 350 °C using the TDPA complex of cadmium. Multiple precursor injections resulted in a more defined shape, although in the 1D–2D ripening phase the rice-shaped particles grew into spherical dots. Multiple injections have also been used to produce nanorods using cadmium acetate as a precursor.[182,183]

During the formation of rods, branched (multipod) structures were always formed. The population of branched particles could be increased by controlling the monomer concentration and by reducing the growth temperature. The prevalent factor was found to be the monomer concentration; by increasing the precursor concentration, the particles grew through short rods, to larger rods, to branched particles, feeding the growth of the structure through the diffusion flux.

Notably, rods were also grown in a non-coordinating growth solution (ODE) with TDPA alone, demonstrating that two ligand systems with differing binding strength on various crystal facets is not necessarily required for anisotropic growth. Also, nuclear magnetic resonance (NMR) experiments revealed that all rods were capped with predominantly phosphonic acids, irrespective of aspect ratio or growth conditions.

The new diffusion model for rod growth suggested that a high monomer concentration is the prevalent factor when attempting to grow anisotropic structures, and that chemical potential, diffusion of monomers and strongly binding monomer ligands (to maintain the high monomer concentration) all need to be taken into account when assessing the complicated mode of growth. It is also noteworthy that the classical Gibbs–Thompson model of crystal growth does not fit with the observed growth phases of CdSe nanoparticles.

Interestingly, the growth of anisotropic CdSe particles has also been achieved simply by changing the chain length on the chalcogen delivery system, without the need for high concentrations, temperatures or phosphonic acids.[184] The use of TBP instead of TOP (along with 2-octenoic acid) increased the diffusion coefficient and hence resulted in CdSe nanorod formation.

1.6.2   Tetrapods

Tetrapods of CdSe, present in most experiments that formed rods, were explained as wurtzite arms growing from the four equal (111) faces out of a zinc blende core crystal, with HPA increasing the growth of the (111) facets in the same manner as a $(00\bar{1})$ facet in a wurtzite-structured particle (Figure 1.4). Notably, the arms were terminated with the $(000\bar{1})$ facet, atomically identical to the zinc blende (111) facet. If the arms were pure wurtzite, they continued to grow upon further addition of precursor; if they possessed stacking faults or had zinc blende layers, up to three additional arms could be grown at the end of the initial arm, forming dendritic structures. The presence of magic clusters has also been used to explain the growth of branched particles, as the small clusters exist in a zinc blende-type geometry.[153]

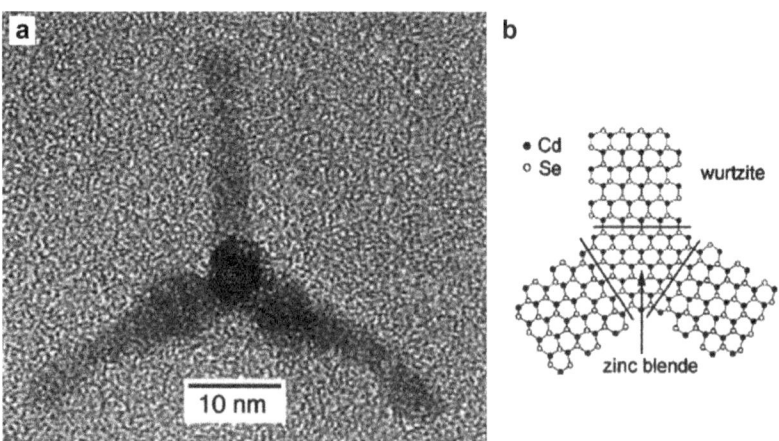

Figure 1.4   a: high-resolution TEM image of CdSe tetrapod looking down the (100) arm. b: two-dimensional cartoon of tetrapod structure. Reprinted with permission from L. Manna, E. C. Scher and A. P. Alivisatos, *J. Am. Chem. Soc.*, 2000, **122**, 12700. Copyright 2000 American Chemical Society.

During attempts to prepare Mn- and Fe-doped CdSe, CdSe tetrapods have serendipitously been produced in large amounts due to the generation of acidic protons by addition of the metal salts and acids to the cadmium precursor prior to TOPSe injection.[185] The *in situ* generated H^+ ions directed the growth of a zinc blende crystalline core, which resulted in wurtzite arms growing on the (111) facets. CdSe tetrapods have also been prepared using quaternary ammonium salts, which were introduced into the reaction with TOPSe in a typical reaction. Of the particles generated by this route, 90% were tetrapods with numerous rods, bipods and tripods being observed alongside.[186]

A key factor identified in the preparation of tetrapods is the energy difference between the two polytypes,[187] which dictates the reaction temperature, as ideally one polytype should form during nucleation and the other during growth. In cases where the energy is more than 10 meV per atom, it is difficult to alternate between the differing structures. In the case of materials such as CdS, CdSe and ZnS the energy gap is significantly smaller, making it difficult to grow one specific phase. CdTe is an intermediate material where the preparation of zinc blende particles is possible at temperatures normally associated with the wurtzite structure. Phosphonic acids used during CdTe growth were found to bind specifically to non-polar faces in wurtzite structures (Figure 1.5),[187] (which have no equivalent in the cubic material) reducing the growth of these facets, facilitating the growth of a cubic core and wurtzite arms resulting in tetrapods. By specifically controlling the kinetics, the ratio of Cd : Te monomers and amount of *n*-octadecylphosphonic acid, high yields (*ca.* 70%) of CdTe tetrapods with arms of almost equal length were obtained, although bipods and tripods were also found. Once the monomer concentration had dropped, the facet with the highest interfacial energy started to dissolve, and form more stable facets. As this facet is the one that grows fastest in solution, this resulted in the round ends on the arms observed in some samples. Tripods were found to be extremely soluble relative to simple rods of an equal length, reinforcing the comparison to dendrimers and polymers. The optics of CdTe tetrapods have been investigated and the properties found to be highly dependant on the particle size and structure.[188] Later work also suggested that an impurity in *n*-octadecylphosphonic acid (the shorter-chained methylphosphonic acid) was responsible for forming the multiple wurtzite twin boundaries which is responsible for driving the branching.[189] Tetrapods have also been prepared at a relatively low temperature of only 180 °C using a non-coordinating solvent and oleic acid as a capping agent, a weaker binding ligand than the phosphonic acids. In this case, the rapid release of monomers from the cadmium oleate complex was suggested to be sufficient to allow diffusion-limited growth.[190] A notable calculation from Thomas and O'Brien has suggested that the formation of branched particles may be in response to minimising the percentage of surface atoms and therefore lowering the internal energy or maximising the lattice energy. This also explains why systems without templates can grow into multipod structures.[191]

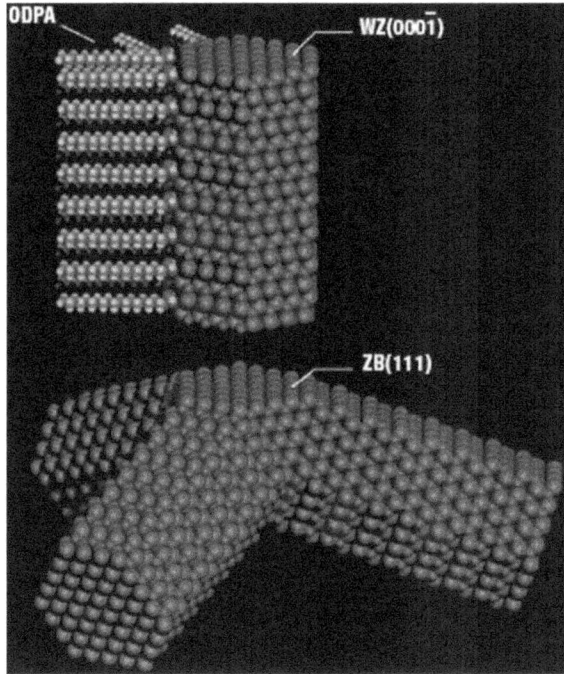

Figure 1.5   Cartoon showing model of a CdTe tetrapod, with the differing crystalline regions. Reprinted by permission from Macmillan Publishers Ltd: *Nat. Mater.*, L. Manna, D. J. Milliron, A. Meisel, E. C. Scher and A. P. Alivisatos, *Nat. Mater.*, 2003, **2**, 382. Copyright 2003.

The selective growth of metal tips on to rods and tetrapods is a notable goal due to the potential to 'wire' nanoparticles into nanoelectronic circuitry. The highly reactive terminal facets of the rods and tetrapods described make them ideal for such structures. This has been achieved by the addition of gold chloride and dodecyldimethyl ammonium bromide (DDAB) to a toluene solution of anisotropic dots.[192] During the reaction, the rods react with the DDAB and become smaller due to a dissolution process. The tipped materials resembled dumbbells and a study of the Au–CdSe interface inferred the presence of covalent bonds between the two phases, essential for efficient electrical connectivity. The electronic structure of CdSe/gold nanorod dumbbells has been investigated.[193] Although there was evidence of growth on other parts of the rods, this could be controlled by altering the amount of gold precursors. The dumbbells could be self-assembled by adding thiolated biotin to the gold terminals, which were then linked together by avidin, a protein which can accommodate up to four biotin units, yielding flower-like and end-to-end assemblies.[194] Gold has also been grown on CdSe/CdS seed/rods, where the gold 'island' was found to grow at the position of the CdSe seed.[195]

Tips could also be grown on the arms of tetrapods and on CdSe/ZnS rods. Coupling between gold tips and CdSe/ZnS rods was evidenced in the absorption spectra where the fine structure was lost and increased absorption in the visible region was observed. Quenching of the emission was also observed, attributable to the electron transfer to the metal.[192] Conductive atomic force microscopy (AFM) experiments confirmed the metal tips were conducting, while the rods remained semiconducting. Linking the rods was achieved by adding a solution of dithiols, which selectively bound the ends of the rods.

Similarly, asymmetric tetrapods have been prepared by depositing a layer of CdTe tetrapods on a substrate in a tetrahedral configuration, followed by partial coating with a polymer. The exposed arms were passivated with dithiols, and then exposed to gold nanoparticles, which coordinated to the thiol linker. As a result, multiple nanoparticles coordinated to the arm. The polymer was dissolved, leaving the structures exposed on the substrate. Removing the tetrapods was problematic, as the deposition process results in the flattening of the three tetrahedral legs. By mechanical interactions, the gold-capped arms could be removed, leaving CdTe rods capped with gold particles.[196] Hyperbranched CdTe particles have also been capped with gold on the tips, using preformed CdTe nanostructures, DDAB and HAuCl$_4$.[197]

Other semiconducting phases have also been grown on the tips of nanorods; lead selenide (PbSe) tips have been grown on both ends of CdSe and CdS rods by addition of the relevant precursors (to be discussed later) to a diphenylether/oleic acid solution of rods at 130 °C.[198,199] Higher temperatures of synthesis resulted in the unselective growth of PbSe on the lateral faces of the crystal. A lower concentration of precursor delivered by a fast injection resulted in the formation of tips on just one end of the rods (65% of the sample—the remaining structures were asymmetric dumbbells, with one tips significantly larger than the other). Growth of PbSe tips on CdS rods required a higher growth temperature and faster precursor injection, although growth was always uniquely on the ends of the rods. Due to the higher temperature of growth, separate PbSe crystals always formed. Unlike the gold tips on CdSe rods described above which had no preferred growth with respect to the wurtzite structure of the rod, the (002) planes of the PbSe rock salt structure were aligned with the (100) planes of either CdSe or CdS.

The preparation of tipped rods was extended to the preparation of highly organised, designed heterostructures with specific topologies. By controlling the chemistry, distinct structures were prepared, often in a single-pot reaction.[200] In a typical reaction, CdS rods were prepared, and tipped with CdSe (extensions) on both ends. This was achieved by adding the anion precursor to the reaction mixture at the required point, controlling the precise length and composition of the rods. Addition of a tellurium precursor resulted in the formation of a branching point *via* a CdTe tripod on the terminal CdSe extension. Attempts to make the same structures in reverse order resulted in CdTe tripods with a shell of CdSe. In a similar manner, CdSe tripods could have CdTe extensions added to the wurtzite arms, which could also form

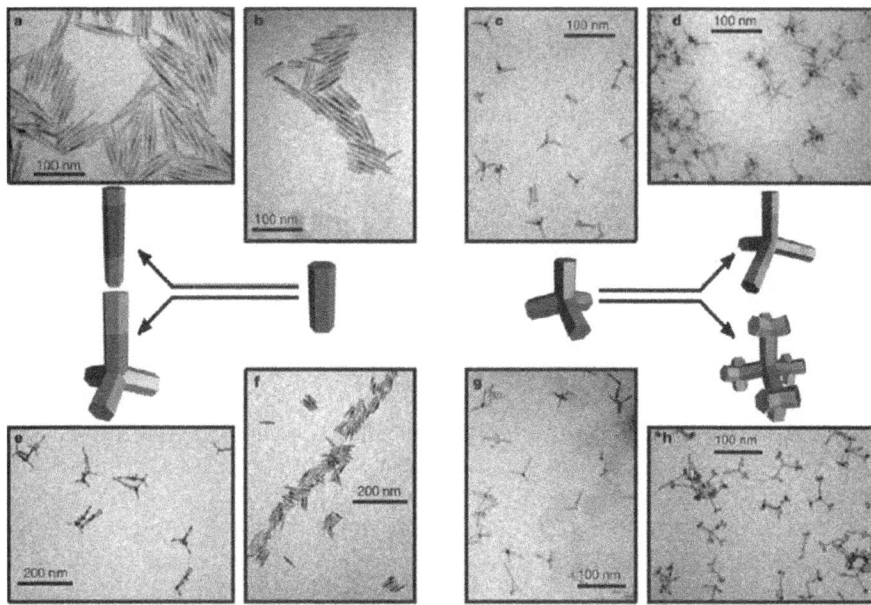

Figure 1.6   Nanocrystal heterostructures: (a) CdS nanorods; (b) CdS nanorods with
CdSe ends; (c) CdSe tetrapods; (d) CdSe tetrapods with CdTe arms; (e)
CdSe nanorods with CdTe tetrapods termination; (f) CdSe nanorods;
(g) CdSe tetrapods; (h) CdTe tetrapods on the ends of the arms of
CdSe tetrapods. Reprinted by permission from Macmillan Publishers
Ltd: *Nature*, D. J. Milliron, S. M. Hughes, Y. Cui, L. Manna, J. Li, L.-W.
Wang and A. P. Alivisatos, *Nature*, 2004, **430**, 190. Copyright 2004.

tetrapods at the end of the arms (Figure 1.6). Similarly, CdTe tetrapods with
CdSe arms have been prepared using the standard chemistry, with on
average 10 arms per crystals being observed. These materials exhibited type II
behaviour and were successfully incorporated into photovoltaic devices.[201]
This high degree of control over heterostructure morphology and composition may make molecular wiring a realistic proposal, and is especially
impressive as it was achieved by a single, simple reaction. The end-to-end
assembly of various structures can also be achieved using gold nanoparticles
grown on specific parts of a nanoparticles as welding points. The gold
particles, destabilised by I_2, coalesced forming larger particles and in the
process linked the particles, including rods and tetrapods, together.[202] The
potential for the use of anisotropic materials is high; CdTe tetrapods have
been suggested to be suitable components in electronic devices.[203]

1.6.3   Seed Growth of Anisotropic Particles

Straight and branched rods of CdSe have also been grown using gold/
bismuth particles as catalysts and seeds for growth in combination with CdO,
TOPO, octanoic acid and TOPSe in a 'geminate' nucleation mechanism.[204]

The origins of this work can be found in the early vapour–liquid–solid (VLS)[205] and solution–liquid–solid (SLS)[206] work where anisotropic growth of a macroscopic wire can be catalysed by a molten metal particle. In CdSe-seeded growth, the actual concentration of precursor does not need to be high: the more dilute the reaction mixture, the straighter the rods, attributed to a decrease of selenium in the vicinity of the Au/Bi spheres. A dilute solution also increased the nanowires' diameter, highlighting the difference between the catalysed longitudinal growth and uncatalysed transverse growth. A growth temperature of 330–350 °C favoured straight rods, whereas growth temperatures starting at 280 °C favoured bend rod growth. By controlling the cadmium : selenium ratio, the diameter of the wires could be controlled; increasing the ratio increased the diameter of the wires (optimum ratio of 7 : 1), and decreasing the ratio by increasing the concentration of TOPSe resulted in branching of the structures. Branching could be avoided by using larger Au/Bi seed particles, although Bi_2Se_3 was observed in systems that utilised thicker bismuth shells. The addition of a small amount of TOP to a system usually expected to produce straight wires produced branched structures, although the mechanism of branched growth from such 'doping' is unknown. The rods also exhibited random regions of both zinc blende and wurtzite structures, as determined by TEM investigations. (Essentially the same reaction has been reported, using Me_2Cd as a precursor, which resulted in much thinner wires when compared to the same reactions using CdO.[207]) Branched structures such as tripods, V-shaped and Y-shaped crystals were formed with a zinc blende crystal core. As well as Y-shaped structures, highly ordered structures were also observed. Notably, the diameter of the branches did not vary, making the structures ideal candidates for molecular wires. In the case of multipods grown by the seeded method, most appear to have three arms, as opposed to the four arms grown by the high-concentration routes, attributable to surface energy consideration of the seed where three arms may be thermodynamically stable.

Grebinski[204] suggested four mechanisms relevant to the growth of seeded particles: uncatalysed growth (diffusion); catalytic fission (which depended on the seed particle to split into either two or three particles and rod growth to occur in different directions to promote V-shaped or Y-shaped rods—however, this is not supported by the variety of experimental observations); collision (which has already been described as an unsuitable mechanism for describing the anisotropic particles); and germinate nanowire nucleation, which has been assigned as the growth mechanism for the seeded particles described.

The germinate nucleation mechanism suggested multiple wires were grown from the surface of a single particle where sufficient precursor is present not only to supersaturate the system, but also to initiate the growth of at least one arm. It was suggested that following nucleation, two nanowires merge to form a branched structure. As the wires grow and become comparable to the size of the dot, the wires join and form a grain boundary, as long as the angle is not too extreme. The wires can then diffuse across grain boundaries to form a single zinc blende junction that connects the arms. Where no zinc blende core is present, two arms join through a high angle grain boundary, resulting in Y-shaped nanowires. It was also noted that all higher structures retained their nanoparticle seed. The number of observed blunt and sharp-tipped nanowires, where the blunt end of a rod is the end associated with the catalyst particle and the sharp end is the terminal point of growth, supports this hypothesis. Similar results, have been obtained using bismuth nanoparticles, where the bandgap size dependence of the rods was monitored.[208] An interesting extension is the preparation of homo- and hetero-branched structures, where bismuth seeds were attached to a poly(ethylenimine) (PEI) passivated ZnSe nanowire backbone and then used as seeds for further growth from the particle.[209] The density of the branching was controlled by the density of seeds along the wire, which was in turn controlled by the amount of particles added during the colloidal work-up. By expanding this method, ZnSe arms could be grown on CdSe wires, with alloy junctions of $Zn_{1-x}Cd_xSe$. The optical properties of these materials showed broad absorption spectra with two features assigned to CdSe and ZnSe. Weak emission was also observed. Similarly, ZnE (E = S, Se, Te) nanowires have also been grown using a SLS-type route using bismuth nanoparticles as seeds, TOPO or trioctylamine (TOA) as solvents and capping agents, and a range of Zn precursors.[210]

Methods for the reproducible synthesis of CdTe have also been reported using the SLS method.[211] In this work, Au/Bi particles were used as seeds, mixed with Me_2Cd and injected into a solution of TOP, TOPO, a long-chain phosphonic acid and a large excess of tributylphosphine telluride (TBPTe) relative to the cadmium, at an elevated temperature, producing either branched wires or straight wires depending on injection temperature and Cd : Te ratio, with lower injection temperatures resulting in wires that displayed no significant branching. The branched wires exhibited Y, V and merged Y structures, the zinc blende core of the wires being attributed to the central branching points. The wires had diameters of 8–10 nm, with lengths often exceeding 10 μm, and composed of a mixture of zinc blende and wurtzite phases. As the diameters of the wires are below the excitonic diameter of CdTe, a shift in the optical band edge was observed, although emission quantum yields were low. An optimisation has been reported, where straight, unbranched, exclusively wurtzite-structured CdTe wires, several micrometres long with diameter 5–11 nm were grown using cadmium octadecylphosphonate and TOPTe as precursors in TOPO.[212,213]

Similarly, multipods of CdSe have been grown using noble metal particles as seeds, although at much lower temperatures than is usual for the SLS-type growth.[214] In these cases, the rods generated had unusually high quantum yields of up to *ca.* 10%. The use of $BiCl_3$ instead of bismuth-based nanoparticles has also been reported as a simpler method of preparing anisotropic II–VI nanomaterials, where an acetone solution of $BiCl_3$ was included with the TOPSe injection into a standard solution of CdO, oleic acid and

TOPO at 250 °C, giving wires with both wurtzite and zinc blende structures, although other bismuth compounds were also found to catalyse the wire formation.[215] It was suggested that the bismuth compounds formed *in situ* bismuth nanoparticles, and bismuth particles were occasionally imaged on the tips of the resulting rods.

Seeds of CdSe with differing crystal structures have also been used to prepare heterostructures of differing types, using the inherent difference in crystal facets.[216] Wurtzite-structured CdSe, when added to a CdS reaction mixture, resulted in the formation of CdSe/CdS nanorods (with the CdSe seed shifted to one end), whereas the use of zinc blende CdSe seeds resulted in tetrapods, as might be expected. Unusually, the tetrapods exhibited remarkably high-emission quantum yields of up to 60%, while the rods exhibited quantum yields of up to 75%. These tetrapods were also shown to be susceptible to applied pressure, with notable shifts in the emission wavelength making them potentially suitable optical strain gauges.[217]

Similar work was reported by Carbone *et al.*, where CdSe particles were used as seeds for the formation of extremely monodispersed and ordered CdSe/CdS rods which self-assembled on substrates.[218] Interestingly, the rods were found not to alloy, even at the high temperatures (up to 380 °C) used during synthesis. Again, the rods were found to have unusually high quantum yields of up to 75% for the shorter rods, although longer rods exhibited a much lower quantum yield. The seeds of CdSe were found to be located at between 1/4 and 1/3 of the rod's length.

1.7 Alloys

The preparation of simple compound semiconductors described above can be extended to the synthesis of related alloy materials, which have novel electronic profiles dictated by the particle composition rather than just the quantum size effects.[219] In some cases, alloys have optical properties unrelated to the parent material: for example, when preparing an alloy of CdSe and CdTe, the effective mass of the excitons of both parent materials need to be considered. In the case of CdTe, the effective exciton mass is substantially lower than that of CdSe. Therefore, the exciton mass of an alloy should be tuneable with alloy composition. In practise, a non-linear effect was observed where the effective exciton mass of the alloy was significantly lower than that of either parent material. This manifested itself as a depression in the absorption and emission energies, hence giving materials with unexpected novel properties.

In the case of CdSeTe, a simple combination of precursors such as TOPSe, TOPTe and CdO thermolysed in TOPO and HDA resulted in the alloy material. Under the growth conditions described, tellurium was found to be more reactive towards cadmium than selenium, allowing the preparation of two differing compositional types of alloy by taking advantage of the difference in kinetics.[220] The amount of tellurium or selenium could be accurately controlled by the amount of either precursor used. By varying the amount of cadmium, either a homogenous (prepared under cadmium-limited conditions) or a graded structure (prepared under cadmium-rich conditions) could be synthesised. In cadmium-limited conditions, the composition of the particle was determined by the relative growth rates of CdSe and CdTe and not by the cadmium concentration, as cadmium was common to both species and cancelled out from the kinetic equations. Therefore the composition remained constant. In the case of the cadmium-rich reaction, a tellurium-rich core was produced during nucleation due to the faster reaction rate of tellurium as compared to selenium. As the tellurium was removed from the reagent mixture, selenium became the predominant species in the subsequent growth step. Once all reagents were consumed, growth stopped and a graded structure was produced. This might be thought of as a CdTe/CdSe core/shell material (to be discussed in Chapter 5) without a discrete boundary. Ostwald ripening, where smaller particles grow on the surface of larger ones, was not considered problematic as the particles formed before this growth mechanism became important.

Both types of alloyed particles consisted of a wurtzite crystal core 2–8 nm in diameter and were extremely fluorescent, with quantum yields of 30–60%. Due to the non-linear optics of alloyed particles, emission was significantly shifted towards the red end of the visible spectrum. This is an excellent example of the tuneability of the optical properties of nanoparticles. It should be stressed that the non-linear effect is not necessarily related to the size quantisation effects that shift the band edge in nanoparticles, and that 'optical bowing' has been previously observed in bulk alloys of CdSeTe.[221] In fact, optical bowing has also been observed in CdS_xTe_{1-x} alloyed particle prepared by similar methods, where the emission was shifted to beyond 700 nm, despite CdS emitting at *ca.* 425 nm and CdTe emitting at *ca.* 575 nm.[222]

In a similar manner, $Zn_xCd_{1-x}S$ (with x between 0.10 and 0.53) alloy nanoparticles have been prepared using cadmium and zinc oleates with sulfur in ODE, with oleic acid as a capping agent.[223] The particle diameter was controlled between 2.4 nm ($x = 0.53$) and 4 nm ($x = 0.10$), while the emission was band edge, extremely narrow (FWHM of up to 18 nm) with reported quantum yields of up to 50%. The emission could be tuned towards the blue end of the spectrum by increasing the zinc content, giving a range from 391 nm ($Zn_{0.53}Cd_{0.47}S$) to 474 nm ($Zn_{0.1}Cd_{0.9}S$). The FWHM of 14 nm ($Zn_{0.1}Cd_{0.9}S$) is one of the narrowest emission profiles for QDs reported at room temperature. All particles exhibited a wurtzite structure. The diffraction pattern displayed a shift to larger angles as zinc content increased due to a decrease in lattice parameter c, consistent with Vergard's law,[224] confirming the alloy composition. The particles were homogenous and did not possess a graded structure despite the faster growth rate of CdS relative to ZnS. The particle homogeneity was attributed to high-temperature-induced internal atom diffusion, which was also used to explain the narrow emission profile. Similar results were observed with the zinc blende analogue of the same system, obtained by using cadmium acetate or stearate as precursors and amines as capping agents.[225] Chloride or sulfate salts of cadmium could also be used, giving wurtzite-structured material; however, no noticeable emission was detected. A related non-injection route has also been described using similar precursors, reporting ZnCdS particles with a cadmium-rich core and an increasingly zinc-rich outer shell. The particles, with quantum yields of up to 23%, exhibited a cubic crystalline phase.[226]

Homogenous $Zn_xCd_{1-x}Se$ particles have been prepared by the reaction between preformed, cooled CdSe nanoparticles and $Me_2Zn/TOPSe$.[227] Conventional synthetic methodologies suggest a core/shell system would result; however, the resulting blue shift in the band edge absorption and emission of the preformed CdSe particles suggested either the particles were getting smaller, which was unlikely, or an alloy system formed. When $x = 0$ (pure CdSe), the band edge emission was found at *ca.* 625 nm. Gradual increase in the zinc content shifted the absorption and emission by *ca.* 125 nm to a band edge of *ca.* 500 nm, tuning the emission from red to blue. The formation of a core/shell system is ruled out by these observations, as a slight red shift is expected in the absorption spectrum of a compatible core/shell material. XRD confirmed the wurtzite structure and the gradual shift to larger angle with the increased zinc content, again confirming the formation of an alloy.

To investigate why the alloy formed rather than the expected core/shell material, pre-prepared core/shell QD structures were subjected to high-temperature annealing. Nanoparticles of CdSe/ZnS and CdSe/CdS were heated to 300 °C for 10 minutes, and the absorption spectra recorded, revealing a slight red shift due to Ostwald ripening. The annealing of CdSe/ZnSe at 300 °C resulted in a blue shift in the absorption spectrum consistent with alloying. CdSe/ZnSe differ only by the cations, and as cations diffuse easily through a semiconductor structure, the formation of an alloy structure is easier than that of the CdSe/ZnS structure, where anion diffusion is more difficult. Three regions of dynamics were observed: up to 270 °C there was a ripening step, with a red shift in the absorption spectrum signifying slight growth; between 270 °C (the alloying point) and 290 °C the alloying process began, signified by the blue shift in the absorption and emission spectra; and heating above 290 °C resulted in the rapid complete formation of stable alloys. The particles prepared had quantum yields of 70–85%, significantly higher than the parent CdSe particles. This was attributed to the larger particle size, the high degree of crystallinity, the hardened lattice structure and the spatial composition fluctuations. A notable property of alloyed nanoparticles prepared by organometallic chemistry is their stability when phase-transferred to water. Usually, the exchange of surfactants damages the particle surface, quenching the luminescence to some degree; hence core/shell particles are used instead. The alloy particles reported here retain their strong emissive properties, making them attractive alternatives to core/shell particles, which are difficult to prepare.

In related work, embryonic nuclei-induced routes have been used to prepare CdZnSe alloy particles, where either CdSe seeds are grown, followed by the immediate addition of Et_2Zn, or by the growth of ZnSe seeds followed by the addition of Me_2Cd.[228] The absorption and emission of alloy particles starting from CdSe seeds gradually blue-shifted over time due to the incorporation of the wide-bandgap ZnSe into the narrow-bandgap CdSe. After 7 hours growth, little change was detected, suggesting the particles had finished growing and formed a homogenous, stable alloy. When CdSe is incorporated into a ZnSe lattice, a red shift in the optical properties was observed. The red shift, although evident, was not as pronounced as when the alloys are prepared from the CdSe seeds. After *ca.* 10 hours growth, the emission started to broaden, attributable to Ostwald ripening and the gradual oxidation and desorption of the surfactants. Emission quantum yields were 45–70%, with the highest efficiencies being obtained when growth continued to the stable phase. Prolonged growth past this stage resulted in the gradual decrease in the quantum yield. Phase-transfer to water resulted in particles with quantum yields of 25–35%.

Likewise, CdSeS nanoparticles were prepared in a one-pot reaction between CdO/oleic acid and a premixed stock solution of TOPSe/TOPS in TOA.[229] The emission could be tuned between *ca.* 460 and 580 nm, with a maximum quantum efficiency of 85% being reported. In this case, selenium was reported to react faster than sulfur with cadmium, resulting, in some cases, in particles with a CdSe-rich core, which was the dominant species when considering the bandgap of the resulting particles. Investigations into the structure using X-ray photoelectron spectroscopy (XPS) confirmed a gradual graded structure that avoided problems with lattice mismatch and dislocations at the interface.[230] Further studies into the role of surfactants in the synthesis of CdSeS alloyed particles highlighted that choosing different surfactants could control the crystalline phase of the particles.[231] These particles have been capped with SiO_2 and used in cell imaging.[232] Other attempts to prepare CdSeS particles using non-coordinating solvents and precursors reported earlier[68] resulted in homogenous alloyed particles with excitonic features in the absorption spectra tuneable from *ca.* 400 nm to *ca.* 575 nm.[233,234] The emission from these particles was notable due to the presence of trap emission. Inclusion of TBP in the reaction mixture resulted in the coordination of sulfur, allowing selenium to react first, followed by the sulfur once the TOPS had been exhausted, resulting in a graded structure. Large amounts of TBP drastically affected the morphology of the particles, yielding nanorods.

Attempts to prepare quaternary alloyed $Cd_{1-x}Zn_xS_ySe_{1-y}$ particles resulted in a CdSe-rich core, with a hybrid CdS and ZnS shell[235] (in contrast to alloys of $HgSe_{1-x}S_x$,[135] where such particles were found to be sulfur rich, attributed to sulfur's increased reactivity towards mercury). ZnS shells have purposely been deposited on $Cd_{1-x}Zn_xSe$ using zinc ethylxanthate and a zinc carboxylate as shell precursors, giving a core/shell system that emitted in the blue region of the spectrum with a quantum efficiency of 67%.[236]

Other quaternary materials include ZnCdSSe, prepared by a phosphine-free synthesis.[237] The particles remained at *ca.* 6 nm diameter, but could be tuned to emit across the entire visible spectrum by altering the composition.

The band edge was tuneable from *ca.* 440 nm to *ca.* 650 nm, with near band edge emission ranging across the same region, with quantum yields of 40–60%. The precursors CdO and ZnO were dissolved in a mixture of oleic acid and 2-ethylhexanoic acid prior to injection into a hot mixture of Se–S dissolved in paraffin liquid. The dissolution of the chalcogens in paraffin liquid reportedly reduced the differences in reactivity, an essential feature of the alloyed particle formation.

Alloys are an excellent example of the tuning of the optical properties of nanoparticles. Depending upon starting parameters, nanoparticle can be prepared in which the band edge emission either red-shifts or blue-shifts from the initial expected position for the parent binary materials. Usually, in simple binary materials, gradual particle growth over time results in a shrinking of the bandgap, resulting in a red shift in the optics, and is often the only option available when growing particles. CdZnSe provides an alternative, where blue-shifted emission is observed in particle growth when starting from CdSe seeds. In the case of CdSeTe, non-linear effects resulted in emission that could not be attributed to either parent material. The preparation of alloys and the associated blue shifts and non-linear effects presents further options when bandgap-engineering small particles.

1.8 Microfluidic Synthesis

The synthesis of nanoparticles by the green chemistry described earlier in this chapter is a major advance. However, in practice, scale-up procedures result in problems with precursor delivery, homogenous mixing in large volumes and greater variations in temperature. One potential way to avoid these problems is the use of microreactors, often termed 'lab-on-a-chip' systems.[238] The small scales of these reactors allow excellent control of temperature and are usually designed as flow processes that allow accurate control of reaction time. The use of reduced spatial dimensions to carry out chemical reactions is attractive because of the potential to screen the process while using the smallest possible amount of precursor. Calculations have shown that 70 reactors in parallel with a flow rate of 0.25 mL min^{-1} can produce 10 L of product in 10 hours.[239]

Initial work on the microfluidic preparation of nanomaterials centred on the preparation of semiconductors QDs by aqueous routes[240] and the preparation of TiO$_2$.[241] The microfluidic preparation of QDs using the green chemistry described earlier was achieved by heating cadmium acetate with stearic acid at 130 °C to give the cadmium stearate precursor.[242,243] This was followed by the addition of TOPO under a nitrogen atmosphere, which was then allowed to cool below 100 °C. Addition of TOPSe provided the reaction mixture, which was loaded into a syringe. The reagents were then injected into a capillary of a known length and diameter, of which a predetermined length was immersed in an oil bath stabilised at temperatures of 230–300 °C. The length of the capillary in the oil bath and the injection rate controlled the residence time. Particles of 2–4.5 nm diameter could be prepared by the reaction in a capillary, with reaction times 7–150 seconds. The particle size–reaction time curve was similar to that of particles grown in the batch processes. The reaction was extremely reproducible, giving materials with quantum yields of *ca.* 1.5% and an emission range of 450–600 nm. The size distribution, as evidenced by the width of the exciton peak in the absorption spectrum, was narrowed using a capillary with a smaller diameter; a capillary with a diameter of 500 µm required 4 seconds to reach reaction temperature, whereas a capillary with a diameter of 200 µm required only 0.4 seconds. Introduction of a 0.5 µL nitrogen bubble into the capillary every 3 seconds, usually used to avoid a velocity (and hence residence time) distribution, was also found to narrow the particle size distribution by inhibiting mixing between segments of the precursors.

A more sophisticated version was reported using a chip-based micro-fluidic reactor, where channels were etched in a 100 mm diameter glass wafer sandwich.[244] In this case, Me_2Cd and TBPSe were mixed with DDA and TOPO in ODE at 60 °C under an inert atmosphere. After being degassed, the reagents were loaded into the sample loop of a HPLC injection valve. The precursors could (if required) be diluted before injection (100 µL plug of precursors) into the reaction channel, which was heated from below to 175–185 °C. After a reaction time of 300 seconds, the product was diluted with ODE and pumped into a capillary flow cell where, 110 seconds after exiting the reaction channels, the emission spectra were recorded using a fibre-optic charge-coupled device (CCD) camera. The nanoparticles produced were approximately 2.5 nm in diameter, although increasing the reaction temperature modestly increased particle size (from 2.44 to 2.69 nm). Decreasing the flow (increasing the residence time in the reactor) also increased the particle size by a similar amount, as did increasing the concentration of precursor. Interestingly, the particle sizes were estimated from the emission wavelengths as opposed to the absorption band edge.

There are however, problems associated with using the organometallic approach to microfluidic-based synthesis; the capping agent, TOPO, is a solid at room temperature and starts to decompose upon prolonged heating, potentially blocking the reaction channels. The cadmium precursor, Me_2Cd, is not ideal because of the hazards associated with its handling and the gas evolution upon its decomposition. Work by Krishnadasan *et al.* in which Me_2Cd was replaced by $Cd(CO_2CH_3)_2$ resulted in particles with quantum yields as high as 10%; however, samples prepared at temperatures of 220 °C displayed emission spectra consistent with surface defects.[245] A detailed investigation into reaction kinetics highlighted the preferable use of high flow rates and high temperatures to minimise the residence time distribution that ultimately led to polydispersed samples.

Another chemistry was developed to be totally compatible with the lab-on-a-chip process; cadmium oleate and TOPSe were chosen as precursors and dissolved in squalane, OAm and TOP, chosen because of their liquid state at room temperature.[246] The reactor was a continuous-flow channel with an internal diameter of 250 μm and a reaction length of 14.6 cm maintained at a temperature of 180–320 °C. The two precursors were delivered in separate channels to a mixer prior to injection into the reaction channel, avoiding the room-temperature formation of clusters that affect the reproducibility in final nanoparticle diameter. Again, particle size could be tuned to between *ca.* 2.2 nm and *ca.* 2.8 nm by careful alteration of flow rate, temperature and precursor composition. By maintaining a constant precursor concentration, it was observed that higher temperatures and residence times increased the particle diameter, although size distribution was compromised at high flow rates or low temperatures. By keeping the concentration of cadmium ions constant and by varying the amount of TOPSe, effective tuning of the particle size was achieved while maintaining an acceptable size distribution. The quantum yields of particles prepared by this procedure (28–51%) were significantly higher than the other particles prepared on a chip; this was attributed to the presence of the amine capping group. Over an 8 hour run, the absorption spectra of all samples were identical, highlighting the stability of the system. The same group also reported the use of supercritical hexane in a continuous-flow process, which overcame problems with solvents such as viscosity, solubility and diffusivity.[247] Reactions were carried out in silicon/Pyrex microreactors, which allowed high-pressure reactions. The reactors consisted of a nitrogen-purged channel 400 μm × 250 μm, with a 0.1 m long mixing zone at room temperature and a 1 m long reaction zone at 350 °C. The CdSe particles were prepared using a similar solvent chemistry as described immediately above, under a pressure of 5 MPa. The use of the supercritical solvent resulted in more nuclei forming due to a higher supersaturation of precursors, which resulted in the narrowing of the size distribution.

The preparation of CdSe/ZnS core/shell particles by a multi-step process in a microcapillary has been reported,[248] by the addition of Et_2Zn and $S(SiMe_3)_2$ in TOP *via* a ceramic micromixer to preformed CdSe particles prepared as described above. In this case, two oil baths and two different temperatures were used to prepare the particles, with the emission intensity being increased fivefold upon capping the particles with the inorganic shell. Slower flow rates resulted in a decrease in the photoluminescence, due to a thicker shell than necessary being deposited. The same group and others have also reported the use of a single-source precursor to deposit the ZnS shell on to preformed CdSe.[249,250] It is also possible to carry out ligand-exchange reactions on chip immediately after shell deposition, to provide water-soluble particles that are potentially useful in biology.[251]

An important advance in the chip-based synthesis was described by Krishnadasan *et al.*, who developed their earlier work by including an in-line spectrometer to feedback the optical properties into an algorithm, which then altered reaction conditions to obtain the required emission wavelength.[252] This intelligent optimisation of the synthetic parameter has obvious benefits in the preparation of nanomaterials and is potentially applicable to other systems. Similarly, a combinatorial approach using several microreactors and an online detector has been developed to allow the optimised synthesis of CdSe QDs, examining reaction time, temperature, and concentration of amine and yielding outcomes on emission quantum yields, particle size, diameter and product yield.[253]

One of the drawbacks of microfluidic synthesis is the use of relatively low temperature because of the polymeric substrates, therefore missing the higher range of temperatures normally associated with the synthesis of high-quality crystalline particles. To avoid this, Chan *et al.* have developed a segmented flow droplet-based glass microreactor that is stable at high temperatures, and developed a synthesis chemistry based on perfluorinated polyethers as carrier solvents, using standard reagents such as cadmium carboxylates, TOPSe and ODE as nanoparticle precursors and solvents.[254] Using this set-up, temperatures as high as 300 °C have been used and the reaction carried out in nanolitre-sized droplets of ODE in the perfluorinated polyether, yielding materials spectroscopically identical to particles prepared by the standard synthetic route. Related to this is the development of chemical aerosol flow synthesis, where precursors and solvent are nebulised into a mist, which was then carried under inert gas flow into a reaction furnace where the QDs formed in sub-micron droplets.[255]

There are numerous problems associated with the typical flask synthesis of semiconductor nanoparticles, such as side reactions, experimental conditions that inadvertently effect reaction kinetics, and less reproducible variables such as stirring rate and precursor injection rate. To overcome this, the synthesis and analysis of CdSe, CdTe and even $NaYF_4$ has been automated using combinatorial chemistry and high-throughput analytical techniques using a combinatorial rig specially designed for high-tempera-ture synthesis required for semiconductor nanomaterials.[256] The set-up included low thermal mass reactor elements that allowed controlled heat-ing, cooling and stirring under an inert atmosphere, and 96-well quartz microplates or XRD microwell plates that allowed particle analysis. Using simple green chemical routes described earlier in this chapter, a range of experimental parameters were explored, allowing the optimisation of particle growth. This degree of control resulted in extremely high repro-ducibility with a 0.2% coefficient of variation over numerous batch runs, while allowing the tuning of the synthesised particles. One notable discovery using the combinatorial approach was that the use of inhomogeneous reactants produced reproducible results; it was suggested that these insol-uble materials acted as precursor reservoirs. This advance in the repro-ducible synthesis of nanomaterials, if combined with intelligent optimisation as described above, offers great potential for the future of synthetic nanomaterial chemistry.

Here, we hope to have shown that the chemistry of II–VI nanomaterials is varied and can be tuned across a wide spectral range, using a variety of techniques, structures and synthetic methodologies.

References

1. K. Kalyanasundaram, E. Borgarello, D. Duonghong and M. Grätzel, *Angew. Chem., Int. Ed.,* 1981, **20**, 987.
2. R. Rossetti and L. Brus, *J. Phys. Chem.,* 1982, **86**, 4470.
3. A. I. Ekimov and A. A. Onushchenko, *Pis'ma Zh. Tekh. Fiz.,* 1981, **34**, 363.
4. A. I. Ekimov, A. L. Efros and A. A. Onushchenko, *Solid State Commun.,* 1985, **56**, 921.
5. A. Henglein, *Ber. Bunsen-Ges. Phys. Chem.,* 1982, **86**, 301.
6. M. Meyer, C. Wallberg, K. Kurihara and J. H. Fendler, *J. Chem. Soc., Chem. Commun.,* 1984, 90.
7. A. Fojtic, H. Weller, U. Koch and A. Henglein, *Ber. Bunsen-Ges. Phys. Chem.,* 1984, **88**, 969.
8. M. Steigerwald, *Polyhedron,* 1994, **13**, 1245.
9. H. M. Manasevit, *Appl. Phys. Lett.,* 1968, **12**, 156.
10. M. L. Steigerwald and C. R. Sprinkle, *Organometallics,* 1988, **7**, 245.
11. R. A. Zingaro, B. H. Stevens and K. Irgolic, *J. Organomet. Chem.,* 1965, **4**, 320.
12. S. M. Stucynski, J. G. Brennan and M. L. Steigerwald, *Inorg. Chem.,* 1989, **28**, 4431.
13. H. M. Manasevit and W. I. Simpson, *J. Electrochem. Soc.,* 1971, **118**, 644.
14. M. L. Steigerwald, A. P. Alivisatos, J. M. Gibson, T. D. Harris, R. Kortan, A. J. Muller, A. M. Thayer, T. M. Duncan, D. C. Douglas and L. E. Brus, *J. Am. Chem. Soc.,* 1988, **110**, 3046.
15. M. G. Bawendi, A. R. Kortan, M. L. Steigerwald and L. E. Brus, *J. Chem. Phys.,* 1989, **91**, 7282.
16. A. R. Kortan, R. Hull, R. L. Opila, M. G. Bawendi, M. L. Steigerwald, P. J. Carroll and L. E. Brus, *J. Am. Chem. Soc.,* 1990, **112**, 1327.
17. C. B. Murray, D. J. Norris and M. G. Bawendi, *J. Am. Chem. Soc.,* 1993, **115**, 8706.
18. C. D. M. Donegá, S. G. Hickey, S. F. Wuister, D. Vanmaekelbergh and A. Meijerink, *J. Phys. Chem. B,* 2003, **107**, 489.
19. S. Monticone, R. Tufeu, A. V. Kanaev, E. Scolan and C. Sanchez, *Appl. Surf. Sci.,* 2000, **162–163**, 565.
20. V. K. La Mer and R. H. Dinegar, *J. Am. Chem. Soc.,* 1950, **72**, 4847.
21. C. B. Murray, PhD thesis, MIT, 1995.
22. E. Jang, S. Jun, Y. Chung and L. Pu, *J. Phys. Chem. B,* 2004, **108**, 4597.
23. J. E. Bowen Katari, V. L. Colvin and A. P. Alivisatos, *J. Phys. Chem.,* 1994, **98**, 4109.
24. The term 'spherical' is loosely applied. The particles described have facets and are usually highly structured. This is often hard to observe with smaller particles, therefore, for the sake of simplicity, a particle with an aspect ratio of *ca.* 1 that appears approximately spherical under TEM examination will be termed spherical. The growth of anisotropic structures, TEM analysis and assignment will be discussed later.

25. X. Peng, J. Wickham and A. P. Alivisatos, *J. Am. Chem. Soc.,* 1998, **120**, 5343.
26. C. R. Bullen and P. Mulvaney, *Nano Lett.,* 2004, **4**, 2303.
27. C. D. Dushkin, S. Saita, K. Yoshie and Y. Yamaguchi, *Adv. Colloid Interface Sci.,* 2000, **88**, 37.
28. D. V. Talapin, A. L. Rogach, M. Haase and H. Weller, *J. Phys. Chem. B,* 2001, **105**, 12278.
29. C. de Mello Donegá, P. Liljeroth and D. Vanmaekelbergh, *Small,* 2005, **1**, 1152.
30. B. D. Dickerson, D. M. Irving, E. Herz, R. O. Claus, W. B. Spillman and K. E. Meisner, *Appl. Phys. Lett.,* 2005, **86**, 171915.
31. J. Park, J. Joo, S. G. Kwon, Y. Jang and T. Hyeon, *Angew. Chem., Int. Ed.,* 2007, **46**, 4630.
32. R. Xie, Z. Li and X. Peng, *J. Am. Chem. Soc.,* 2009, **131**, 15457.
33. J. van Embden and P. Mulvaney, *Langmuir,* 2005, **21**, 10226.
34. D. Tonti, M. B. Mohammed, A. Al-Salman, P. Pattison and M. Chergui, *Chem. Mater.,* 2008, **20**, 1331.
35. D. R. Lide, *Handbook of Chemistry and Physics*, CRC Press, 73rd edn, 1992.
36. F. V. Mikulec, PhD thesis, MIT, 1999.
37. D. V. Talapin, S. Haubold, A. L. Rogach, A. Kornowski, M. Haase and H. Weller, *J. Phys. Chem. B,* 2001, **105**, 2260.
38. D. V. Talapin, A. L. Rogach, I. Mekis, S. Haubold, A. Kornowski, M. Haase and H. Weller, *Colloids Surf., A,* 2002, **202**, 145.
39. S. F. Wuister, F. van Driel and A. Meijerink, *Phys. Chem. Chem. Phys.,* 2003, **5**, 1253.
40. S. F. Wuister, F. van Driel and A. Meijerink, *J. Lumin.,* 2003, **102–103**, 327.
41. P. Dagtepe, V. Chikan, J. Jasinski and V. J. Leppert, *J. Phys. Chem. C,* 2007, **111**, 14977.
42. N. Gaponik, D. V. Talapin, A. L. Rogach, K. Hoppe, E. V. Shevchenko, A. Kornowski, A. Eychmüller and H. Weller, *J. Phys. Chem. B,* 2002, **106**, 7177.
43. H. Bao, Y. Gong, Z. Li and M. Gao, *Chem. Mater.,* 2004, **16**, 3853.
44. A. L. Rogach, *Mater. Sci. Eng., B,* 2000, **69–70**, 435.
45. I. L. Garzón, C. Rovira, K. Michaelian, M. R. Beltrán, P. Ordejón, J. Junquera, D. Sánchez-Portal, E. Artacho and J. M. Soler, *Phys. Rev. Lett.,* 2000, **85**, 5250.
46. S. F. Wuister, C. de Mello Donegá and A. Meijerink, *J. Phys. Chem. B,* 2004, **108**, 17393.
47. A. C. Jones, D. J. Cole-Hamilton, A. K. Holliday and M. M. Ahmad, *J. Chem. Soc., Dalton Trans.,* 1983, 1047.
48. M. Green and P. O'Brien, *Adv. Mater. Opt. Electron.,* 1997, **7**, 277.
49. J. Hambrock, A. Birkner and R. A. Fischer, *J. Mater. Chem.,* 2001, **11**, 3197.

50. T. J. Boyle, S. D. Bunge, T. M. Alam, G. P. Holland, T. J. Headley and G. Avilucea, *Inorg. Chem.*, 2005, **44**, 1309.
51. M. Lazell and P. O'Brien, *J. Mater. Chem.*, 1999, **9**, 1381.
52. S. M. Stuczynski, Y.-U. Kwon and M. L. Steigerwald, *J. Organomet. Chem.*, 1993, **449**, 167.
53. P. Reiss, S. Carayon, J. Bleuse and A. Pron, *Synth. Met.*, 2003, **139**, 649.
54. X. Peng, *Chem.–Eur. J.*, 2002, **8**, 334.
55. Z. A. Peng and X. Peng, *J. Am. Chem. Soc.*, 2001, **123**, 183.
56. H.-S. Chen and R. V. Kumar, *Cryst. Growth Des.*, 2009, **9**, 4235.
57. H.-S. Chen and R. V. Kumar, *J. Phys. Chem. C*, 2009, **113**, 12236.
58. X. Chen, A. C. S. Samia, Y. Lou and C. Burda, *J. Am. Chem. Soc.*, 2005, **127**, 4372.
59. M. S. Wong and G. D. Stucky, *Mater. Res. Soc. Symp. Proc.*, 2001, **676**, Y2.3.1.
60. L. Qu, Z. A. Peng and X. Peng, *Nano Lett.*, 2001, **1**, 333.
61. S. Kumar and T. Nann, *J. Chem. Soc., Chem. Commun.*, 2003, 2478.
62. H. Liu, J. S. Owen and A. P. Alivisatos, *J. Am. Chem. Soc.*, 2007, **129**, 305.
63. J. S. Steckel, B. K. H. Yen, D. C. Oertel and M. G. Bawendi, *J. Am. Chem. Soc.*, 2006, **128**, 13032.
64. O. S. Oluwafemi and N. Revaprasadu, *New J. Chem.*, 2008, **10**, 1432.
65. N. Mntungwa, P. V. S. R. Rajasekhar and N. Revaprasadu, *Mater. Chem. Phys.*, 2011, **126**, 500.
66. A. Hoshino, K. Fujioka, T. Oku, M. Suga, Y. F. Sasaki, T. Ohta, M. Yasuhara, K. Suzuki and K. Yamamoto, *Nano Lett.*, 2004, **4**, 2163.
67. F. Zezza, R. Comparelli, M. Sticcoli, M. L. Curri, R. Tommasi, A. Agostiano and M. Della Monica, *Synth. Met.*, 2003, **139**, 597.
68. W. W. Yu and X. Peng, *Angew. Chem., Int. Ed.*, 2002, **41**, 2368.
69. C. Bullen, J. van Embden, J. Jasieniak, J. E. Cosgriff, R. J. Mulder, E. Rizzardo, M. Gu and C. L. Raston, *Chem. Mater.*, 2010, **22**, 4135.
70. M. B. Mohamed, D. Tonti, A. Al-Salman, A. Chemseddine and M. Chergui, *J. Phys. Chem. B*, 2005, **109**, 10533.
71. J. Jasieniak, C. Bullen, J. van Embden and P. Mulvaney, *J. Phys. Chem. B*, 2005, **109**, 20665.
72. Z. Deng, L. Cao, F. Tang and B. Zou, *J. Phys. Chem. B*, 2005, **109**, 16671.
73. R. K. Čapek, K. Lambert, D. Dorfs, P. F. Smet, D. Poelman, A. Eychmüller and Z. Hens, *Chem. Mater.*, 2009, **21**, 1743.
74. Y. A. Yang, H. Wu, K. R. Williams and Y. C. Cao, *Angew. Chem., Int. Ed.*, 2005, **44**, 6712.
75. B. Mahler, N. Lequeux and B. Dubertret, *J. Am. Chem. Soc.*, 2010, **132**, 953.
76. J. Zhang, X. Zhang and J. Y. Zhang, *J. Phys. Chem. C*, 2010, **114**, 3904.
77. B. Mahler, P. Spinicelli, S. Buil, X. Quelin, J.-P. Hermier and B. Dubertret, *Nat. Mater.*, 2008, **7**, 659.
78. Y. Chen, J. Vela, H. Htoon, J. L. Casson, D. J. Werder, D. A. Bussian, V. I. Klimov and J. A. Hollingsworth, *J. Am. Chem. Soc.*, 2008, **130**, 5026.

79. R. K. Čapek, I. Moreels, K. Lambert, D. De Muynck, Q. Zhao, A. Van Tomme, F. Vanhaecke and Z. Hens, *J. Phys. Chem. C*, 2010, **114**, 6371.

80. L. Liu, Z. Zhuang, T. Xie, Y.-G. Wang, J. Li, Q. Peng and Y. Li, *J. Am. Chem. Soc.*, 2009, **131**, 16423.

81. W. W. Yu, Y. A. Wang and X. Peng, *Chem. Mater.*, 2003, **15**, 4300.

82. J. W. Cho, H. S. Kim, Y. J. Kim, S. Y. Jang, J. Park, J.-G. Kim, Y.-J. Kim and E. H. Cha, *Chem. Mater.*, 2008, **20**, 5600.

83. V. Kloper, R. Osovsky, J. Kolny-Olesiak, A. Sashchiuk and E. Lifshitz, *J. Phys. Chem. C*, 2007, **111**, 10336.

84. R. Osovsky, V. Kloper, J. Kolny-Olesiak, A. Sashchiuk and E. Lifshitz, *J. Phys. Chem. C*, 2007, **111**, 10841.

85. J. Joo, H. Bin Na, T. Yu, Y. Woon Kim, F. Wu, J. Z. Zhang and T. Hyeon, *J. Am. Chem. Soc.*, 2003, **125**, 11100.

86. U. K. Gautam, R. Seshadri and C. N. R. Rao, *Chem. Phys. Lett.*, 2003, **375**, 560.

87. D. Pan, S. Jiang, L. An and B. Jiang, *Adv. Mater.*, 2004, **16**, 982.

88. D. Pan, Q. Wang, S. Jiang, X. Ji and L. An, *Adv. Mater.*, 2005, **17**, 176.

89. D. Pan, Q. Wang, S. Jiang, X. Ji and L. An, *J. Phys. Chem. C*, 2007, **111**, 5661.

90. Q. Wang, D. Pan, S. Jiang, X. Ji, L. An and B. Jiang, *Chem.–Eur. J.*, 2005, **11**, 3843.

91. Y. C. Cao and J. Wang, *J. Am. Chem. Soc.*, 2004, **126**, 14336.

92. S. Sun and C. B. Murray, *J. Appl. Phys.*, 1999, **85**, 4325.

93. S. Asokan, K. M. Krueger, A. Alkhawaldeh, A. R. Carreon, Z. Mu, V. L. Colvin, N. V. Mantzaris and M. S. Wong, *Nanotechnology*, 2005, **16**, 2000.

94. M. Green, P. Rahman and D. Smyth-Boyle, *Chem. Commun.*, 2007, 574.

95. P. J. Newman and D. R. MacFarlane, *Z. Phys. Chem.*, 2006, **220**, 1473.

96. J. Jasieniak and P. Mulvaney, *J. Am. Chem. Soc.*, 2007, **129**, 2841.

97. K. Biswas and C. N. R. Rao, *Chem.–Eur. J.*, 2007, **13**, 6123.

98. C.-Q. Zhu, P. Wang, X. Wang and Y. Li, *Nanoscale Res. Lett.*, 2008, **3**, 213.

99. B. Xing, W. Li, H. Dou, P. Zhang and K. Sun, *J. Phys. Chem. C*, 2008, **112**, 14318.

100. S. Sapra, A. L. Rogach and J. Feldmann, *J. Mater. Chem.*, 2006, **16**, 3391.

101. J. A. Gerbec, D. Magana, A. Washington and G. F. Strouse, *J. Am. Chem. Soc.*, 2005, **127**, 15791.

102. A. L. Washington II and G. F. Strouse, *J. Am. Chem. Soc.*, 2008, **130**, 8916.

103. A. L. Washington II and G. F. Strouse, *Chem. Mater.*, 2009, **21**, 3586.

104. Q. Song, X. Ai, T. Topuria, P. M. Rice, F. H. Alharbi, A. Bagabas, M. Bahattab, J. D. Bass, H.-C. Kim, J. C. Scott and R. D. Miller, *Chem. Commun.*, 2010, **46**, 4871.

105. M. A. Hines and P. Guyot-Sionnest, *J. Phys. Chem. B*, 1998, **102**, 3655.

106. W. K. Woo, PhD thesis, MIT, 2002.

107. C. A. Tolman, *Chem. Rev.*, 1977, **77**, 313.

108. P. D. Cozzoli, L. Manna, M. L. Curri, S. Kudera, C. Giannini, M. Striccoli and A. Agostiano, *Chem. Mater.*, 2005, **17**, 1296.

109. L. S. Li, N. Pradhan, Y. Wang and X. Peng, *Nano Lett.,* 2004, **4**, 2261.
110. P. Reiss, G. Quemard, S. Carayon, J. Bleuse, F. Chandezon and A. Pron, *Mater. Chem. Phys.,* 2004, **84**, 10.
111. H.-S. Chen, B. Lo, J.-Y. Hwang, G.-Y. Chang, C.-M. Chen, S.-J. Tasi and S.-J. J. Wang, *J. Phys. Chem. B,* 2004, **108**, 17119.
112. C.-S. Yang and K.-L. Ku, *J. Chin. Chem. Soc.,* 2004, **51**, 65.
113. A. B. Panda, S. Acharya, S. Efrima and Y. Golan, *Langmuir,* 2007, **23**, 765.
114. J. H. Yu, J. Joo, H. M. Park, S.-I. Baik, Y. W. Kim, S. C. Kim and T. Hyeon, *J. Am. Chem. Soc.,* 2005, **127**, 5662.
115. Q. Dai, N. Xiao, J. Ning, C. Li, D. Li, B. Zou, W. W. Yu, S. Kan, H. Chen, B. Liu and G. Zou, *J. Phys. Chem. C,* 2008, **112**, 7567.
116. J. Zhang, K. Sun, A. Kumbhar and J. Fang, *J. Phys. Chem. C,* 2008, **112**, 5454.
117. X. W. Zhang and J. B. Xia, *J. Phys. D: Appl. Phys.,* 2006, **39**, 1815.
118. A. L. Rogach, A. Eychmüller, S. G. Hickey and S. V. Kershaw, *Small,* 2007, **3**, 536.
119. M. Green, P. Prince, M. Gardener and J. Steed, *Adv. Mater.,* 2004, **16**, 994.
120. A. Delin, *Phys. Rev. B: Condens. Matter Mater. Phys.,* 2002, **65**, 153205.
121. K. U. Gawlik, L. Kipp, M. Skibowski, N. Orlowski and R. Manske, *Phys. Rev. Lett.,* 1997, **78**, 3165.
122. M. von Truchseß, A. Pfeuffer-Jeschke, C. R. Becker, G. Landwehr and E. Batke, *Phys. Rev. B: Condens. Matter Mater. Phys.,* 2000, **61**, 1666.
123. N. Orlowski, J. Augustin, Z. Golacki, C. Janowitz and R. Manzke, *Phys. Rev. B: Condens. Matter Mater. Phys.,* 2000, **61**, R5058.
124. Landolt-Börnstein, *Numerical Data and Functional Relationships in Science and Technology: New Series*, Springer, Berlin, 1982, vol. 17b: *Semiconductors*.
125. M. T. Harrison, S. V. Kershaw, M. G. Burt, A. L. Rogach, A. Kornowski, A. Eychmüller and H. Weller, *Pure Appl. Chem.,* 2000, **72**, 295.
126. A. Rogach, S. Kershaw, M. Burt, M. Harrison, A. Kornowski, A. Eychmüller and H. Weller, *Adv. Mater.,* 1999, **11**, 552.
127. M. T. Harrison, S. V. Kershaw, M. G. Burt, A. L. Rogach, A. Kornowski, A. Eychmüller and H. Weller, *Adv. Mater.,* 2000, **12**, 123.
128. M. T. Harrison, S. V. Kershaw, M. G. Burt, A. Eychmüller, H. Weller and A. L. Rogach, *Mater. Sci. Eng., B,* 2000, **69–70**, 355.
129. M. T. Harrison, S. V. Kershaw, M. G. Burt, A. L. Rogach, A. Eychmüller and H. Weller, *J. Mater. Chem.,* 1999, **9**, 2721.
130. É. O'Connor, A. O'Riodan, H. Doyle, S. Moynihan, A. Cuddihy and G. Redmond, *Appl. Phys. Lett.,* 2005, **86**, 201114.
131. H. Kim, K. Cho, H. Song, B. Min, J.-S. Lee, G.-T. Kim and S. Kim, *Appl. Phys. Lett.,* 2003, **83**, 4619.
132. M. L. Steigerwald and C. R. Sprinkle, *J. Am. Chem. Soc.,* 1987, **109**, 7200.
133. J. G. Brennan, T. Siegrist, P. J. Carroll, S. M. Stuczynski, P. Reynders, L. E. Brus and M. L. Steigerwald, *Chem. Mater.,* 1990, **2**, 403.
134. K. A. Higginson, M. Kuno, J. Bonevich, S. B. Qadri, M. Yousuf and H. Mattoussi, *J. Phys. Chem. B,* 2002, **106**, 9982.

135. M. Kuno, K. A. Higginson, S. B. Qadri, M. Yousuf, S. H. Lee, B. L. Davis and H. Mattoussi, *J. Phys. Chem. B,* 2003, **107**, 5758.
136. S. B. Qadri, M. Kuno, C. R. Feng, B. B. Rath and M. Yousef, *Appl. Phys. Lett.,* 2003, **83**, 4011.
137. M. Green, G. Wakefield and P. J. Dobson, *J. Mater. Chem.,* 2003, **13**, 1076.
138. S. Keuleyan, E. Lhuillier, V. Brajuskovic and P. Guyot-Sionnest, *Nat. Photonics,* 2011, **5**, 489.
139. M.-O. M. Piepenbrock, T. Stirner, S. M. Kelly and M. O'Neill, *J. Am. Chem. Soc.,* 2006, **128**, 7087.
140. L. S. Li, H. Wang, Y. Liu, S. Lou, Y. Wang and Z. Du, *J. Colloid Interface Sci.,* 2007, **308**, 254.
141. S. Kim, T. Kim, S. H. Im, S. I. Seok, K. W. Kim, S. Kim and S.-W. Kim, *J. Mater. Chem.,* 2011, **21**, 15232.
142. S. Keuleyan, E. Lhuillier and P. Guyot-Sionnest, *J. Am. Chem. Soc.,* 2011, **133**, 16422.
143. S. Keuleyan, E. Lhuillier, V. Brajuskovic and P. Guyot-Sionnest, *Nat. Photonics,* 2011, **5**, 489.
144. E. Lhuillier, S. Keuleyan and P. Guyot-Sionnest, *Nanotechnology,* 2012, **23**, 175705.
145. P. Howes, M. Green, C. Johnston and A. Crossley, *J. Mater. Chem.,* 2008, **18**, 3474.
146. E. A. Turner, H. Rösner, Y. Huang and J. F. Corrigan, *J. Cluster Sci.,* 2007, **18**, 764.
147. W. Wichiansee, M. N. Nordin, M. Green and R. J. Curry, *J. Mater. Chem.,* 2011, **21**, 7331.
148. W. Xu, S. Lou, S. Li, H. Wang, H. Shen, J. Z. Niu, Z. Du and L. S. Li, *Colloids Surf., A,* 2009, **341**, 68.
149. S. Kumar and T. Nann, *Small,* 2006, **2**, 316.
150. J.-W. Jun, J.-H. Lee, J.-S. Choi and J. Cheon, *J. Phys. Chem. B,* 2005, **109**, 14795.
151. Y. Yu, P. V. Kamat and M. Kuno, *Adv. Funct. Mater.,* 2010, **20**, 1464.
152. X. Peng, L. Manna, W. Yang, J. Wickham, E. Scher, A. Kadanavich and A. P. Alivisatos, *Nature,* 2000, **404**, 59.
153. L. Manna, E. C. Scher and A. P. Alivisatos, *J. Am. Chem. Soc.,* 2000, **122**, 12700.
154. E. C. Scher, L. Manna and A. P. Alivisatos, *Philos. Trans. R. Soc. London, Ser. A,* 2003, **361**, 241.
155. F. Shieh, A. E. Saunders and B. A. Korgel, *J. Phys. Chem. B,* 2005, **109**, 8538.
156. J. Joo, J. S. Son, S. G. Kwon, J. H. Yu and T. Hyeon, *J. Am. Chem. Soc.,* 2006, **128**, 5632.
157. T. Nann and J. Riegler, *Chem.–Eur. J.,* 2002, **8**, 4791.
158. Y. Chen and L. Gao, *Chem. Lett.,* 2002, **31**, 557.
159. S. H. De Paoli Lacerda, J. F. Douglas, S. D. Hudson, M. Roy, J. M. Johnson, M. L. Becker and A. Karim, *ACS Nano,* 2007, **1**, 337.

160. A. G. Kanaras, C. Sonnichsen, H. Liu and A. P. Alivisatos, *Nano Lett.,* 2005, **5**, 2164.
161. Y.-W. Jun, S.-M. Lee, N. J. Kang and J. Cheon, *J. Am. Chem. Soc.,* 2001, **123**, 5150.
162. P. Sreekumari Nair, T. Radhakrishnan, N. Revaprasadu, G. A. Kolawole and P. O'Brien, *Chem. Commun.,* 2002, 584.
163. P. Christian and P. O'Brien, *J. Mater. Chem.,* 2008, **18**, 1689.
164. P. Christian and P. O'Brien, *Chem. Commun.,* 2005, 2817.
165. R. Li, Z. Luo and F. Papadimitrakopoulos, *J. Am. Chem. Soc.,* 2006, **128**, 6280.
166. J. D. Doll, G. Pilania, R. Ramprasad and F. Papadimitrakopoulos, *Nano Lett.,* 2010, **10**, 680.
167. S. Kumar, M. Ade and T. Nann, *Chem.–Eur. J.,* 2005, **11**, 2200.
168. Z. A. Peng and X. Peng, *J. Am. Chem. Soc.,* 2001, **123**, 1389.
169. X. Peng, *Adv. Mater.,* 2003, **15**, 459.
170. S. A. Blanton, R. L. Leheny, M. A. Hines and P. Guyot-Sionnest, *Phys. Rev. Lett.,* 1997, **79**, 865.
171. D. V. Talapin, R. Koeppe, S. Götzinger, A. Kornowski, J. M. Lupton, A. L. Rogach, O. Benson, J. Feldmann and H. Weller, *Nano Lett.,* 2003, **3**, 1677.
172. Z. A. Peng and X. Peng, *J. Am. Chem. Soc.,* 2002, **124**, 3343.
173. A. Fojtik, H. Weller, U. Koch and A. Henglein, *Ber. Bunsen-Ges. Phys. Chem.,* 1984, **88**, 969.
174. A. Kasuya, R. Sivamohan, Y. A. Barnakov, I. M. Dmitruk, T. Nirasawa, V. R. Romanyuk, V. Kumar, S. V. Mamykin, K. Tohji, B. Jeyadevan, K. Shinoda, T. Kudo, O. Terasaki, Z. Liu, R. V. Belosludov, V. Sundararajan and Y. Kawazoe, *Nat. Mater.,* 2004, **3**, 99.
175. Z.-J. Jiang and D. F. Kelley, *ACS Nano,* 2010, **4**, 1561.
176. S. Kudera, M. Zanella, C. Giannini, A. Rizzo, Y. Li, G. Gigli, R. Cingolani, G. Ciccarella, W. Spahl, W. J. Parak and L. Manna, *Adv. Mater.,* 2007, **19**, 548.
177. E. Kuçur, J. Ziegler and T. Nann, *Small,* 2008, **4**, 883.
178. M. Li, J. Ouyang, C. I. Ratcliffe, L. Pietri, X. Wu, D. M. Leek, I. Moudrakovski, Q. Lin, B. Yang and K. Yu, *ACS Nano,* 2009, **3**, 3832.
179. J. Ouyang, M. B. Zaman, F. J. Yan, D. Johnston, G. Li, X. Wu, D. Leek, C. I. Ratcliffe, J. A. Ripmeester and K. Yu, *J. Phys. Chem. C,* 2008, **112**, 13805.
180. K. Yu, M. Z. Hu, R. Wang, M. Le Piolet, M. Frotey, M. B. Zaman, X. Wu, D. M. Leek, Y. Tao, D. Wilkinson and C. Li, *J. Phys. Chem. C,* 2010, **114**, 3329.
181. M. Zanella, A. Z. Abbasi, A. K. Schaper and W. J. Parak, *J. Phys. Chem. C,* 2010, **114**, 6205.
182. P. Sreekumari Nair, K. P. Fritz and G. D. Scholes, *Chem. Commun.,* 2004, 2084.
183. P. Sreekumari Nair, K. P. Fritz and G. D. Scholes, *Small,* 2007, **3**, 481.
184. S. Sapra, J. Poppe and A. Eychmüller, *Small,* 2007, **3**, 1886.

185. Q. Pang, L. Zhao, Y. Cai, D. P. Nguyen, N. Regnault, N. Wang, S. Yang, W. Ge, R. Ferreira, G. Bastard and J. Wang, *Chem. Mater.*, 2005, **17**, 5263.
186. S. Asokan, K. M. Krueger, V. L. Colvin and M. S. Wong, *Small*, 2007, **3**, 1164.
187. L. Manna, D. J. Milliron, A. Meisel, E. C. Scher and A. P. Alivisatos, *Nat. Mater.*, 2003, **2**, 382.
188. D. Tarì, M. De Giorgi, F. Della Sala, L. Carbone, R. Krahne, L. Manna and R. Cingolani, *Appl. Phys. Lett.*, 2005, **87**, 224101.
189. L. Carbone, S. Kudera, E. Carlino, W. J. Parak, C. Giannini, R. Cingolani and L. Manna, *J. Am. Chem. Soc.*, 2006, **128**, 748.
190. A. Sugunan, S. H. M. Jafri, J. Qin, T. Blom, M. S. Toprak, K. Leifer and M. Muhammed, *J. Mater. Chem.*, 2010, **20**, 1208.
191. P. J. Thomas and P. O'Brien, *J. Am. Chem. Soc.*, 2006, **128**, 5614.
192. T. Mokari, E. Rothenberg, I. Popov, R. Costi and U. Banin, *Science*, 2004, **304**, 1787.
193. D. Steiner, T. Mokari, U. Banin and O. Millo, *Phys. Rev. Lett.*, 2005, **95**, 056805.
194. A. Salant, E. Amitay-Sadovsky and U. Banin, *J. Am. Chem. Soc.*, 2006, **128**, 10006.
195. G. Menagen, D. Mocatta, A. Salant, I. Popov, D. Dorfs and U. Banin, *Chem. Mater.*, 2008, **20**, 6900.
196. H. Liu and A. P. Alivisatos, *Nano Lett.*, 2004, **4**, 2397.
197. Y. Khalavka and C. Sönnichsen, *Adv. Mater.*, 2008, **20**, 588.
198. S. Kudera, L. Carbone, M. Francesca Casula, R. Cingolani, A. Falqui, E. Snoeck, W. J. Parak and L. Manna, *Nano Lett.*, 2005, **5**, 445.
199. L. Carbone, S. Kudera, C. Giannini, G. Ciccarella, R. Cingolani, P. D. Cozzoli and L. Manna, *J. Mater. Chem.*, 2006, **16**, 3952.
200. D. J. Milliron, S. M. Hughes, Y. Cui, L. Manna, J. Li, L.-W. Wang and A. P. Alivisatos, *Nature*, 2004, **430**, 190.
201. H. Zhong, Y. Zhou, Y. Yang, C. Yang and Y. Li, *J. Phys. Chem. C*, 2007, **111**, 6538.
202. A. Figuerola, I. R. Franchini, A. Fiore, R. Mastria, A. Falqui, G. Bertoni, S. Bals, G. van Tendeloo, S. Kudera, R. Cingolani and L. Manna, *Adv. Mater.*, 2009, **21**, 550.
203. S. L. Teich-McGoldrick, M. Bellanger, M. Caussanel, L. Tsetseris, S. T. Pantelides, S. C. Glotzer and R. D. Schrimpf, *Nano Lett.*, 2009, **9**, 3683.
204. J. W. Grebinski, K. L. Hull, J. Zhang, T. H. Kosel and M. Kuno, *Chem. Mater.*, 2004, **16**, 5260.
205. R. S. Wagner and W. C. Ellis, *Appl. Phys. Lett.*, 1964, **4**, 89.
206. T. J. Trentler, K. M. Hickman, S. C. Goel, A. M. Viano, P. C. Gibbons and W. E. Buhro, *Science*, 1995, **270**, 1791.
207. Z. Li, A. Kornowski, A. Myalitsin and A. Mews, *Small*, 2008, **4**, 1698.
208. H. Yu, R. A. Loomis, P. C. Gibbons, L.-W. Wang and W. E. Buhro, *J. Am. Chem. Soc.*, 2003, **125**, 16168.
209. A. Dong, R. Tang and W. E. Buhro, *J. Am. Chem. Soc.*, 2007, **129**, 12254.

210. D. D. Fanfair and B. Korgel, *Cryst. Growth Des.,* 2008, **8**, 3246.
211. M. Kuno, O. Ahmad, V. Protasenko, D. Bacinello and T. H. Kosel, *Chem. Mater.,* 2006, **18**, 5722.
212. J. Sun, L.-W. Wang and W. E. Buhro, *J. Am. Chem. Soc.,* 2008, **130**, 7997.
213. J. Sun, W. E. Buhro, L.-W. Wang and J. Schrier, *Nano Lett.,* 2008, **8**, 2913.
214. K.-T. Yong, Y. Sahoo, M. T. Swihart and P. N. Prasad, *Adv. Mater.,* 2006, **18**, 1978.
215. J. Puthussery, T. H. Kosel and M. Kuno, *Small,* 2009, **10**, 1112.
216. D. V. Talapin, J. H. Nelson, E. V. Shevchenko, S. Aloni, B. Sadtler and A. P. Alivisatos, *Nano Lett.,* 2007, **7**, 2951.
217. C. L. Choi, K. J. Koski, S. Sivasankar and A. P. Alivisatos, *Nano Lett.,* 2009, **9**, 3544.
218. L. Carbone, C. Nobile, M. De Giorgi, F. D. Sala, G. Morello, P. Pompa, M. Hytch, E. Snoeck, A. Fiore, I. R. Franchini, M. Nadasan, A. F. Silvestre, L. Chiodo, S. Kudera, R. Cingolani, R. Krahne and L. Manna, *Nano Lett.,* 2007, **7**, 2942.
219. Y. F. Zhu, X. Y. Lang and Q. Jiang, *Adv. Funct. Mater.,* 2008, **18**, 1422.
220. R. E. Bailey and S. Nie, *J. Am. Chem. Soc.,* 2003, **125**, 7100.
221. S.-H. Wei, S. B. Zhang and A. Zunger, *J. Appl. Phys.,* 2000, **87**, 1304.
222. N. P. Gurusinghe, N. N. Hewa-Kasakarage and M. Zamkov, *J. Phys. Chem. C,* 2008, **112**, 12795.
223. X. Zhong, Y. Feng, W. Knoll and M. Han, *J. Am. Chem. Soc.,* 2003, **125**, 13559.
224. L. Vergard and H. Schjelderup, *Phys. Z.,* 1917, **18**, 93.
225. X. Zhong, S. Liu, Z. Zhang, L. Li, Z. Wei and W. Knoll, *J. Mater. Chem.,* 2004, **14**, 2790.
226. J. Ouyang, C. I. Ratcliffe, D. Kingston, B. Wilkinson, J. Kuijper, X. Wu, J. A. Ripmeester and K. Yu, *J. Phys. Chem. C,* 2008, **112**, 4908.
227. X. Zhong, M. Han, Z. Dong, T. J. White and W. Knoll, *J. Am. Chem. Soc.,* 2003, **125**, 8589.
228. X. Zhong, Z. Zhang, S. Liu, M. Han and W. Knoll, *J. Phys. Chem. B,* 2004, **108**, 15552.
229. E. Jang, S. Jun and L. Pu, *Chem. Commun.,* 2003, 2964.
230. D. D. Sarma, A. Nag, P. K. Santra, A. Kumar, S. Sapra and P. Mahadevan, *J. Phys. Chem. Lett.,* 2010, **1**, 2149.
231. N. Al-Salim, A. G. Young, R. D. Tilley, A. J. McQuillan and J. Xia, *Chem. Mater.,* 2007, **19**, 5185.
232. R. Han, M. Yu, Q. Zheng, L. Wang, Y. Hong and Y. Sha, *Langmuir,* 2009, **25**, 12250.
233. L. A. Swafford, L. A. Weigand, M. J. Bowers II, J. R. McBride, J. L. Rapaport, T. L. Watt, S. K. Dixit, L. C. Feldman and S. J. Rosenthal, *J. Am. Chem. Soc.,* 2006, **128**, 12299.
234. M. D. Garrett, A. D. Dukes III, J. R. McBride, N. J. Smith, S. J. Pennycook and S. J. Rosenthal, *J. Phys. Chem. C,* 2008, **112**, 12736.
235. A. Nag, A. Kumar, P. Prem Kiran, S. Chakraborty, G. Ravindra Kumar and D. D. Sarma, *J. Phys. Chem. C,* 2008, **112**, 8229.

236. M. Protiére and P. Reiss, *Small,* 2007, **3**, 399.

237. Z. Deng, H. Yan and Y. Liu, *J. Am. Chem. Soc.,* 2009, **131**, 17744.

238. J. C. de Mello and A. J. de Mello, *Lab Chip,* 2004, **4**, 11N.

239. H. Nakamura, Y. Yamaguchi, M. Miyazaki, H. Maeda, M. Uehara and P. Mulvaney, *Chem. Commun.,* 2002, 2844.

240. J. B. Edel, R. Fortt, J. C. de Mello and A. J. de Mello, *Chem. Commun.,* 2002, 1136.

241. H. Wang, H. Nakamura, M. Uehara, M. Miyazaki and H. Maeda, *Chem. Commun.,* 2002, 1462.

242. H. Nakamura, A. Tahiro, Y. Yamaguchi, M. Miyazaki, T. Watari, H. Shimizu and H. Maeda, *Lab Chip,* 2004, **4**, 237.

243. H. Nakamura, Y. Yamaguchi, M. Miyazaki, M. Uehara, H. Maeda and P. Mulvaney, *Chem. Lett.,* 2002, 1072.

244. E. M. Chan, R. A. Mathies and A. P. Alivisatos, *Nano Lett.,* 2003, **3**, 199.

245. S. Krishnadasan, J. Tovilla, R. Vilar, A. J. de Mello and J. C. de Mello, *J. Mater. Chem.,* 2004, **14**, 1655.

246. B. K. H. Yen, N. E. Stott, K. F. Jensen and M. G. Bawendi, *Adv. Mater.,* 2003, **15**, 1858.

247. S. Marre, J. Park, J. Rempel, J. Guan, M. G. Bawendi and K. F. Jensen, *Adv. Mater.,* 2008, **20**, 4830.

248. H. Wang, X. Li, M. Uehara, Y. Yamaguchi, H. Nakamura, M. Miyazaki, H. Shimizu and H. Maeda, *Chem. Commun.,* 2004, 48.

249. H. Wang, H. Nakamura, M. Uehara, Y. Yamaguchi, M. Miyazaki and H. Maeda, *Adv. Funct. Mater.,* 2005, **15**, 603.

250. H. Yang, W. Luan, Z. Wan, S.-T. Tu, W.-K. Yuan and Z. M. Wang, *Cryst. Growth Des.,* 2009, **9**, 4807.

251. R. Kikkeri, P. Laurino, A. Odedra and P. H. Seeberger, *Angew. Chem., Int. Ed.,* 2010, **49**, 2054.

252. S. Krishnadasan, R. J. C. Brown, A. J. deMello and J. C. deMello, *Lab Chip,* 2007, 7, 1434.

253. A. Toyota, H. Nakamura, H. Ozono, K. Yamashita, M. Uehara and H. Maeda, *J. Phys. Chem. C,* 2010, **114**, 7527.

254. E. M. Chan, A. P. Alivisatos and R. A. Mathies, *J. Am. Chem. Soc.,* 2005, **127**, 13854.

255. Y. T. Didenko and K. S. Suslick, *J. Am. Chem. Soc.,* 2005, **127**, 12196.

256. E. M. Chan, C. Xu, A. W. Mao, G. Han, J. S. Owen, B. E. Cohen and D. J. Milliron, *Nano Lett.,* 2010, **10**, 1874.

CHAPTER 2

The Preparation of III–V Semiconductor Nanomaterials

So far, we have described the synthesis of II–VI quantum dots (QDs), as these materials are the easiest to prepare and analyse (having optical properties mainly in the visible region) with a wide varieties of structures accessible from simple precursors. There are, however, numerous other families of semiconductors that can be prepared as QDs using related chemistry, again inspired by the seminal paper describing trioctylphosphine oxide (TOPO) as a capping agent.[1] These new materials are often used to access regions of the optical spectrum not normally obtainable using typical CdSe-based QDs, although the chemistry is quite clearly similar.

2.1   Properties and Applications of III–V Materials

III–V materials are a common family of semiconductors with numerous optoelectronic applications. InP has a bandgap of 1.27 eV and an excitonic diameter of 21 nm, whereas indium arsenide (InAs) is a narrow-bandgap material with a bandgap of 0.36 eV and an excitonic diameter of 74 nm. InSb has a bandgap of 0.165 eV and an excitonic diameter of an estimated 138 nm, while GaSb has a bandgap of 0.67 eV and an excitonic diameter of 40 nm.[2] After the II–VI family of QDs, the III–V family might be considered the next most popular materials, with numerous publications describing the optical properties and physical characteristics of TOPO-capped InP and InAs, including Raman spectra,[3] exciton recombination kinetics,[4] anti-Stokes luminescence from surface states,[5] excited-state spectroscopy,[6] coupling in particle lattices[7] and fluorescence intermittency.[8] Key studies include the

RSC Nanoscience & Nanotechnology No. 33
Semiconductor Quantum Dots: Organometallic and Inorganic Synthesis
By Mark Green
© Mark Green 2014
Published by the Royal Society of Chemistry, www.rsc.org

observation of atom-like states in III–V QDs,[9] and the production of hydrogen by a nanophotocathode containing InP.[10] The III–V family of materials are of immense interest because of the large excitonic diameters,[2] which should allow size quantisation effects to manifest at much larger particle sizes than in the analogous II–VI materials, allowing size-dependent properties to be explored in more detail. The chemistry is somewhat more difficult with this group of materials; problems with the synthesis of III–V QDs have been highlighted by Heath and Shiang,[11] notably the separation of the nucleation and growth steps, attributable to both the strongly complexed precursors used and the relatively covalent nature of the resulting materials when compared to II–VI materials.

Reports describing the synthesis of capped III–V materials emerged shortly after the first papers on TOPO-capped CdSe, and the use of QDs in biology has ensured the continued development of such materials, providing alternatives to cadmium-containing biolabels. Whereas simple precursors are available for II–VI materials, notably the anionic components which can be provided by the simple elemental species, there are few suitable group V precursors; elemental precursors have not been successfully utilised in the same manner. The origins of III–V nanomaterials might be traced back to researchers examining novel organometallic compounds,[12] with the emphasis not on the end products but on new precursors, which were usually thin films prepared by chemical vapour deposition (CVD); such precursors could also be used in the solution synthesis of simple nanomaterials.[13]

The key reaction for the synthesis of III–V QDs was described in 1989 by both the Wells[14] and the Barron[15] groups, for the synthesis of GaAs, InAs and InP respectively. The reaction, termed dehalosilylation, was based on previous work by Wells using silylated compounds to form Ga–As bonds[16] and was initially envisaged as a possible alternative to the use of gaseous group V precursors. The work described a simple solution route to III–V bulk materials using a group III salt, such as $GaCl_3$ or $InCl_3$, with a silylated pnictogen such as $As(SiMe_3)_3$ or $P(SiMe_3)_3$ in solution at low temperatures. Although the reaction between $InCl_3$ and $P(SiMe_3)_3$ yielded $[Cl_2InP(SiMe_3)_2]_x$, heating the intermediate at 650 °C under vacuum yielded bulk InP of high purity. The analogous reaction with $GaCl_3$ or $InCl_3$ and $As(SiMe_3)_3$ proceeded *via* an unidentified intermediate stage, yielding the bulk material of at least 84% purity upon thermolysis. These reactions are extremely important as they provide the first simple solution routes to high-quality III–V materials using precursors that are readily available and easy to handle. It is worth noting at this point that a similar reaction was reported a couple of years later, describing the use of an indium metal alkyl and $P(SiMe_3)_3$ yielding InP,[17] although this has proved less popular, possibly due to the air sensitivity and the difficulties associated with handling metal alkyls.

This was the starting point for numerous explorations into III–V chemistry, by variations on the dehalosilylation reaction. Reports included the alcoholysis of single-source precursors such as $[R(Cl)In(\mu\text{-}P(SiMe_3)_2)]_2$ (prepared from R_2InCl and $P(SiMe_3)_3$), giving a mixture of InP and In_2O_3, although there was little discussion regarding optical properties of the particles.[18] Likewise, the reaction of metal alkyls with silylated phosphines produced adduct compounds that were used as precursor to metal phosphides, although some metal impurities were also observed.[19] The group of Wells explored new routes to GaAs using precursors using precursors such as $AsCl_3Ga_2$, prepared by reacting $GaCl_3$ with $As(SiMe_3)_3$, which was then washed in hot hydrocarbons followed by heating up to 410 °C under an inert atmosphere.[20] Similarly, the single-source precursor $[X_2GaP(SiMe_3)_2]_2$ (X = Cl, Br, I) was prepared by the reaction of $GaCl_3$ with $P(SiMe_3)_3$, which could then be thermolysed to give bulk GaP.[21,22] The same reactions using indium halides with either $As(SiMe_3)_3$ or $P(SiMe_3)_3$ gave similar precursors to those described above, which could all be used to prepare bulk III–V compounds by thermolysis.[23]

The first solution route to III–V QDs using the dehalosilylation reaction was reported by Olshavsky *et al.*,[24] who carried out the reaction between $GaCl_3$ and $As(SiMe_3)_3$ in quinoline at 240 °C for 3 days, followed by flame-annealing the product under vacuum, which improved the crystallinity but reduced the solubility of the material presumably by removing the surface species. The particles, *ca.* 4.5 × 3.5 nm in size, were zinc blende in structure with an excitonic peak in the absorption spectra at *ca.* 500 nm, shifting from the bulk band edge of *ca.* 850 nm. This report is notable as the first to explore the effects of quantum confinement using the theoretical framework provided by Brus, exploring the effects of quantum confinement of the optical properties. Interestingly, the presence of molecular species in solution and on the particle surface was reported to interfere with the optics,[25] which is significant as this was highlighted again in later years as a problem with III–V QD synthesis, while confirming the solvent molecules coordinated to the particle surface. Another report[26] on the same reaction described the use of differing solvents, such as trioctylamine (TOA) and hexadecane which provided higher synthesis temperatures. A similar reaction using $In(OH)_3$ and $As(SiMe_3)_3$ in triglyme gave 4 nm InAs particles with a band edge of approximately 500 nm, and broad emission at approximately 550 nm.[27]

2.2   Group III–Phosphides

Mićić *et al.* were the first to report the incorporation of the dehalosilylation reaction with the high-temperature,[28] inert-atmosphere reaction using TOPO as a surfactant, inspired by the work of Murray *et al.* described earlier. In this case, chloroindium oxalate and $P(SiMe_3)_3$ of varying molar ratios were mixed at room temperature to form an unidentified precursor, which was then heated to 270 °C with either TOPO or a mix of TOPO and trioctylphosphine (TOP) for several days, giving slightly phosphorus-rich InP QDs of approximately 2.5 nm diameter with a narrow size distribution, passivated with the surfactant(s) in at least two bonding configurations.[29] The particles are notable for being reportedly the first examples of InP to exhibit an excitonic peak in the absorption spectra, at *ca.* 500 nm, with the onset of absorption being approximately 600–700 nm. Broad photoluminescence was also reported, between 600 nm and 900 nm, and was attributed to sub-gap surface states.

This initial report was extended to describe the properties of InP in more depth (notably describing the inability to prepare crystalline particles using InCl$_3$), while also reporting the preparation of GaP and GaInP$_2$ QDs.[30,31] In the case of GaP particles, the amorphous particles were initially prepared by mixing GaCl$_3$ and P(SiMe$_3$)$_3$ with TOPO and TOP, giving particles with a 3 nm diameter. These particles could be heated further with a large amount of TOPO at 360 °C for 3 days giving a crystalline nanomaterial. The synthesis of 2.5 nm diameter GaInP$_2$ required a mixture of both precursors and a slight molar excess of P(SiMe$_3$)$_3$, followed by heating in the presence of a small amount of TOPO and tris(2-diphenylphospinoethyl)phosphine at 400 °C for 3 days. All particles could be isolated using methanol as a non-solvent, followed by redispersion in toluene.

The size of the zinc blende InP particles could be controlled by varying the amount of precursor used, from 2.6 nm with a band edge of *ca.* 600 nm and a clear excitonic peak, to 4.6 nm with a band edge at *ca.* 800 nm. Emission from the InP particles showed a band in the region of 400–600 nm, and a second beyond 800 nm. GaP particles 2–3 nm diameter displayed band edges between *ca.* 400 and 450 nm, complicated by the fact that below 3 nm the bandgap is predicted to decrease with decreasing size rather than increase. GaInP$_2$ particles 2.5–6.5 nm in diameter were reported, with the 2.5 nm particles having a band edge at *ca.* 2.7 eV (*ca.* 460 nm). Importantly, the photoluminescent properties of the particles could not be fully ascertained because of interference from a luminescent decomposition product from heated TOPO, and this remains a significant factor when exploring the optical characteristics of nanoparticles. An in-depth exploration of the optical properties of capping agent decomposition products has highlighted that TOA and hexadecylamine (HDA) also exhibit emission that may well be mistakenly assigned as III–V QD luminescence.[32] The broad emission was found to be significantly reduced in 99% pure TOPO (as opposed to technical-grade, 90% pure TOPO), indicating the luminescent materials may well be the thermal decomposition products of the impurities outlined in Chapter 6.

Guzelian *et al.* extended the synthesis using essentially the same chemistry, but adding InCl$_3$ directly to TOPO before heating, forming an In–TOPO complex.[33] After prolonged stirring and gentle heating, an equimolar amount of P(SiMe$_3$)$_3$ was injected into the In–TOPO complex and the temperature increased to 265 °C, where it was maintained for several days, with nucleation and growth occurring simultaneously in direct contrast to the discrete separation of nucleation and growth easily achieved in the preparation of II–VI nanomaterials. The temperature was lowered, and another capping agent added if required, followed by several days further heating at a lower temperature yielding crystalline (zinc blende) InP QDs. The long period of growth was essential for high-quality particles, although crystals were obtained after 2 days. Size-selective precipitation allowed particles between 2 and 5 nm to be isolated with a size distribution of ±20%, capped with both TOPO and the second capping agent, such as dodecylamine (DDA), the presence of which improved the solubility of the particles. The particles exhibited bandgaps of 1.7–2.4 eV (*ca.* 730 nm and *ca.* 520 nm) as shown in Figure 2.1. It was also reported that when exposed to air, both surface indium and phosphorus sites oxidised to a depth of more than one monolayer. This is important, as the particles did not exhibit any photoluminescence prior to surface oxidation, attributed to the surface oxide blocking trapping sites. The emission spectra consisted of both band edge emission, and weaker deep

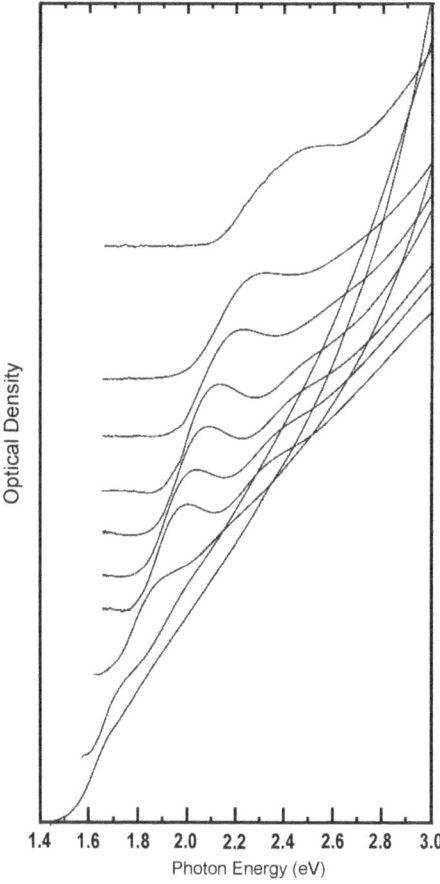

Figure 2.1 Absorption edges of a range of InP QDs. Reprinted with permission from A. A. Guzelian, J. E. B. Katari, A. V. Kadavanich, U. Banin, K. Hamad, E. Juban, R. H. Wolters, C. C. Arnold and J. R. Heath, *J. Phys. Chem.*, 1996, **100**, 7212. Copyright 1996 American Chemical Society.

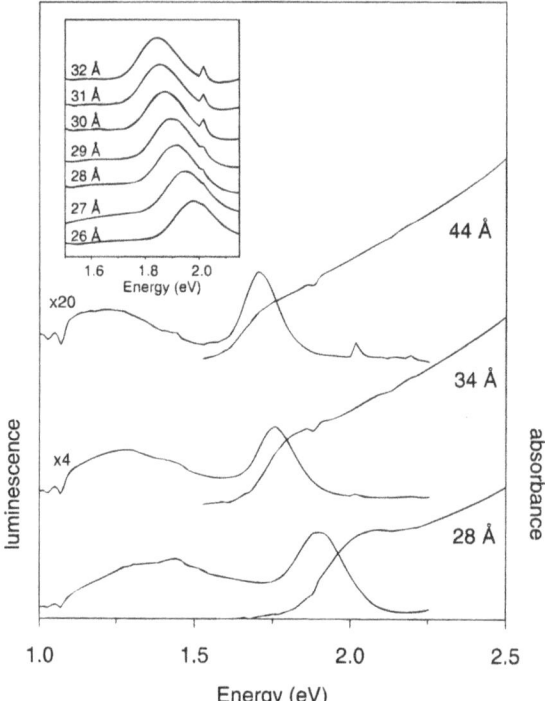

Figure 2.2 Absorption and emission spectra of various sizes of InP QDs, showing a mixture of band edge emission and deep trap emission. Reprinted with permission from A. A. Guzelian, J. E. B. Katari, A. V. Kadavanich, U. Banin, K. Hamad, E. Juban, R. H. Wolters, C. C. Arnold and J. R. Heath, *J. Phys. Chem.*, 1996, **100**, 7212. Copyright 1996 American Chemical Society.

trap emission at lower energies (Figure 2.2), with the smallest particles (2.8 nm in diameter) having the highest quantum yield of up to 0.12%. The oxide layer on InP particles has also been shown to limit size-selective precipitation.[34] The oxidation of InP is significant as the oxidation has been shown to extend beyond the surface; the storage of InP in ambient condition for a few days reportedly resulted in the total oxidation to In_2O_3, and this also occurred in high-quality commercially available InP/ZnS samples, giving evidence of the formation of both In_2O_3 and ZnO.[35]

2.2.1 Increasing the Emission Quantum Yield of III–V Materials

The low quantum yield for TOPO-capped InP QDs was the major limiting factor in the use of these materials, and this was attributed to phosphorus vacancies on the particle surface[36] (although several excellent 'one-pot' routes to highly luminescent core/shell structures, such as InP/ZnS are described in

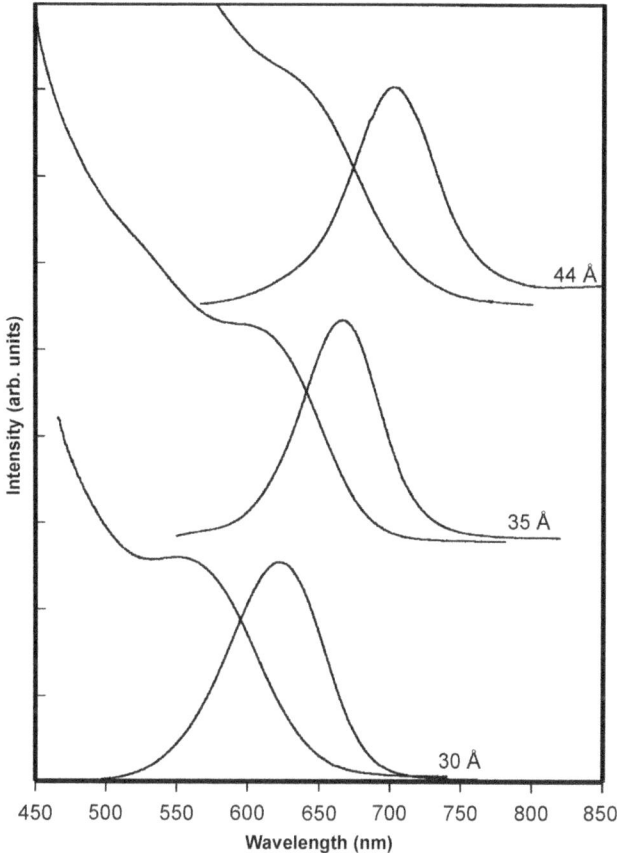

Figure 2.3 Absorption and emission spectra of different-sized InP QDs after treatment with HF. Reprinted with permission from O. I. Mićić, J. Sprague, Z. Lu and A. J. Nozik, *Appl. Phys. Lett.*, 1996, **68**, 3150. Copyright 1996, AIP Publishing LLC.

Chapter 5.) This was resolved by Mićić *et al.*, who described a simple etching process of the InP QDs with either HF or NH$_4$F, in which the fluoride ions removed phosphorus vacancies (V$_P$) and dangling bonds (as confirmed by electron paramagnetic resonance[37]) which are known electron traps. The fluoride ions replaced oxygen in the surface oxide layer,[38] although some defects were known to still exist in the core of the particle.[39] This etching process resulted in a blue shift in the absorption spectra of the particles as they became smaller, while the emission became entirely band edge, losing the low trap emission feature as shown in Figure 2.3.[38] Importantly, the emission quantum yields increased massively to 30% at 300 K. It was reported that deep trap emission could be avoided if the InP particles were prepared with an excess of phosphorus, although the band edge emission was extremely weak, and the absorption spectra did not have an excitonic feature, attributed to a wider size distribution due to a lack of indium sites for the TOPO to coordinate to. Talapin *et al.* extended the study to report that the etching process was predominately photochemical in nature, and required illumination from a xenon lamp to generate reliable and reproducible materials.[36,40] Combining the etching process with size-selective precipitation allowed a wide range of luminescent particles to be isolated in the size range *ca.* 1.7–6.5 nm, giving materials emitting from the green part of the visible spectrum through to the near infrared, with quantum yields of 20–40%. An elegant extension is the *in situ* etching of InP QDs during particle growth.[41] In this reaction, indium palmitate and $P(SiMe_3)_3$ were dissolved in decane and injected into a microwave reactor containing a microwave-absorbing fluorine-based ionic liquid which generated F^- ions during the reaction. By varying the ionic liquid and reaction conditions, zinc blende InP QDs *ca.* 3 nm in diameter were obtained with emission quantum yields of up to 47% without the need of further processing. Interestingly, the ionic liquid is suggested to be a spectator solvent and takes no place in the reaction other than generating the etchant. Another method of blocking dangling bonds on the surface of InP is the inclusion of $Zn(COOC_{10}H_{19})_2$ in the reaction, which blocks surface trapping sites, resulting in impressive emission quantum yields of up to 30% without further processing.[42]

Interestingly, extremely small (1.5–2.3 nm diameter) particles of InP have also been reported using essentially the same chemistry, but with a much shorter synthesis time, lower temperature and different capping group such as trioctylamine (TOA).[43] The particles of InP 1.5 nm in diameter were reported to have a very clear excitonic peak at *ca.* 420 nm and a broad band edge emission spectrum with a quantum yield of 20% without any etching, shelling or processing. Interestingly, the high quantum yields cannot be explained by the formation of an In_2O_3 shell as the tertiary amine capping agent would not undergo the relevant condensation reaction. The role of amines in InP formation is still unresolved. The majority of papers[43,44] suggest amines (and protic reagents such as alcohols) activate the reaction, leading to faster nucleation and improved particle quality. This role has been disputed and will be discussed later.

2.2.2   The Growth of Anisotropic Particles

Most of the work on III–V nanomaterials has resulted in small spherical particles, possibly because of the difficulty of controlling the nucleation and growth steps. Anisotropic particles have, however, been prepared by controlling the surfactant chemistry. The thermolysis of a single-source precursor, $Ga(P^tBu_2)_3$, in a mixture of HDA and TOA resulted in wurtzite-structured GaP nanorods, 8 nm × 45 nm in size.[45] The crystallinity of the rods could be controlled by altering the surfactant ratio, with HDA directing anisotropic particle growth and the formation of the hexagonal phase, whereas the use of just TOA resulted in zinc blende spherical particles. The optical characteristics were similar, with rods emitting at *ca.* 3.46 eV and spheres at 3.48 eV.

Notably, nanorods of zinc blende InP up to 100 nm long, with twin boundaries and stacking faults, could also be prepared using an amendment to the popular solution–liquid–solid route (SLS) to anisotropic III–V materials[46] using two single-source precursors, prepared *in situ* by dehalosilylation chemistry.[47] To either a 2% DDA in TOA solution, or a 2% TOPO in TOP solution containing $InCl_3$ and $In(^tBu)_2Cl$, was added $P(SiMe_3)_3$ followed by a small amount of methanol to hydrolyse the P–Si bond, which was then stirred at room temperature for a day. The reagents were then heated at various temperatures and for various times, and isolated by precipitation using methanol. The resulting rods were air sensitive, and the metal impurities removed by addition of mercury to form alloys, which could also be etched with HF. The use of the *in situ* formed $[(^tBu)_2InP(SiMe_3)_2]_2$ allowed the formation of metallic indium particles which catalysed the growth of the rods, while $[Cl_2InP(SiMe_3)_2]_2$ acted as the InP precursor. As the long-chain amine capping agents bound strongly to indium, the seed particles prepared were small, and the resulting rods therefore had diameters as small as 2.8 nm with a length of up to 33 nm, with the small rods showing, in some cases zigzag and tripod-type morphologies. In comparison, TOPO/TOP-capped rods 5 nm in diameter of have been prepared under similar conditions. The optical properties of quantum rods (2.8 nm × 10–30 nm) were similar in spectral position to 3 nm spherical InP dots, although the absorption and emission spectra were notably broader, with a larger Stokes shift of *ca.* 400 meV.

InP rods, with average diameter of 4.2–7.8 nm and length of 10–39 nm have also been grown by a similar SLS method with bismuth particles,[48] using $In(COOC_{13}H_{27})$, HDA, TOP, TOPO and polydecene (and notably no phosphonic acid) into which was injected $P(SiMe_3)_3$ at temperatures between *ca.* 240 and 265 °C. After a certain length of time, a stock solution of bismuth particles in dioctyl adipate (DOA) was also injected into the reaction flask, the source of heat removed after 0.5–3 minutes and the product purified by typical solvent/non-solvent interactions. The resulting rods were capped with HDA and contained the bismuth seed, which could be removed by sonicating a toluene solution of the rods with oleic acid for several hours, although this also reduced the thickness of the structures. The rods exhibited two excitonic features, with the onset of absorption at *ca.* 1.5 eV (*ca.* 825 nm), with no detectable emission from the as-prepared rods. Etching the rods with HF resulted in band edge emission with quantum yields of 0.16%, whereas etched rods with the tips removed exhibited quantum yields of up to 0.6%. These rods were used to determine the 3D–2D rod–wire transition which was found to occur at 25 nm, and a more in-depth study exploring 2D and 3D confinement, which differentiates between wire and dot behaviour, has also been reported.[49] In this case, the rods were grown using the single-source precursor $[Me_2InP(SiMe_3)_2]_2$, $P(SiMe_3)_3$ in TOP using indium nanoparticles as the seeds, although the use of bismuth particles as seed catalysts for SLS-type growth of III–V materials has arguably become the most popular, and has been extended to most materials.[50]

The SLS growth of InP wires, with diameters of 4.3–19 nm and lengths 50–500 nm for thin wires and up to 10 μm for the thicker wires, has been achieved[51] using essentially the same method but including 1-octylphosphonic acid with HDA; the phosphonic acid is essential for high-quality wires and the inclusion of TOPO and TOP altered the wire's width. All four surfactants were found to be necessary for the growth of high-quality wires longer than 10 nm. The absorption spectra displayed band edges from *ca.* 700 to 850 nm with few discernible excitonic features. Attempts to obtain emission from the sample by photochemical etching using HF resulted in the thinning of the wires, and intense irradiation resulted in the oxidation of the wires, giving InPO$_4$ with luminescent InP domains.

Another SLS method of preparing InP rods involved the use of indium nanoparticles, which acted as an indium source rather than a catalyst.[44] The monodispersed indium particles, prepared by the thermolysis of indium alkyls in TOP and had, were reacted with phosphine ions generated from the hydrolysis of P(SiMe$_3$)$_3$. This route was notable as the resulting rods were composed entirely of InP, as the seed reacted with the phosphorous precursor giving materials with no catalyst contamination. Similar work has been reported where InP and GaP nanowires have been produced, using metal alkyls and P(SiMe$_3$)$_3$ in octadecene (ODE) at 305 °C with myristic acid as a capping agent, where the metal alkyls form the metal seed catalysts.[52]

2.2.3 Other Phosphorus Precursors

Few suitable solution-based precursors exist for phosphide-based nano-materials, with P(SiMe$_3$)$_3$ clearly the most suitable starting material. Despite the effectiveness of silylated precursors, they still have limitations. They are difficult to prepare, expensive (where commercially available) and extremely air sensitive. Although silylated phosphines are the solution functional equivalent of phosphine gas (PH$_3$), it is still possible to produce InP QDs by generating PH$_3$ *in situ* by adding HCl to Ca$_3$P$_2$ under an argon atmosphere, then bubbling the nascent gas through a reaction flask containing ODE, InCl$_3$ and myristic acid at 250 °C.[53] Notably, the phosphine delivery was strongest in the first few minutes, but was maintained throughout the reaction, although at a lower output, allowing size-focusing of the particle size and yielding particles with excellent optical properties and narrow size distributions of *ca.* 10%. This continuous delivery also overcame the phosphorus depletion problem associated with P(SiMe$_3$)$_3$. The particles, 3–6 nm in diameter, exhibited a zinc blende core, with no evidence of oxide side products. The resulting nanoparticles exhibited a clear excitonic peak which could be tuned between 650 and 700 nm, with emission between 675 and 720 nm by varying precursor ratios. The emission quantum yield was low, less than 1%, but could be improved using a shelling technique to give InP/ZnS QDs. An interesting alternative is the use of solid hydrogen phosphide (PH)$_x$, generated by the reaction between PBr$_3$ and LiAlH$_4$.[54] The solid, which can be handled in air, was prepared *in situ* during the formation of InP wires using bismuth or indium seed particles in an SLS-type reaction with TOP as a capping agent. A very similar reaction was reported, where PCl_3 and $LiBH(C_2H_5)_3$ were added sequentially to a solution of indium stearate at 40 °C, followed by heating at 250 °C for InP particle growth.[55] The resulting particles were approximately 3 nm in diameter and zinc blende in structure. The absorption spectra showed a clear excitonic feature, indicating the narrow size distribution, although the excitonic peaks were not as sharp as particles grown in non-coordinating solvents, which was attributed to the reaction being a slow and continued nucleation process. The emission spectra consisted of weak band edge emission and lower energy deep trap emission. After etching with HF, the deep trap emission was removed and the band edge quantum yield was increased from 0.25% to *ca.* 20%.

The lack of other obvious precursors has limited the amount of novel chemistry reported; however, other safer, more readily available suitable inorganic starting materials have been explored. It is worth noting that a common capping agent, TOP, has been used as a phosphorus precursor for InP by catalytic cleavage of the P–C bond, although the mixed product included indium metal and the particles were large in size with no detailed optical characteristics reported.[56] Interestingly, TOP has been used to prepare In/InP nanoneedles in a one-pot reaction, by initially forming indium nanodroplets that catalysed the decomposition of TOP.[57]

The use of Na_3P as a phosphorus precursor has also been reported with 4-ethylpyridine and TOP as solvents, and $InCl_3$ as the indium precursor.[58] The resulting material was reportedly 5 nm in diameter and zinc blende in structure, exhibiting a clear excitonic peak in the absorption spectrum at approximately 580 nm, although no description of the emissive properties was provided. Possibly the most successful alternative for silylated phosphines is tris(dimethylamino)phosphine, $P(NMe_2)_3$, which has been utilised as a precursor in the synthesis of TOPO-capped InP at 365 °C, using $InCl_3$ as a starting material.[59] The resulting materials, approximately 6 nm in diameter with a large standard deviation of 50%, exhibited a zinc blende core, with optical properties consistent with quantum confinement but without any excitonic feature in the absorption spectra and with broad emission. The use of $P(NMe_2)_3$ has been extended to the synthesis of InP, 2–4 nm in diameter, in an autoclave, using $InCl_3$ and DDA (as a capping agent), and toluene as a solvent.[60] The reaction proceeded at 180 °C for 24 hours before the sample was isolated by size-selective precipitation and etched with HF as described above, giving particles with emission quantum yields of up to 58%, before the addition of a wide-bandgap shell. Notably, trap emission was observed after size-selective precipitation, which was attributed to the formation of a surface oxide. This explanation is in contrast with earlier reports, described above, which suggested the oxide layer was actually essential to forming an emissive species. In this case, the ZnS shell was, unusually, added after the phase transfer step, in water. This was improved by adding the shell in a second autoclave step.[61] In an interesting amendment to the synthesis of anisotropic InP, Dorn *et al.*

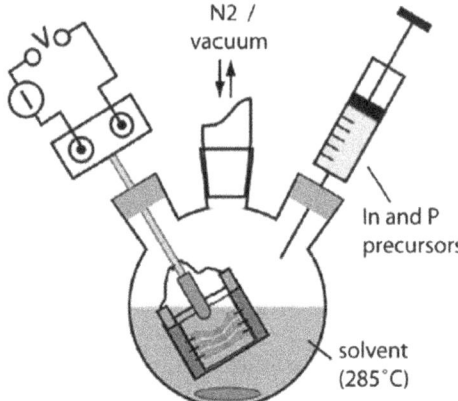

Figure 2.4 Experimental set-up for electrically controlled solution–liquid–solution growth of InP Nanowires. Reprinted with permission from A. Dorn, P. M. Allen and M. G. Bawendi, *ACS Nano*, 2009, **3**, 3260. Copyright 2009 American Chemical Society.

prepared nanowires using InI_3 and $P(NMe_2)_3$ in OAm, which was injected into a flask of TOP (285 °C) under nitrogen into which electrodes coated with bismuth catalysts had been placed; this was termed electrically controlled solution–liquid–solution (EC-SLS) growth (Figure 2.4).[62] The polycrystalline nanowires, with an amorphous 5 nm oxide layer, were rapidly grown between the gaps on the electrode, with the degree of growth controlled by the bias. Notably, the bridging of the wires could be monitored by measuring the conductivity. Other solution-based phosphines have been reported as potential precursors for InP nanofibre synthesis, such as tri(*m*-tolyl)phosphine ($P(PhMe)_3$), although this precursor was utilised in a sealed ampoule synthesis at 370 °C, with no capping agent and using indium nanoparticles as precursors.[63] White phosphorus, P_4, has also been used as a precursor and reacted with indium particles yielding InP particles, although no optical properties were reported.[64]

2.3 Group III–Arsenides

The success of the dehalosilylation reaction in producing phosphide-based QDs was extended to the synthesis of arsenide-based materials. These materials are of interest as the bandgap of InAs is 0.36 eV, and when blue-shifted by quantum confinement potentially gives optical properties in the near infrared, a region of the electromagnetic spectrum with numerous applications. InAs QDs have therefore been explored in some depth, including examining the shifting in energy levels due to surface ligand exchange,[65] exchange interactions,[66] absorption cross-section and oscillator strength,[67] size-selective spectroscopy,[68] high-resolution electron microscopy studies on conducting surfaces[69] and, importantly, the assignment of atom-like electronic states[70,71] and the preparation of InAs-based near infrared light-emitting devices.[72]

InAs QDs capped with TOP were prepared in an analogous manner to the InP particles mentioned above.[73,74] The zinc blende particles, 2.5–6 nm in diameter were prepared using $InCl_3$ and $As(SiMe_3)_3$ (As : In molar ratio 1.15–1.2) which were mixed together in TOP forming a stock solution, a portion of which was then injected into TOP at a temperature of 240–265 °C. The particles showed near band edge emission at *ca.* 1.5 eV (*ca.* 830 nm) for 3.4 nm diameter particles, to *ca.* 1.0 eV (*ca.* 1240 nm) for 6 nm diameter particles, with quantum yields of 0.5–2.5%. Interestingly, the particles were insensitive to oxidation, unlike the InP particles described above.

Zinc blende InAs nanorods could also be prepared using gold nanoparticles as catalysts for anisotropic growth in an SLS-inspired reaction, as described above.[75] In these examples, the nanorods were prepared by a rapid injection of a precursor stock solution containing dodecanethiol-stabilised gold particles into TOPO at 360 °C, followed by almost immediate reaction quenching. The reaction solution was then repeatedly purified using solvent/non-solvent interactions, initially removing nanowires (20 nm width up to 1 μm in length) from the reaction solution, followed successively by quantum rods of progressively smaller lengths, until spherical InAs dots remained. The rods varied from 22.7 nm × 4.4 nm to a length of 9.4 nm while keeping the same diameter. In some cases, the gold particle was visible. The rods exhibited a significant red shift in bandgap with increasing length, unlike the CdSe rods described in Chapter 1 which displayed minimal change in the optical spectra. The red shift in bandgap was also accompanied by a reduction in emission intensity, although no quantum yields were given. Similarly, InAs rods with similar optical properties could be grown using either indium or silver particles or gold clusters as catalysts.[76] The use of indium catalysts resulted in slightly narrower rods (*ca.* 3–4 nm), whereas gold clusters gave significantly narrower rods (*ca.* 2 nm), and the use of 3 nm diameter silver particles gave rods of relatively poor quality. The use of 4 nm silver particles resulted in the formation of spherical particles. Interestingly, simple GaAs spherical particles could not be grown using the dehalosilylation reaction; this was attributed to the ligand binding too strongly to the gallium precursor. However, using gold nanoparticles as catalysts, slightly anisotropic GaAs particles were observed, although the yield was reportedly low. This same methodology has also been applied to the synthesis of GaAs wires using bismuth particles as catalysts, where $(^tBu)_3Ga$ and $As(SiMe_3)_3$ were added to an ODE solution containing TOPO/TOP, followed by rapid but brief heating and the injection of the catalyst particles.[77] After a further short period of growth, the rods could be isolated by centrifugation. The use of just TOPO as a surfactant did not result in wire growth, but the use of ODE resulted in zinc blende GaAs wires 500–1000 nm in length. InAs rods were reported by the same group, who used essentially the same reaction, with $In(COO(CH_2)_{12}CH_3)$, TOP, HDA and polydecene in the reaction flask, to which was added $As(SiMe_3)_3$ and bismuth simultaneously at 240–330 °C.[78]

Interestingly, emission spectra were recorded as far into the red region of the electromagnetic spectrum as 2500 nm.

Again, the chemistry of arsenide-containing particles has changed relatively little, with few advances being made, possibly due to the lack of simple alternative precursors. Other precursors and solvent systems have been explored, such as the use of $As(NMe_2)_3$ in the synthesis of InAs nanoparticles using 4-ethylpyridine as solvent and capping agent.[79] In this case, the $InCl_3$ was found to form an air-stable complex with the solvent, yielding *mer*-[In(4-ethylpyridine)$_3$Cl$_3$], which gave small (2 nm) poorly crystalline InAs particles when slowly heated to reflux with $As(NMe_2)_3$. The particles could be isolated using petroleum spirits as a non-solvent, although size-selective precipitation could not be achieved because of the formation of a surface oxide, as described above for InP. The particles exhibited substantial shifts in the bandgap to between 2.63 and 2.96 eV after several days of growth, with emission profiles between *ca.* 475 and 550 nm, although no quantum yield was reported. Further growth could be effected by the introduction of additional precursors, which resulted in particle growth and a red shift in the optical spectra, with the emergence of an excitonic feature in the absorption spectra along with a narrowing of the emission profile, consistent with growth focusing as described in Chapter 1. Nanoparticles of GaAs *ca.* 4 nm in diameter have also been prepared using the same precursor and GaCl$_3$.[80] Although slightly larger, the particles were still poorly crystalline as determined by X-ray powder diffraction (XRD), and the presence of Ga_2O_3 and As_2O_3 was also observed. When annealed in air the diffraction pattern for $GaAsO_4$ was obtained. The absorption spectra exhibited an excitonic shoulder at *ca.* 475 nm, with emission varying at *ca.* 450 nm. Interestingly, the capping agent was found to emit in the same spectral region, although GaAs was thought to be the predominant emitting species.

2.3.1 Non-Coordinating Solvent Route

In all the cases described so far, prolonged synthesis is required to obtain materials that are ultimately of a lower quality than the II–VI analogues as the nucleation and growth steps are hard to separate. The materials are poorly crystalline, polydispersed with, more often than not, poor optical properties. An important advance has been reported by Battaglia and Peng,[81] where non-coordinating solvents have been used to prepare monodispersed, high-quality crystalline samples of zinc blende InP and InAs in a similar manner to the green route used by the same research group, as described in Chapter 1. In this route, rather than using the capping agent as the solvent, a high-boiling point solvent, ODE, was used into which could be added precise amounts of the required capping agent. The use of non-coordinating solvents allowed controllable nucleation and hence improved the growth process. In contrast to the work described above, common passivating agents such as phosphines, phosphine oxides and amines were initially suggested to be ineffective, but fatty acids of an intermediate chain length were found to be excellent capping agents although the concentration of the ligand was crucial to the quality of the particles produced. The unsuitability of amines is surprising because of their previous successful use in InP synthesis, especially as they have previously been noted as activating the phosphorus precursor.[43,44] The reaction utilised the same precursors as those described above; $In(COOCH_3)_3$ and $E(SiMe_3)_3$, (E = As, P) as—unlike the non-coordinating solvent route to II–VI materials—metal oxides or elemental phosphorus were found to be insoluble and unreactive respectively. The indium precursor was mixed in ODE with the capping agent and heated to 120 °C to affect dissolution, which was followed by 2 hours of evacuation and back-flushing three times with an inert gas. This was followed by injection of an ODE solution of $E(SiMe_3)_3$ (E = As, P) at 300 °C, and growth at 270 °C. Successive precursor additions could be performed, dropwise, at 250 °C, where the precursors were added separately to maintain the indium-rich reaction and the tight size distribution. For the best-quality results, the indium : pnictide ratio was found to be 2 : 1 for InP, and 8 : 1 for InAs. The solvent also had to be rigorously dried before use, unlike the synthesis of II–VI, some of which could even be prepared in air. The absorption spectra of both InP and InAs displayed clear excitonic shoulders at 500–600 nm for InP and 600–900 nm for InAs, which suggested a limited size range of materials with an excellent low size distribution. Unfortunately, the emission quantum yield of InP was described as low and the emission of InAs was not reported. The synthesis of InAs was reported in more depth elsewhere,[82] which described the synthesis of extremely small (*ca.* 2 nm) InAs QDs with TOP in ODE at 150 °C; their small size was due to the high reactivity of $As(SiMe_3)_3$. The particles exhibited absorption spectra with features in the blue region consistent with magic cluster formation, and the particles grew through self-focusing by interparticle diffusion, shifting the absorption spectra well into the red with a maximum excitonic shoulder at *ca.* 850 nm.

The use of ODE is not without problems. It was suggested that the high temperatures used with long-chain carboxylate precursors resulted in hydrolysis leading to In_2O_3 particle formation unless thorough degassing was employed.[83] This oxidation could reportedly be avoided and the need for a degassing step removed completely by using low temperatures (<200 °C) and long-chain amines to activate the reaction, even though previously these ligands were found to be unsuitable. Using this reaction, where a mixture of $P(SiMe_3)_3$ and octylamine (OA) were injected into a hot solution (178 °C) of indium acetate and myristic acid in ODE, InP particles that emitted across the entire visible region of the electromagnetic spectrum could be obtained by varying reaction conditions, notably the concentration of myristic acid. Clear excitonic peaks could be obtained in the absorption spectra from 500 to 700 nm, suggesting that the particles were monodispersed, which was confirmed by electron microscopy. The particles were shown to oxidise rapidly in air, with the absorption spectra changing over 12 hours. The emission quantum yields were typically low (<1%) but were markedly improved to a maximum of 40% by simple passivation with a ZnS shell.

The InP/ZnS materials could also be transferred to water using thiols, as described in Chapter 6, without losing any brightness.

This improvement in the synthesis of InP is significant as the materials are highly monodispersed, crystalline and importantly cover the entire visible region; an even wider range than CdSe particles, without size-selective precipitation or harsh synthesis conditions. The use of long-chain amines to avoid oxide formation is, however, in direct contrast to a report that offers a differing explanation of the role of the amine; while the presence of an oxide layer has been reported as essential for luminescent InP QD, this is normally the result of exposure of the particle to ambient conditions. However, the amine-induced growth of an oxide layer on InP has been reported, which increased the emission quantum yield to 6–8%.[84] The introduction of OAm into the reaction flask with the indium precursor before the addition of the phosphorus reagent reportedly resulted in the growth of an In_2O_3 shell. In this example, the use of indium myristate as the indium precursor resulted in the generation of myristic acid anhydride, which underwent a condensation reaction with the amine to generate an amide and water, which hydrolysed excess indium precursor generating the wide-bandgap oxide shell, which has a lattice mismatch of 14%. The use of the amine altered the size distribution of the particles, again leading to the suggestion that the amine activated the phosphorus precursor.

This activation claim is again in contrast to the report by Allen *et al.*,[85] who reported a detailed study into the mechanism of InP formation and showed that the phosphorus precursors are completely depleted after injection, suggesting the growth step was due to ripening from non-InP species. Investigations (using NMR) into InP formation using long-chain amines at relatively low temperatures have shown that in the case of OA-capped particles, the indium precursor coordinates with the capping agent, before one $P(SiMe_3)_3$ reversibly coordinates to the solvation sphere, giving complex **1** (Figure 2.5). The complex then lost one ligand from the indium precursor, forming a stable In–P bond, while the $P(SiMe_3)_3$ lost a silylmethyl group, forming intermediate **2** in Figure 2.5, which then formed InP clusters and QDs. This convincingly suggested, contrary to prior results which indicated that protic reagents such as primary amines induced the decomposition of $P(SiMe_3)_3$ and hence acted as activators, that amines are actually inhibitors for InP QD formation, restricting precursor decomposition due to steric factors where the bulky silylated precursor is unlikely to approach the indium to allow the formation of reaction intermediate **1** (Figure 2.5).

Another study showed that when using indium carboxylate precursors, further surfactants were not actually necessary to produce capped nano-particles of InP.[86] In this case, the carboxylates were made prior to the reaction, from $In(C_5H_5)_3$ and the carboxylic acid directly, followed by puri-fication. The carboxylate was then added to ODE, which was then degassed and stabilised at an elevated temperature before the injection of an ODE solution of $P(SiMe_3)_3$. After several minutes, the nucleation stage was terminated by the injection of room-temperature ODE, allowing a discrete

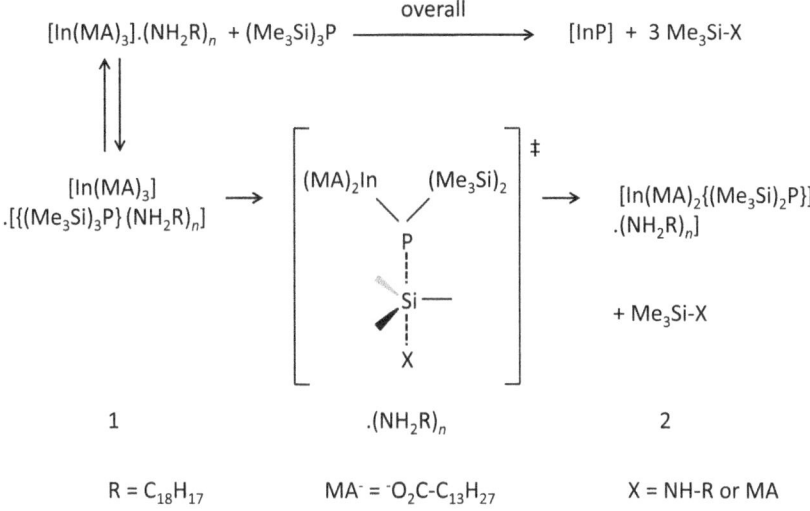

Figure 2.5 Mechanism of InP formation. Reprinted with permission from P. M. Allen, B. J. Walker and M. G. Bawendi, *Angew. Chem., Int. Ed.*, 2010, **49**, 760. Copyright Wiley-VCH Verlag GmbH & Co. KGaA.

separate growth stage at a slightly lower temperature for up to 2 hours. In this case, the growth of the nanoparticles did not significantly shift the absorption edge from the initial position, although by varying reaction temperatures, time and carboxylate chain length, the absorption excitonic shoulder could be altered from 476 nm to 574 nm, with associated near band edge emission tuneable from 545 nm to 661 nm and quantum yields of up to 2%.

One method of obtaining controlled nucleation is the use of weakly coordinating solvents such as high-boiling esters, as described by Xu *et al.*[87] Compounds such as methyl myristate and dibutylsebacate were combined with long-chain carboxylic acids (myristic acid was found to be the most favourable) and amines and heated to 260 °C, followed by the injection of a mixture of $(CH_3)_3In$ and $P(SiMe_3)_3$ in an ester. The reaction was then cooled to 200 °C for growth, after which the InP particles could be isolated by precipitation using acetone/methanol and redispersed in chloroform. Comparison of the optical properties of InP grown in esters to InP grown in ODE suggested that the nucleation of ester-grown materials was slower than those grown in the truly non-coordinating solvent, although faster than those grown in TOPO. The use of the metal alkyl rather than the metal carboxylate gave InP of a higher quality with higher emission quantum yields. Although these values were not reported, the emission was shown to be predominantly band edge in origin. The use of protic amines, again suggested to accelerate the nucleation, was found to improve the optical properties of the particles.

Although most of the work on the synthesis of III–V materials in non-coordinating solvents has focused on InP, high-quality InAs can also be made by a different precursor. An interesting report[88] describes the generation of arsine gas (AsH_3) from the addition of HCl to Zn_3As_2, which was then passed through a hot (220 °C) solution of ODE and indium acetate and myristic acid over a 20 minute period to result in zinc blende InAs particles, in a similar reaction to that described by the Reiss group.[53] In a similar manner to InP grown using PH_3, the generation of AsH_3 flow was strongest in the first 5 minutes of particle growth, and was then slower for the remaining duration of the reaction. Addition of further ODE was also used to drop the solution temperature and stop the nucleation, allowing a separate and discrete growth step, although prolonged reaction times resulted in the precipitation of elemental arsenic. Varying reaction conditions improved the size distribution and photoluminescence, with the emission being particularly sensitive to the indium : capping agent ratio. The absorption profile showed a clear excitonic peak at about 700 nm, with emission between *ca.* 750 and 800 nm although the quantum yield was still below 1% and decreased upon prolonged oxidation, in contrast to InAs prepared by $As(SiMe_3)_3$. The emission could be improved by addition of a ZnSe shell, as described in Chapter 5.

2.4   Group III–Antimonides

There are few reports of solution routes to antimonide-based QDs, although the dehalosilylation reaction is suitable for the synthesis of GaSb nanoparticles.[89] Addition of $Sb(SiMe_3)_3$ to a toluene solution of $GaCl_3$ at room temperature resulted in an immediate reaction, giving GaSb and Me_3SiCl after 24 hours reflux. The resulting materials could be purified by sublimation at 400 °C, giving an antimony-rich sample of nanocrystalline GaSb with dimensions of 30–40 nm. Although this route does not utilise a capping agent and no optical properties were reported, it does confirm that the dehalosilylation reaction is a viable route to these materials.

This route was also extended to the preparation of InSb particles using $In(COOCH_3)_3$ and stearic acid in ODE.[90] After degassing the indium precursor and capping agents at elevated temperatures, followed by cooling to 100 °C, a solution of $Sb(SiMe_3)_3$ in TOP was injected into the reaction flask, followed by 5 minutes growth and then cooling using an ice bath. Isolation of the InSb particles was achieved using solvent/non-solvent interactions. The particles had an absorption edge at *ca.* 700–800 nm, a significant shift from the bulk band edge at 0.17 eV (7300 nm) although no excitonic feature was observed. Near band edge emission was also observed with a maximum at 800 nm. The particles were 3–13 nm in diameter and crystalline, although no discussion of the crystal phase was reported.

Zinc blende-structured alloys of $InAs_xSb_{1-x}$ could be prepared by combining arsenic and antimony precursors.[91] The particles, *ca.* 3 nm in diameter, were prepared by injections of varying amounts of $Sb(SiMe_3)_3$ and $As(SiMe_3)_3$ into an ODE solution of $In(COOCH_3)_3$ and oleic acid at 300 °C, followed by growth for only 30 minutes at 270 °C, resulting in alloyed particles of $InAs_xSb_{1-x}$ of varying compositions. Precursor ratios (As : Sb) of 90 : 10, 66 : 33 and 50 : 50 resulted in composition of 97 : 3, 90 : 10 and

86 : 14 due to the difference in reactivity of the precursors. Increasing the antimony content resulted in a red shift in the optical properties due to the small bandgap on InSb relative to InAs, with emission tuneable between 770 nm and 880 nm. Attempts to prepare InP_xSb_{1-x} were less successful because of the large size difference between phosphorus and antimony atoms.

2.5 Group III–Nitrides

Group III–nitrides are a class of wide-bandgap semiconductors with numerous applications in optoelectronics. Although several chemical precursor routes exist to nitrogen-containing nanoparticles,[2] these have not expanded into organometallic solution routes as easily as other III–V materials. Unlike most other III–V QDs, which are easily accessible using the dehalosilylation reaction, nitride-containing QDs are hard to prepare because there are few obvious, simple or available precursors. The use of silylated precursors is not possible because of the relatively strong N–Si bond (when compared to the P–Si bond) although the successful use of $HN(SiMe_3)_2$ to prepare GaN in an autoclave has been reported; however, no optical properties were discussed.[92] The use of simple compounds, such as $InBr_3$, has been found to be unsuitable for the synthesis of InN, for example, when using Li_3N.[93] Interestingly, indium metal formation has been reported as an unwanted side product in similar reactions.[94]

In some of the earliest work towards solution synthesis of nitride-based QDs, zinc blende-structured GaN QDs, 3.0 $\pm$ 1.2 nm in diameter were prepared by the thermolysis of the poly(imidogallane)$\{Ga(NH)_{3/2}\}_n$ in TOA at 360 °C over 24 hours, under an ambient flow of ammonia, followed by cooling to 220 °C and the addition of a further surface capping agent (HDA).[95] GaN reportedly has an excitonic diameter of just 5 nm, so the materials prepared can be described as weakly confined, exhibiting an excitonic feature at *ca.* 330 nm, shifted from the bulk band edge at 390 nm with weak broad emission centred at *ca.* 380 nm. This route was improved by using a dimeric version of the poly(imidogallane), $Ga_2[N(Me)]_6$, avoiding the need for ammonia, which was thermolysed in TOA and HDA for up to 3 days.[96] The resulting GaN, 2–4 nm in diameter, with an average size of 2.4 nm, had a clear zinc blende structure as determined by XRD and an optical band edge at *ca.* 290 nm after 1 day of heating, and *ca.* 300 nm after 60 hours. Impressively, the 60 hour sample appeared to have a slight excitonic shoulder. Emission was dominated by broad deep trap emission between 400 and 600 nm, although importantly, band edge emission was observed at 305 nm. Interestingly, the use of HDA was found to be essential for the formation of GaN. This amendment to the earlier route by Mićić produced a larger yield of materials, although solubility in the desired solvent was still an issue.

The use of TOA has been explored in the synthesis of InN materials using $InBr_3$ and NaN_3 as precursors,[97] although it was suggested that the ligand was not entirely suitable as it induced the reduction of the indium precursor to its elemental form and promoted the transformation of the azide to ammonia through hydrogen abstraction. InN was, however, reportedly prepared using TOPO. A stabilised azide, $In(CH_2)_3N(CH_3)_2N_3$ was thermolysed in TOPO at 230 °C for several hours, yielding InN, as determined by electron microscopy and XPS.[98] The origin of the optical characteristics was unclear, with bright blue emission being observed and assigned as emission from TOPO, with absorption from the nanoparticles measured at 570 nm and weak emission at 690 nm, consistent with InN particles *ca.* 4.5 nm in diameter.

The same surfactant was again successfully used in the production of relatively monodispersed sample of GaN, *ca.* 5 nm in diameter, using $GaCl_3$ and Li_3N as precursors in dibenzofuran containing TOPO at 280 °C.[99] The particle size could be tuned by altering the dibenzofuran : precursor ratio. The particles were analysed by XRD; however, a crystalline phase was not assigned despite the presence of clear reflections. Few optical properties were discussed, although a broad emission spectrum was reported which suggested, when referring to the known difference in conduction band between zinc blende and wurtzite-structured materials, that a mix of crystalline materials might have been present.

In contrast to earlier reports by Wells,[94] a similar method using $InBr_3$ and $NaNH_2$ has been reported, where the two precursors were reacted at 250 °C under ammonia, and the resulting powder (InN) isolated by centrifugation.[100] The black powder was then washed to remove excess indium precursor. At this point, the product consisted of aggregated materials, which, when sonicated with nitric acid, released the nitride nanoparticles and removed indium metal contaminants. These particles could then be sonicated with OAm, forming a stable colloidal solution of InN. The particle shape was polydispersed and irregular, with an average particle size of 6.2 ± 2.0 nm and an optical room-temperature band edge of 1.29 eV (*ca.* 960 nm), although no emission was observed and the actual band edge was estimated to be approximately 0.8 eV (*ca.* 1550 nm), a result of the Moss–Burstein shift. This method is notable as the nanoparticles could be obtained from the large aggregates obtained initially after the reaction by simple processing, and the capping agent was added after the reaction and processing was finished.

In conclusion, it might be assumed that III–V QDs are poor relations to the more popular II–VI materials. This is not the case; although III–V materials are harder to prepare, recent advances such as the use of non-coordinating solvents have resulted in high-quality monodispersed samples that are structurally the equal of, for example, CdSe. The optical properties are also of high quality, with the exception of quantum yield, although simple procedures such as fluoride etching produce materials that are actually more emissive. It is also simple to prepare brightly emitting high-quality core/shell materials, of, for example InP/ZnS in a single-pot reaction which will be described in Chapter 5. These materials can be tuned to emit across the entire visible region, with the added benefit of avoiding cadmium, which is important in applications where the particles may come into contact with the environment or biological materials.

References

1. C. B. Murray, D. J. Norris and M. G. Bawendi, *J. Am. Chem. Soc.,* 1993, **115**, 8706.
2. M. Green, *Curr. Opin. Solid State Mater. Sci.,* 2002, **6**, 355.
3. M. J. Seong, O. I. Mićić, A. J. Nozik and A. Mascarenhas, *Appl. Phys. Lett.,* 2003, **82**, 185.
4. S.-H. Kim, R. H. Wolters and J. R. Heath, *J. Chem. Phys.,* 1996, **105**, 7957.
5. E. Poles, D. C. Selmarten, O. I. Mićić and A. J. Nozik, *Appl. Phys. Lett.,* 1999, **75**, 971.
6. D. Bertram, O. I. Mićić and A. J. Nozik, *Phys. Rev. B: Condens. Matter Mater. Phys.,* 1998, **57**, R4265.
7. M. C. Beard, G. M. Turner, J. E. Murphy, O. I. Micic, M. C. Hanna, A. J. Nozik and C. A. Schmuttenmaer, *Nano Lett.,* 2003, **3**, 1695.
8. M. Kuno, D. P. Fromm, A. Gallagher, D. J. Nesbitt, O. I. Micic and A. J. Nozik, *Nano Lett.,* 2001, **1**, 557.
9. U. Banin, Y. Cao, D. Katz and O. Millo, *Nature,* 1999, **400**, 542.
10. T. Nann, S. K. Ibrahim, P.-M. Woi, S. Xu, J. Ziegler and C. J. Pickett, *Angew. Chem., Int. Ed.,* 2010, **49**, 1574.
11. J. R. Heath and J. J. Shiang, *Chem. Soc. Rev.,* 1998, **27**, 65.
12. A. H. Cowley and R. A. Jones, *Angew. Chem., Int. Ed. Engl.,* 1989, **28**, 1215.
13. W. E. Buhro, *Polyhedron,* 1994, **13**, 1131.
14. R. L. Wells, C. G. Pitt, A. T. McPhail, A. P. Purdy, S. Shafieezad and R. B. Hallcock, *Chem. Mater.,* 1989, **1**, 4.
15. M. D. Healy, P. E. Laibinis, P. D. Stupik and A. R. Barron, *Chem. Commun.,* 1989, 359.
16. C. G. Pitt, A. P. Purdy, K. T. Higa and R. L. Wells, *Organometallics,* 1986, **5**, 1266.
17. S. M. Stuczynksi, R. L. Opila, P. Marsh, J. G. Brennan and M. L. Steigerwald, *Chem. Mater.,* 1991, **3**, 379.
18. T. Douglas and K. H. Theopold, *Inorg. Chem.,* 1991, **30**, 596.
19. S. T. Barry, S. Belhumeur and D. S. Richeson, *Organometallics,* 1997, **16**, 3588.
20. R. L. Wells, R. B. Hallock, A. T. McPhail, C. G. Pitt and J. D. Johansen, *Chem. Mater.,* 1991, **3**, 381.
21. R. L. Wells, M. F. Self, A. T. McPhail, S. R. Aubuchon, R. C. Woudenberg and J. P. Jasinski, *Organometallics,* 1993, **12**, 2832.
22. S. R. Aubuchon, A. T. McPhail, R. L. Wells, J. A. Giambra and J. R. Bowser, *Chem. Mater.,* 1994, **6**, 82.
23. R. L. Wells, S. R. Aubuchon, S. S. Kher, M. S. Lube and P. S. White, *Chem. Mater.,* 1995, **7**, 793.
24. M. A. Olshavsky, A. N. Goldstein and A. P. Alivisatos, *J. Am. Chem. Soc.,* 1990, **112**, 9438.
25. H. Uchida, C. J. Curtis and A. J. Nozik, *J. Phys. Chem.,* 1991, **95**, 5382.
26. L. D. Potter, A. A. Guzelian, A. P. Alivisatos and Y. Wu, *J. Chem. Phys.,* 1995, **103**, 4834.

27. H. Uchida, T. Matsunaga, H. Yoneyama, T. Sakata, H. Mori and T. Sasaki, *Chem. Mater.,* 1993, **5**, 717.
28. O. I. Mićić, C. J. Curtis, K. M. Jones, J. R. Sprague and A. J. Nozik, *J. Phys. Chem.,* 1994, **98**, 4966.
29. M. Tomaselli, J. L. Yarger, M. Bruchez, R. H. Havlin, D. deGraw, A. Pines and A. P. Alivisatos, *J. Chem. Phys.,* 1999, **110**, 8861.
30. O. I. Mićić, J. R. Sprague, C. J. Curtis, K. M. Jones, J. L. Machol, A. J. Nozik, H. Giessen, B. Fluegel, G. Mohs and N. Peyghambarian, *J. Phys. Chem.,* 1995, **99**, 7754.
31. O. I. Mićić and A. J. Nozik, *J. Lumin.,* 1996, **70**, 95.
32. M. Furis, Y. Sahoo, D. J. MacRae, F. S. Manciu, A. N. Cartwright and P. N. Prasad, *J. Phys. Chem. B,* 2003, **107**, 11622.
33. A. A. Guzelian, J. E. B. Katari, A. V. Kadavanich, U. Banin, K. Hamad, E. Juban, R. H. Wolters, C. C. Arnold and J. R. Heath, *J. Phys. Chem.,* 1996, **100**, 7212.
34. O. I. Mićić, K. M. Jones, A. Cahill and A. J. Nozik, *J. Phys. Chem. B,* 1998, **102**, 9791.
35. J. Jasinski, V. J. Leppert, S.-T. Lam, G. A. Gibson, K. Nauka, C. C. Yang and Z.-L. Zhou, *Solid State Commun.,* 2007, **141**, 624.
36. S. Adam, D. V. Talapin, H. Borchert, A. Lobo, C. McGinley, A. R. B. De Castro, M. Haase, H. Weller and T. Möller, *J. Chem. Phys.,* 2005, **123**, 084706.
37. O. I. Mićić, A. J. Nozik, E. Lifshitz, T. Rajh, O. G. Poluektov and M. C. Thurnauer, *J. Phys. Chem. B,* 2002, **106**, 4390.
38. O. I. Mićić, J. Sprague, Z. Lu and A. J. Nozik, *Appl. Phys. Lett.,* 1996, **68**, 3150.
39. L. Langof, E. Ehrenfreund, E. Lifshitz, O. I. Micic and A. J. Nozik, *J. Phys. Chem. B,* 2002, **106**, 1606.
40. D. V. Talapin, N. Gaponik, H. Borchert, A. L. Rogach, M. Haase and H. Weller, *J. Phys. Chem. B,* 2002, **106**, 12659.
41. D. D. Lovingood and G. F. Strouse, *Nano Lett.,* 2008, **8**, 3394.
42. S. Xu, J. Ziegler and T. Nann, *J. Mater. Chem.,* 2008, **18**, 2653.
43. O. I. Mićić, S. P. Ahrenkiel and A. J. Nozik, *Appl. Phys. Lett.,* 2001, **78**, 4022.
44. J. M. Nedeljković, O. I. Mićić, S. P. Ahrenkiel, A. Miedaner and A. J. Nozik, *J. Am. Chem. Soc.,* 2004, **126**, 2632.
45. Y.-H. Kim, Y.-W. Jun, B.-H. Jun, S.-M. Lee and J. Cheon, *J. Am. Chem. Soc.,* 2002, **124**, 13656.
46. T. J. Trentler, K. M. Hickman, S. C. Goel, A. M. Viano, P. C. Gibbons and W. E. Buhro, *Science,* 1995, **270**, 1791.
47. S. P. Ahrenkiel, O. I. Mićić, A. Miedaner, C. J. Curtis, J. M. Nedeljković and A. J. Nozik, *Nano Lett.,* 2003, **3**, 833.
48. F. Wang and W. E. Buhro, *J. Am. Chem. Soc.,* 2007, **129**, 14381.
49. H. Yu, J. Li, R. A. Loomis, L.-W. Wang and W. E. Buhro, *Nat. Mater.,* 2003, **2**, 517.
50. D. D. Fanfair and B. A. Korgel, *Cryst. Growth Des.,* 2005, **5**, 1971.

51. F. Wang, H. Yu, J. Li, Q. Hang, D. Zemlyanov, P. C. Gibbons, L.-W. Wang, D. B. Janes and W. E. Buhro, *J. Am. Chem. Soc.,* 2007, **129**, 14327.
52. Z. Liu, D. Xu, J. Fang, K. Sun, W.-B. Jian and Y.-F. Lin, *Chem.–Eur. J.,* 2009, **15**, 4546.
53. L. Li, M. Protière and P. Reiss, *Chem. Mater.,* 2008, **20**, 2621.
54. T. H. Lim, S. Ravi, C. W. Bumby, P. G. Etchegoin and R. D. Tilley, *J. Mater. Chem.,* 2009, **19**, 4852.
55. Z. Liu, A. Kumbhar, D. Xu, J. Zhang, Z. Sun and J. Fang, *Angew. Chem., Int. Ed.,* 2008, **47**, 3540.
56. P. K. Khanna, K.-W. Jun, K. B. Hong, J.-O. Baeg and G. K. Mehrotra, *Mater. Chem. Phys.,* 2005, **92**, 54.
57. T. Strupeit, C. Klinke, A. Kornowski and H. Weller, *ACS Nano,* 2009, **3**, 668.
58. K.-W. Jun, P. K. Khanna, K. B. Hong, J.-O. Baeg and Y.-D. Suh, *Mater. Chem. Phys.,* 2006, **96**, 494.
59. T. Matsumoto, S. Maenosono and Y. Yamaguchi, *Chem. Lett.,* 2004, **33**, 1492.
60. C. Li, M. Ando, H. Enomoto and N. Murase, *J. Phys. Chem. C,* 2008, **112**, 20190.
61. H.-J. Byun, W.-S. Song and H. Yang, *Nanotechnology,* 2011, **22**, 235605.
62. A. Dorn, P. M. Allen and M. G. Bawendi, *ACS Nano,* 2009, **3**, 3260.
63. J. Wang, Q. Yang, Z. Zhang, T. Li and S. Zhang, *Dalton Trans.,* 2010, **39**, 227.
64. S. Carenco, M. Demange, J. Shi, C. Boissière, C. Sanchez, P. Le Floch and N. Mézailles, *Chem. Commun.,* 2010, **46**, 5578.
65. M. Soreni-Harari, N. Yaacobi-Gross, D. Steiner, A. Aharoni, U. Banin, O. Millo and N. Tessler, *Nano Lett.,* 2008, **8**, 678.
66. U. Banin, J. C. Lee, A. A. Guzelian, A. V. Kadavanich and A. P. Alivisatos, *Superlattices Microstruct.,* 1997, **22**, 559.
67. P. Yu, M. C. Beard, R. J. Ellingson, S. Ferrere, C. Curtis, J. Drexler, F. Luiszer and A. J. Nozik, *J. Phys. Chem. B,* 2005, **109**, 7084.
68. U. Banin, C. J. Lee, A. A. Guzelian, A. V. Kadavanich, A. P. Alivisatos, W. Jaskolski, G. W. Bryant, A. L. Efros and M. Rosen, *J. Chem. Phys.,* 1998, **109**, 2306.
69. A. konkar, S. Lu, A. Madhukar, S. M. Hughes and A. P. Alivisatos, *Nano Lett.,* 2005, **5**, 969.
70. U. Banin, Y. W. Cao, D. Katz and O. Millo, *Nature,* 1999, **400**, 542.
71. O. Millo, D. Katz, Y. W. Cao and U. Banin, *Phys. Rev. B: Condens. Matter Mater. Phys.,* 2000, **61**, 16733.
72. N. Tessler, V. Medvedev, M. Kazes, S. Kan and U. Banin, *Science,* 2002, **295**, 1506.
73. A. A. Guzelian, U. Banin, A. V. Kadavanich, X. Peng and A. P. Alivisatos, *Appl. Phys. Lett.,* 1996, **69**, 1432.
74. G. Schmid, *Nanoparticles*, Wiley-VCH, Weinheim, 2nd edn, 2010, p. 106.
75. S. Kan, T. Mokari, E. Rothenberg and U. Banin, *Nat. Mater.,* 2003, **2**, 155.
76. S. Kan, A. Aharoni, T. Mokari and U. Banin, *Faraday Discuss.,* 2004, **125**, 23.

77. A. Dong, H. Yu, F. Wang and W. E. Buhro, *J. Am. Chem. Soc.,* 2008, **130**, 5954.
78. F. Wang, H. Yu, S. Jeong, J. M. Pietryga, J. A. Hollingsworth, P. C. Gibbons and W. E. Buhro, *ACS Nano,* 2008, **2**, 1903.
79. M. Green, S. Norager, P. Moriarty, M. Motevalli and P. O'Brien, *J. Mater. Chem.,* 2000, **10**, 1939.
80. M. A. Malik, P. O'Brien, S. Norager and J. Smith, *J. Mater. Chem.,* 2003, **13**, 2591.
81. D. Battaglia and X. Peng, *Nano Lett.,* 2002, **2**, 1027.
82. R. Xie and X. Peng, *Angew. Chem., Int. Ed.,* 2008, **47**, 7677.
83. R. Xie, D. Battaglia and X. Peng, *J. Am. Chem. Soc.,* 2007, **129**, 15432.
84. M. Protière and P. Reiss, *Chem. Commun.,* 2007, 2417.
85. P. M. Allen, B. J. Walker and M. G. Bawendi, *Angew. Chem., Int. Ed.,* 2010, **49**, 760.
86. D. W. Lucey, D. J. MacRae, M. Furis, Y. Sahoo, A. N. Cartwright and P. N. Prasad, *Chem. Mater.,* 2005, **17**, 3754.
87. S. Xu, S. Kumar and T. Nann, *J. Am. Chem. Soc.,* 2006, **128**, 1054.
88. J. Zhang and D. Zhang, *Chem. Mater.,* 2010, **22**, 1579.
89. S. Schulz, L. Martinez and J. L. Ross, *Adv. Mater. Opt. Electron.,* 1996, **6**, 185.
90. C. M. Evans, S. L. Castro, J. J. Worman and R. P. Raffaelle, *Chem. Mater.,* 2008, **20**, 5727.
91. S.-W. Kim, S. Sujith and B. Y. Lee, *Chem. Commun.,* 2006, 4811.
92. K. Sardar and C. N. R. Rao, *Adv. Mater.,* 2004, **16**, 425.
93. R. L. Wells and J. F. Janik, *Eur. J. Solid State Inorg. Chem.,* 1996, **33**, 1079.
94. R. W. Cumberland, R. G. Blair, C. H. Wallace, T. K. Reynolds and R. B. Kaner, *J. Phys. Chem. B,* 2001, **105**, 11922.
95. O. I. Mićić, S. P. Ahrenkiel, D. Bertram and A. J. Nozik, *Appl. Phys. Lett.,* 1999, **75**, 478.
96. G. Pan, M. E. Kordesch and P. G. Van Petten, *Chem. Mater.,* 2006, **18**, 3915.
97. J. Choi and E. G. Gillan, *J. Mater. Chem.,* 2006, **16**, 3774.
98. P. S. Schofield, W. Zhou, P. Wood, I. D. W. Samuel and D. J. Cole-Hamilton, *J. Mater. Chem.,* 2004, **14**, 3124.
99. C. C. Chen and C.-H. Liang, *Tamkang J., Sci. Eng.,* 2002, **5**, 223.
100. J. C. Hsieh, D. S. Yun, E. Hu and A. M. Belcher, *J. Mater. Chem.,* 2010, **20**, 1435.

CHAPTER 3

The Preparation of IV–VI Semiconductor Nanomaterials

3.1 PbE (E = S, Se, Te)

Bulk lead chalcogenides are narrow-bandgap semiconductors with face-centred cubic structures, lending themselves to several potential applications, notably in thermoelectrics due to the large figure of merit (ZT) associated with the materials, and photovoltaics due to the ideal position of the quantum-confined bandgap.[1] Bulk PbSe exhibits a bandgap of 0.26 eV and an excitonic diameter of 92 nm.[2] PbTe has a bandgap of 0.25 eV and will reportedly exhibit quantum confinement effects below 152 nm diameter [3] (which is the longitudinal Bohr radius; the transverse Bohr radius is reportedly 12.9 nm), whereas PbS has a bandgap of 0.37 eV and an excitonic diameter of 40 nm.[4] Upon quantum confinement, the blue shift of the bandgap results in materials with optical properties in the infrared region, making these particles the materials of choice for most infrared-dependent applications,[5] with numerous reports of PbE (E = S, Se, Te)-based photodetectors,[6–9] solar cells,[10–12] LEDS,[13,14] lasers,[15] and field-effect transistors.[16] Quantum dots (QDs) of lead chalcogenides have even been used as biological labels,[17] where phase transfer of PbS particles using simple thiolated ligands was successful and maintained emission quantum yields of up to 26%.[18] PbSe particles have also been capped with SiO_2 and used in cellular imaging without any obvious toxicity issues.[19,20] Unlike the CdSe family of nanoparticles, which have analogous organic compounds with similar optical properties, there are currently no organic equivalents for strongly luminescent QD infrared emitters such as PbSe.

   The most common method of preparing QDs is the hot injection method where, for example, a metal salt is dissolved in solution with a capping agent

RSC Nanoscience & Nanotechnology No. 33
Semiconductor Quantum Dots: Organometallic and Inorganic Synthesis
By Mark Green
© Mark Green 2014
Published by the Royal Society of Chemistry, www.rsc.org

and heated, followed by injection of a chalcogen precursor solution into the reagents. The first organometallic synthesis of capped Pb chalcogenide QDs was actually achieved by the thermolysis of single-source precursors.[21,22] In these cases, lead dithio- and diselenocarbamates were thermolysed in trioctylphosphine oxide (TOPO) at relatively low temperatures (<200 °C), giving lead sulfide and lead selenide particles of varying sizes and shapes. The electronic spectra clearly showed absorption in the red/near infrared region, although no excitonic peak or emission was observed. Similar work, reported later, also described the synthesis of PbS using $Pb(S_2CNEt_2)_2$, with nanoparticles exhibiting a wide range of morphologies.[23]

The first hot injection method using two precursors was reported by Murray *et al.*,[24] who importantly showed the first absorption spectra with clear features (Figure 3.1), highlighting the first excitonic transition which

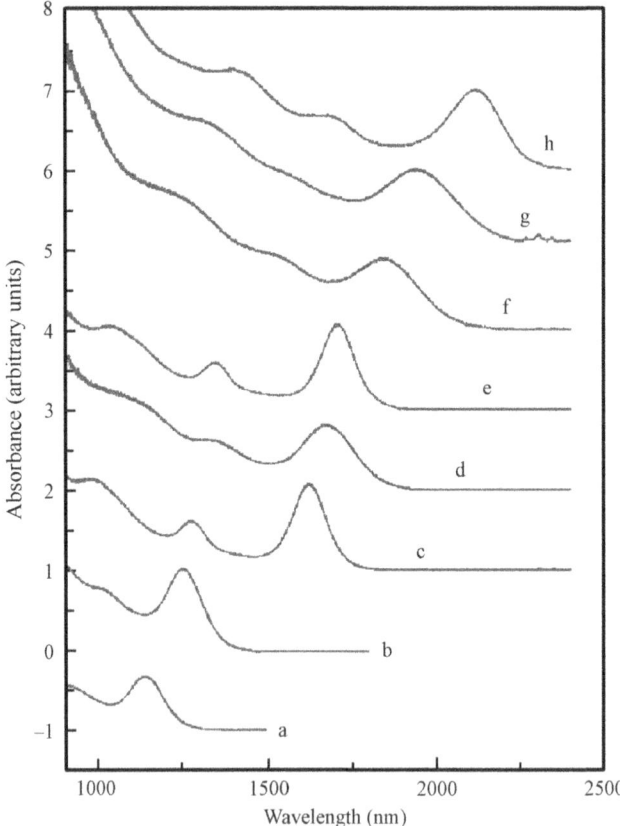

Figure 3.1 Absorption spectra of PbSe QDs of various diameters: (a) 3.0 nm, (b) 3.5 nm, (c) 4.5 nm, (d) 5.0 nm, (e) 5.5 nm, (f) 7.0 nm, (g) 8.0 nm, (h) 9.0 nm. Reprinted with permission from C. B. Murray, S. Sun, W. Gaschler, H. Doyle, T. A. Betley and C. R. Kagan, *IBM J. Res. Dev.*, 2001, **45**, 47. Copyright 2001 IBM.

was assigned as an interband transition for electrons and holes with a 1S envelope function (termed $1S_h1S_e$) and notably the second excitonic feature which was eventually assigned as $1P_h1P_e$.[25] In this seminal report, mono-dispersed quasi-spherical particles of PbSe were prepared by the injection of $Pb(CO_2(CH_2)_7(CH=CH)(CH_2)_7CH_3)_2$ and TOPSe into diphenylether at 150 °C, a notably lower synthesis temperature than those used for the preparation of II–VI materials, a consistent feature of the synthesis of IV–VI nano-materials. The particle size of 3–15 nm diameter was tuned by altering the reaction temperature, with higher temperatures giving larger particles. The route was then used to explore the optical properties of PbSe in some depth, demonstrating the tuneable band edge emission with extremely high quantum yields of up to 85%, ranging typically from 1.3 to 2.0 μm in size.[26,27] The samples exhibited small Stokes shifts of only 22 meV, assigned to the simple exciton structure. Larger particles (>8 nm in diameter) were prepared using the same method with either a higher synthesis temperature (250 °C) or a lower temperature (200 °C) route with essentially secondary precursor additions.[28] Larger particles, which typically favoured cubic morphologies, displayed emission spectra up to 4 μm (Figure 3.2), although the quantum yield dropped noticeably to 0.5% due to electronic coupling with the capping agent. This drop in quantum yield as the particle increased in size might

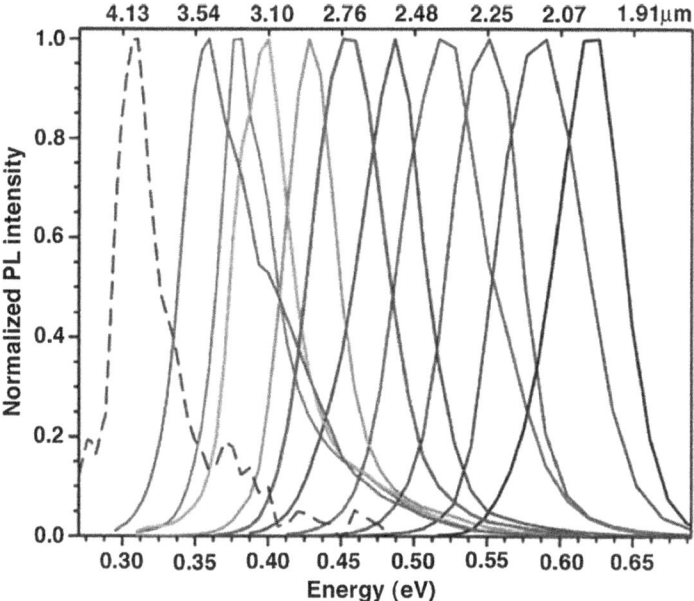

Figure 3.2 Room-temperature emission spectra of a series of PbSe QDs (dashed line spectrum obtained at 77 K). Reprinted with permission from J. M. Pietyga, R. D. Schaller, D. Werder, M. H. Stewart, V. I. Klimov and J. A. Hollingsworth, *J. Am. Chem. Soc.*, 2004, **126**, 11752. Copyright 2004 American Chemical Society.

appear counter-intuitive as larger particles have less surface defects. This unusual size dependence has been investigated using X-ray photoemission spectroscopy, which revealed a complex range of lead and selenium environments, essentially a core PbSe system, with a shell of $Pb_{1-x}Se$ and a final capping layer of selenium ions bound to trioctylphosphine (TOP), essentially a layer of TOPSe. Interestingly, two environments for TOP were observed, suggesting a second layer of TOP above the TOPSe layer. The smaller particles were found to have a thinner $Pb_{1-x}Se$ layer, which was suggested as a source of defects.[29] In contrast, investigations by NMR have reported that the passivation is almost entirely due to oleic acid coordinating to a uniquely lead surface, with just 0–5% of the surface ligands being TOP.[30] Although different, these results are clearly dependent upon reaction conditions. Other noteworthy investigation revealed important optical data, such as the extinction coefficient and the oscillator strength of QD PbSe.[31]

QDs of PbSe have been shown to be extremely air sensitive, spontaneously oxidising in ambient conditions to give PbO, $PbSeO_3$ and SeO_2 on the particle surface.[32] The resulting oxidation reduces the particle size significantly, blue-shifting and broadening the optical properties. The emission quantum yield initially quenches in minutes, attributed to oxygen on the particle surface enhancing non-radiative carrier trapping, although this reverses in a matter of hours to enhance the emission, often beyond the original quantum yield, attributable to the lowering of the rate of non-radiative recombination due to the widening bandgap in oxidised samples. This obviously impacts on the durability of devices fabricated with such materials.[33–35] Similar results were observed by Stouwdam *et al.*[36] and Dai *et al.*,[37] who also observed complete oxidation and dissolution of PbSe over a 12 day period under ultraviolet excitation.

Despite the inherent air sensitivity, highly luminescent magic-sized clusters (MSCs) of PbSe have been grown in ambient conditions using very simple, related chemistry.[38] Heating PbO with oleic acid, followed by cooling to 20 °C produced the lead precursor in a similar manner to the hot injection method. The precursor was then dissolved in octadecene (ODE), to which was added TOPSe in ambient conditions, and left stirring for a few hours while the solution darkened notably. The resulting PbSe particles exhibited reproducible, very specific electronic transitions consistent with magic cluster formation of various sizes. The clusters exhibited absorption excitonic peaks between *ca.* 625 and 880 nm, with broad emission between *ca.* 800 and 1000 nm and quantum yields of up to 90% despite being prepared in the presence of oxygen. Even when phase-transferred to water, the particles maintain relatively high quantum yields (up to 30%) over a matter of weeks. The inorganic core of the particles was suggested to be less than 2 nm, with PEGylated samples displaying hydrodynamic diameters of *ca.* 2.4 nm, extremely small when taking into account that a significant portion of the size could be attributed to the capping agent. This makes PbSe magic clusters potentially relevant to biological imaging applications.

Sashchiuk *et al.* also reported the synthesis of PbSe using Pb(OOCCH(C$_2$H$_5$)C$_4$H$_9$)$_2$ and TBPSe, with no further capping agent added.[39] Again, this reaction was carried out in toluene at room temperature, and apart from the synthesis of mercury selenide and telluride, are the only published examples of the rapid synthesis of QDs by organometallic-related methods at room temperature. The particle size (1.2–5 nm diameter) could be controlled by altering the ratio of reagents, and placing the resulting nanoparticle solution in the refrigerator inhibited reactions for months. The particles displayed absorption spectra consistent with PbSe with several excitonic features, although luminescence was dominated by trap emission. This could easily be improved by addition of a shell, which was easily achieved by the addition of a tributyl phosphine (TBP) solution of sulfur (TBPS), giving PbSe/PbS core/shell materials.

The same group reported the synthesis of PbSe-based assemblies based on individual QDs by increasing reagent concentrations, reaction temperature and time.[40] In these reactions, the structures were prepared by the rapid injection of TBP solutions of Pb(CO$_2$(CH$_2$)$_3$C$_6$H$_{11}$)$_2$ and TBPSe in TOPO at elevated temperatures of 150 °C, followed by relatively long growth periods of 15 minutes, or by gradual heating to 150 °C for a prolonged period of up to 2.5 hours. The assemblies grew from individual particles into large spheres of up to 450 nm diameter, but prolonged growth resulted in wire-like assemblies with a helical structure, with widths increasing from 20 nm to 150 nm over time. The resulting wires were composed of individual nanoparticles, stabilised by dipole–dipole interactions. Similar results were reported by Cho *et al.* who reported the use of the standard method of making PbSe particles using diphenylether as a solvent, but using a much higher synthesis temperature than usual (250 °C), with a range of differing capping agents.[41] The use of tetradecylphosphonic acid (TDPA) and oleic acid resulted in PbSe wires up to 30 μm in length, whereas the use of hexadecylamine (HDA) resulted in PbSe nanorings, all of which consisted of individual particles ordered by oriented attachment and fusion. Wires can also be grown in the presence of primary amines, which exhibited a zigzag morphology, possibly consistent with previous observations of helical structures. This is impressive when one consider that to form a 10 μm wire, 10^3 particles need to assemble, all through the (100) axis *via* dipole interactions. Interestingly, the spherical particles (termed quasi-spherical because of the faceted nature of the particles) have the highest dipole, whereas cubic particles—the preferred morphology for the larger materials—have an almost non-existent dipole moment. This is reinforced by the observation that cubes do not form wires, but individual spheres can be fused into wires upon mild heating. Nanowires of PbSe were also prepared using SeP(NEt$_2$)$_3$ as a chalcogen source rather than TOPSe, allowing synthesis at a lower temperature and for a long period of time. The rods synthesised had well-defined optical properties and exhibited near band edge emission (quantum yields of up to 15%), although the rods exhibited a larger Stokes shift than isotropic particles.[42] Theoretical investigations into the surface energies of

PbSe nanocrystals confirmed the dipole-driven interactions through high-energy surfaces that lead to oriented attachment, although these studies suggested that the interaction range is theoretically shorter than experimental observations.[43] Assembly through the (100) axis has also been achieved post-synthesis by the treatment of small oleic acid-capped PbSe with pyridine.[44]

The growth of these anisotropic materials is unusual as it appears that the mechanisms are solely oriented attachment of small particles. However, a range of straight and branched PbSe nanowires could also be prepared using Au/Bi core/shell particles as seeds, by injecting TOPSe and the seeds into a weakly coordinating capping agent (e.g. TOP or phenyl ether) and a lead carboxylate at temperatures of 150–200 °C.[45] The growth conditions and resulting structures were analogous to the growth of CdSe rods described in Chapter 1 (the 'geminate' mechanism), and were clearly not formed by oriented attachment of smaller particles. Although the growth temperature did not appear to favour either straight or branched structures, the Pb : Se ratio clearly affected product morphology, with Pb : Se ratios of 2 : 1 favouring straight wires whereas a 4 : 1 ratio favoured branched structures. The diameters of the seed particles were found to dictate the width of the wire. As mentioned, PbSe branched wires with similar structure to branched CdSe could be grown, exhibiting Y-shaped morphologies, although in the case of cubic PbSe T-shapes could also be observed, and also right angles which were actually T-shaped particles which had not fully grown. Shape control of PbSe particles was also achieved using noble metal nanoparticles as seeds. A variety of shapes, from rods and diamonds to stars, were all prepared rapidly at low temperatures using typical precursors by varying the metal and reagent concentrations.[46]

The role of the lead precursor counterion in particle morphology has been explored and found to be significant. Spherical (or quasi-spherical) particles were obtained when the precursor (commonly lead acetate and a carboxylic acid) was dried vigorously, removing acetate and acetic acid. If these side products were allowed to remain, then octahedral and star-shaped particles were predominant, due to acetic acid coordinating to the surface; the small size of this facilitated oriented attachment of the particles.[47] Similarly, the use of $Pb(NO_3)_2$ as a precursor resulted in hollow PbSe nanoboxes, due to the decomposition of the nitrate group, which resulted in gas bubbles that assisted the Ostwald ripening process.[48] A more in-depth investigation into the general reaction pathway has been undertaken by Steckel et al., who suggested two mechanisms occur during PbSe formation.[49] In the first mechanism, TOPSe acts as a Se^{2-} source, which reacts with the lead species yielding TOPO, oleate anhydride and PbSe monomers. In the second, TOPSe acts as a Se^0 source, reacting with Pb^0 which is provided by the reduction of the lead oleate by an in situ generated dialkylphosphine (which might also be present as an impurity), yielding PbSe QDs. Based on these results, the introduction of diphenylphosphine into PbSe reactions has been carried out in order to increase the yield of the reaction,[50,51] which is generally significantly lower than the synthesis yields of II–VI nanomaterials. The introduction of dialkylphosphines to improve reactions yield unfortunately also has an adverse effect on emission quantum yield and particle size control due to the fast reaction rate. The use of an alternative reducing agent, 1,2-hexadecanediol, introduced into the reaction with the ODE during chalcogen injection, resulted in an increase in reaction yield while maintaining the high-emission quantum yields, due to the controlled growth rate.[52]

The use of a non-coordinating solvent such as ODE has also been employed in the synthesis of PbSe by Yu *et al.*[53] In these reactions, PbO was complexed with oleic acid in ODE at 150 °C, followed by further heating to 180 °C. This was followed by injection of TOPSe and growth at 150 °C. Seconds after injection, small (3.5 nm diameter) particles could be isolated in low yields, although after 13 minutes large (9 nm) particle could be isolated in almost 100% yields. The particle exhibited a narrow size distribution (5–7% standard deviation on materials <10 nm, above which the standard deviation increased to 10–15%) without the need for size-selective precipitation. The particles were crystalline (rock salt) and spherical in shape; in contrast to other reports, the larger particles (13 nm) still maintained their spherical morphology. The particles displayed evidence of size-focusing growth as determined by the half width at half maximum (HWHM) of the emission profile, which had a maximum quantum yield of 89%. The particles displayed band edge emission, with first absorption peaks between 1100 nm and 2520 nm. Similarly, PbSe has been made in ODE using $Se(SiMe_3)_2$ as a precursor under comparable experimental conditions.[34]

An interesting method of preparing PbSe has been reported by cation exchange.[54] In this example, a lead precursor ($PbCl_2$) is injected into the reaction solution containing precursors for SnSe particle formation. As the tin precursor is extremely reactive, SnSe formation occurs rapidly, which then exchanged to provide PbSe particles in less than 1 second, allowing full growth in 5 minutes. The growth can be tuned by varying the injection and growth temperatures, and by multiple injections of precursor for particles above 7.5 nm diameter, resulting in QDs with emission from *ca.* 1250 nm to 3000 nm.

Although PbSe was the first lead chalcogenide to be investigated in depth, other materials have also been reported. Lead sulfide, PbS, with a cubic crystalline core has also been synthesised from an amendment of the Murray route to CdS.[55] In this case, PbO was heated to 150 °C with oleic acid for 1 hour, under vacuum or an inert gas, until the oleate complex was formed. To this was added rapidly *via* injection a solution of $S(SiMe_3)_2$ in either TOP or ODE with a Pb : S ratio of 2 : 1. The flask was then removed from the heat, and either allowed to cool or returned to the heating source (80–140 °C) after nucleation to allow slow particle growth. The particles exhibited absorption edges between 1200 and 1400 nm, although much smaller particles could be obtained by drastically reducing the oleic acid concentration, giving particles with an excitonic feature at *ca.* 800 nm. Emission was band edge, with quantum yields of up to 20% and a full width half maximum (FWHM) of approximately 100 meV. The particles could be isolated just seconds after precursor injection, and were found to be irregular in size and shape. Prolonged mild heating resulted in more monodispersed spherical particles, which could be isolated by the usual solvent/non-solvent interactions. The excitonic peak of the isolated particles suggested a relatively wide size distribution; however, but unusually, when isolated in a solvent, the excitonic peak and emission peak blue-shifted slightly and sharpened, indicating continued growth (digestive ripening) yielding spherical monodispersed particles. The oleic acid was found to coordinate to the surface lead sites, whereas TOP was found to passivate the surface sulfur sites.[56] This improved passivation was the origin of the increased emission quantum yields (up to 80%), along with the improved optical profiles and stability in TOP-coated PbS.[57] QDs of PbS prepared by this method have successfully been incorporated into light-emitting devices.[58] QDs of PbS have also been found to be air sensitive, in a similar manner to PbSe.[59] Smaller PbS particles, with less sulfur on the surface and denser capping agent concentrations exhibited trap defects introduced by the formation of $PbSO_3$, which merely extended carrier lifetime. The larger faceted particles, with sulfur-rich crystal planes such as (111) which were prone to oxidation, exhibited trapping species from $PbSO_4$ formation which acted as recombination centres resulting in device deterioration. In contrast, a size-dependent oxidation was also found by Peterson and Krauss[60] who found that smaller particles were more prone to oxidation, and suggested that charge trapping to an optically dark state along with photo-oxidation resulted in the decrease in emission. Nitrogen purging was also reported to result in brief photobrightening of the particles.

Related to the initial synthetic route is a method briefly described by Joo *et al.*, where monodispersed cubic particles of PbS 6–13 nm in diameter were prepared by the injection of sulfur in OAm into a solution of $PbCl_2$ complexed to OAm at 90 °C, followed by growth for 1 hour at 220 °C.[61] A further similar report described the synthesis of crystalline, monodispersed PbS using $PbCl_2$ suspended in OAm in more depth,[62] with the addition of sulfur in OAm at 120 °C. The particles were again isolated by non-solvent interactions, and the reaction could be used to produce up to 1.5 g of purified product. Interestingly, the material was found to be lead-rich, and X-ray photoelectron spectroscopy (XPS) confirmed the presence of a Pb–Cl monolayer shell on the surface. The size of the particles was controlled by altering the viscosity of the solution by amending the Pb : amine ratio. The increase in viscosity resulted in a decrease in the mass transfer coefficient, resulting in the particles growing in a diffusion-controlled reaction, essential for focusing the growth of the nanoparticles. The materials again emitted at *ca.* 1200–1600 nm, with a FWHM of approximately 62 meV and quantum yields of up to 40%; further studies revealed the extinction coefficient.[63] The use of elemental sulfur rather than the toxic and noxious-smelling $S(SiMe_3)_2$ is a clear benefit. Reports exist regarding the use of differing chalcogen precursors, such as thioacetamide, CH_3NH_2S, although safety issues with this material may limit its use.[64] Alloyed particles of PbS_xSe_{1-x} have been prepared, and used in photovoltaic devices.[65] In this synthesis, PbO was heated and degassed in oleic acid and ODE at 150 °C for a 1 hour, followed by a single injection of an ODE solution of TOPSe, $S(SiMe_3)_2$ and diphenylphosphine for 90 seconds, followed by cooling and dilution with hexane, and precipitation. The resulting particles were rich in sulfur due to the increased reactivity of the sulfur precursor and displayed emission between the typical values for PbS (which emits at shorter wavelengths than PbSe QDs) and PbSe QDs.

Lead telluride (PbTe) is an interesting material because of its large excitonic diameter, and has been prepared by similar routes to those described above. The first report on the synthesis of PbTe QDs was described by Lu *et al.* where particles were prepared by injecting a cold solution of $Pb(OAc)_3$ and TOPTe into diphenylether with oleic acid at 200 °C.[66] Upon growth for 5 minutes, spherical particles could be isolated, while extended growth for 25 minutes resulted in cubic particles, although size-selective precipitation was required for both spherical and cubic particles in order to obtain a mono-dispersed sample (Figure 3.3). Further studies[67] in which the ratio of precursors was widely varied with the use of a long-chain amine as a capping agent resulted in the formation of differing morphologies, with ratios of Pb : Te of 5 : 1 giving particles with a predominantly octahedral shape, while a ratio of 1 : 5 gave cubic particles. The use of phosphonic acids as capping agents was also found to significantly slow nucleation.

A similar method was reported by Urban *et al.* who described the synthesis of PbTe QDs.[4] In this method, $Pb(OAc)_3$ and oleic acid were dissolved in squalene, and heated under vacuum at 70 °C for 3 hours to remove water and acetic acid, a key factor in preparing shape-controlled particles as described earlier for PbSe. TOPTe was then injected at 180 °C, followed by growth at between 155 and 160 °C for 2 minutes before cooling to room temperature and isolation using solvent/non-solvent interactions. By altering the ratio of $Pb(OAc)_3$ and oleic acid, the particle size could be tuned, with lower concentration of oleic acid resulting in smaller particles due to the less stable lead acetate reacting quickly, yielding more nuclei. Allowing a slightly longer growth time also resulted in slightly larger particles. Using this method, cuboctahedral particles between 4 and 10 nm could be routinely produced, with a size distribution of about 5%. Increasing the temperature above 200 °C produced polydispersed cubic crystals and increasing the reaction time also resulted in an increase in polydispersity. The optical properties of spherical particles between 4 and 8 nm were also explored, with excitonic features observed between 1800 and 2400 nm. Simultaneously, Murphy *et al.* published a very similar synthesis[3] in which PbO and oleic acid (or a mixture of oleic and erucic acid) were mixed with ODE and heated under argon for 30 minutes, after which TOPTe was injected into the solution at 140–170 °C, with growth temperature of 80–130 °C over 6 minutes. The lower injection and growth temperatures resulted in smaller particles than those reported by Urban *et al.*, of 2.6–18 nm. As a result, the exciton was observed in the optical spectra between *ca.* 1000 nm and 2000 nm. Altering the precursor ratio also

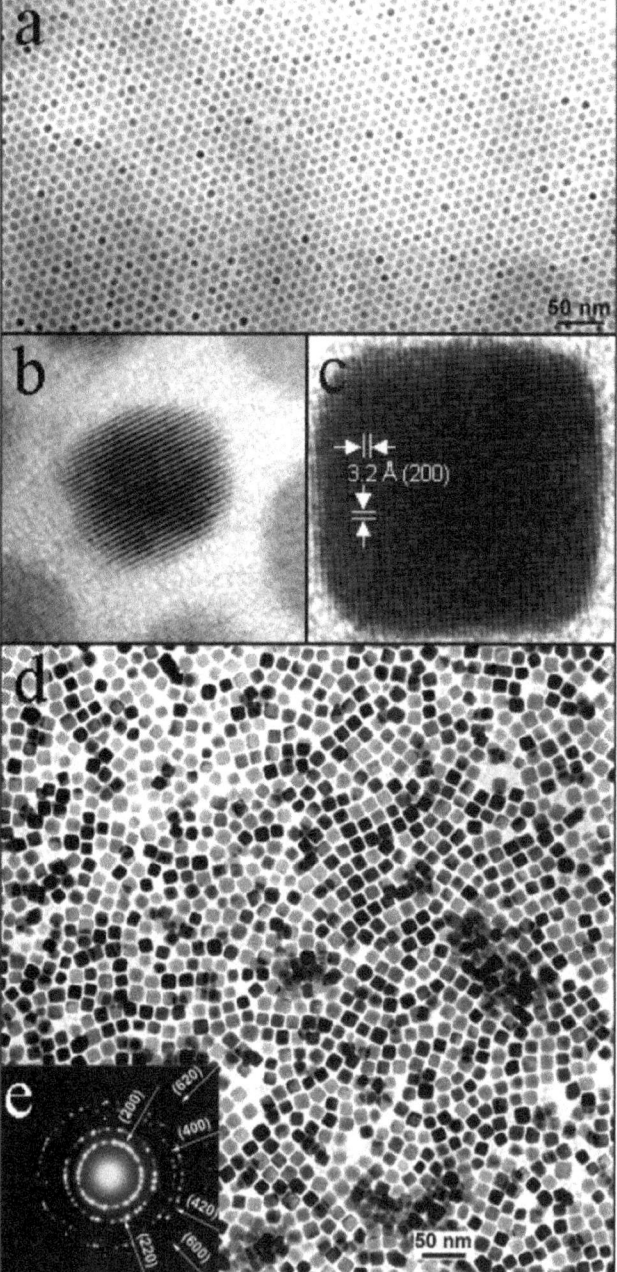

Figure 3.3 (a) Electron microscope image of spherical PbTe QDs. (b) High-resolution electron microscope image of a spherical PbTe QD. (c) High-resolution electron microscope image of a cubic PbTe QD. (d) Electron microscope image of cubic PbTe QDs monolayer. (e) Selected area electron diffraction pattern of 40 nm diameter cubic

changed particle size and prolonged heating at relatively elevated temperatures (193 °C) again resulted in cubic particles being formed, with a sphere/cube transition occurring at approximately 9.3 nm. Importantly, the emission spectra of small (2.9 nm diameter) PbTe particles were also measured for the first time, and found to be band edge with quantum yields of up to 52%, significantly higher than PbS, although not as high as some reports for PbSe. Although most synthetic routes resulted in either quasi-spherical, cubic or octahedral particles, PbTe rods could also be prepared by solution synthesis.[68] In this example, $PbCO_3$ and NaHTe were reacted together in water, the product isolated, dispersed in TOP and injected into HDA at 190–270 °C. The precursor was found to contain PbTe, the carbonate salt and tellurium which resulted in further growth once injected into the amine.

3.2 Other Group IV Chalcogenides

Other group IV–VI nanoscale compounds have been prepared by similar chemistries; GeTe nanowires have been prepared using $Ge(OCH(CH_3)_2)_4$ as a precursor.[69] In this case, the germanium precursor was dissolved in TOP, and placed in a flask containing ODE as the reaction solvent, oleic acid and OAm in the presence of a fluorine-doped tin oxide glass slide covered with a thin film of bismuth. The flask was then heated to 220 °C, resulting in the formation of bismuth particles in solution. Injection of TOPTe into the flask then resulted in the formation of GeTe nanowires, using the bismuth particles as nucleation sites. Interestingly, the wires, 50–100 nm in diameter and up to 2 μm long, had no trace of bismuth in the structure, suggesting that the usual solution–liquid–solid (SLS) growth mentioned in Chapter 1 was not the growth mechanism. The presence of tellurium particles 200–400 nm in diameter were also detected.

A similar method, using a GeI_2–TOP complex as a precursor (with no bismuth particles), and using TOPO and dodecanethiol as capping agents has been reported.[70] The dodecanethiol and TOPTe were injected rapidly and sequentially, resulting in amorphous GeTe particles. The dodecanethiol was found to be essential, as no germanium particles formed unless the ligand was present. It was deduced by NMR that the thiol displaced TOP, forming a Ge–thiol precursor, which reacted with TOPTe to yield GeTe particles 2.5–4.5 nm in diameter. The amorphous particles could be crystallised by heating to 240 °C. Larger nanoparticles of GeTe have also been reported using bis[bis(trimethylsilyl)-amino]Ge(II) as a germanium precursor.[71]

Tin(II) sulfide (α-SnS) has also been synthesised by heating a mixture of $Sn[N[(CH_3)_3Si]_2]_2$, TOP, oleic acid and ODE to 170 °C.[72] It is noteworthy that the use of $Sn(CO_2CH_3)_2$ did not yield the oleate complex, hence the need for the silylated precursor. To the hot precursor solution was injected a solution

---

of $CH_3C(S)NH_2$ in TOP/OAm, which caused particle formation. The reaction was allowed to proceed at 125 °C for 5 minutes, whereupon the particles were isolated by solvent/non-solvent interactions. The particles exhibited an indirect band edge at *ca.* 770 nm. SnTe has also been prepared from Sn $[N[(CH_3)_3Si]_2]_2$, which was dissolved in ODE and injected into a hot a solution of TOPTe and OAm at 150 °C, followed by growth for 1–2 minutes at 120 °C.[73] The resulting particles were approximately spherical, monodispersed and able to close-pack on a microscope grid. Bandgaps were found between *ca.* 3200 nm (14 nm diameter particles) and *ca.* 2300 nm (7.2 nm diameter particles), significantly into the infrared region.

Similarly, SnSe and $SnSe_2$ particles could be prepared by using $(CH_3)_3CSeSeC(CH_3)_3$ and $SnCl_2$ as precursors, attaining different stoichiometries by varying the Sn : Se ratio of the precursors.[74] The selenium precursor was injected into a solution of the $SnCl_2$ in dodecylamine (DDA) and dodecanethiol at 95 °C, followed by growth at 180 °C for 4 minutes. The resulting SnSe particles were clearly anisotropic, 19 nm wide ($\pm$5.1 nm) but varied widely in length. The particles exhibited a direct band edge at *ca.* 730 nm, a blue shift from the bulk value of *ca.* 950 nm.

Here, we demonstrate that the although the II–VI family of materials (Chapter 1) cover the visible region of the electromagnetic spectrum, other materials have been developed that exploit the near infrared region using similar synthetic chemistry.

References

1. J. Tang and E. H. Sargent, *Adv. Mater.,* 2011, **23**, 12.
2. F. W. Wise, *Acc. Chem. Res.,* 2000, **33**, 773.
3. J. E. Murphy, M. C. Beard, A. G. Norman, S. P. Ahrenkiel, J. C. Johnson, P. Yu, O. I. Mićić, R. T. Ellingson and A. J. Nozik, *J. Am. Chem. Soc.,* 2006, **128**, 3241.
4. J. J. Urban, D. V. Talapin, E. V. Shevchenko and C. B. Murray, *J. Am. Chem. Soc.,* 2006, **128**, 3248.
5. E. Sargent, *Adv. Mater.,* 2005, **17**, 515.
6. G. Konstantatos, J. Clifford, L. Levina and E. H. Sargent, *Nat. Photonics,* 2007, **1**, 531.
7. K. Szendrei, F. Cordella, M. V. Kovalenko, M. Böberl, G. Hesser, M. Yarema, D. Jarzab, O. V. Mikhnenko, A. Gocalinska, M. Saba, F. Quocji, A. Mura, G. Bongiovanni, P. W. M. Blom, W. Heiss and M. A. Loi, *Adv. Mater.,* 2009, **21**, 683.
8. V. Sukhovatkin, S. Hinds, L. Brzozwski and E. H. Sargent, *Science,* 2009, **324**, 1542.
9. S. A. McDonald, G. Konstantatos, S. Zhang, P. W. Cyr, E. J. D. Klem, L. Levina and E. H. Sargent, *Nat. Mater.,* 2005, **4**, 138.
10. N. Zhao, T. P. Osedach, L.-Y. Chang, S. M. Geyer, D. Wanger, M. T. Binda, A. C. Arango, M. G. Bawendi and V. Bulovic, *ACS Nano,* 2010, **4**, 3743.

11. J. J. Choi, Y.-F. Lim, M. B. Santiago-Berrios, M. Oh, B.-R. Hyun, L. Sun, A. C. Bartnik, A. Goedhart, G. G. Malliaras, H. D. Abruňa, F. W. Wise and T. Hanrath, *Nano Lett.,* 2009, **9**, 3749.
12. K. S. Leschkies, T. J. Beatty, M. S. Kang, D. J. Norris and E. S. Aydil, *ACS Nano,* 2009, **3**, 3638.
13. K. N. Bourdakos, D. M. N. M. Dissanayake, T. Lutz, S. R. P. Silva and R. J. Curry, *Appl. Phys. Lett.,* 2008, **92**, 153311.
14. G. Konstantos, C. Huang, L. Levina, Z. Lu and E. H. Sargent, *Adv. Funct. Mater.,* 2005, **15**, 1865.
15. R. D. Schaller, M. A. Petruska and V. I. Klimov, *J. Phys. Chem. B,* 2003, **107**, 13765.
16. D. V. Talapin and C. B. Murray, *Science,* 2005, **310**, 86.
17. B.-R. Hyun, H. Chen, D. A. Rey, F. W. Wise and C. A. Batt, *J. Phys. Chem. B,* 2007, **111**, 5726.
18. S. Hinds, S. Myrskog, L. Levina, G. Koleilat, J. Yang, S. O. Kelley and E. H. Sargent, *J. Am. Chem. Soc.,* 2007, **129**, 7218.
19. T. T. Tan, S. Tamil Selvan, L. Zhao, S. Gao and J. Y. Ying, *Chem. Mater.,* 2007, **19**, 3112.
20. M. Darbandi, W. Lu, J. Fang and T. Nann, *Langmuir,* 2006, **22**, 4371.
21. T. Trindade, P. O'Brien, X.-M. Zhang and M. Motevalli, *J. Mater. Chem.,* 1997, **7**, 1011.
22. T. Trindade, O. C. Monteiro, P. O'Brien and M. Motevalli, *Polyhedron,* 1999, **18**, 1171.
23. S.-M. Lee, Y.-W. Jun, S.-M. Cho and J. Cheon, *J. Am. Chem. Soc.,* 2002, **124**, 11244.
24. C. B. Murray, S. Sun, W. Gaschler, H. Doyle, T. A. Betley and C. R. Kagan, *IBM J. Res. Dev.,* 2001, **45**, 47.
25. M. Y. Trinh, A. J. Houtepen, J. M. Schins, J. Piris and L. D. A. Siebbeles, *Nano Lett.,* 2008, **8**, 2112.
26. B. L. Wehrenberg, C. Wang and P. Guyot-Sionnest, *J. Phys. Chem. B,* 2002, **106**, 10634.
27. H. Du, C. Chen, R. Krishnan, T. D. Krauss, J. M. Harbold, F. W. Wise, M. G. Thomas and J. Silcox, *Nano Lett.,* 2002, **2**, 1321.
28. J. M. Pietyga, R. D. Schaller, D. Werder, M. H. Stewart, V. I. Klimov and J. A. Hollingsworth, *J. Am. Chem. Soc.,* 2004, **126**, 11752.
29. S. Sapra, J. Nanda, J. M. Pietryga, J. A. Hollingsworth and D. D. Sarma, *J. Phys. Chem. B,* 2006, **110**, 15244.
30. I. Moreels, B. Fritzinger, J. C. Martins and Z. Hens, *J. Am. Chem. Soc.,* 2008, **130**, 15081.
31. I. Moreels, K. Lambert, D. De Muyunck, F. Vanhaecke, D. Poelman, J. C. Martins, G. Allan and Z. Hens, *Chem. Mater.,* 2007, **19**, 6101.
32. M. Sykora, A. Y. Koposov, J. A. McGuire, R. K. Schulze, O. Tretiak, J. M. Pietryga and V. I. Klimov, *ACS Nano,* 2010, **4**, 2021.
33. J. M. Luther, M. Law, M. C. Beard, Q. Song, M. O. Reese, R. J. Ellingson and A. J. Nozik, *Nano Lett.,* 2008, **8**, 3488.

34. G. I. Koleilat, L. Levina, H. Shukla, S. H. Myrskog, S. Hinds, A. G. Pattantyus-Abraham and E. H. Sargent, *ACS Nano,* 2008, **2**, 833.
35. B. Sun, A. T. Findikoglu, M. Sykora, D. J. Werder and V. I. Klimov, *Nano Lett.,* 2009, **9**, 1235.
36. J. W. Stouwdam, J. Shan, F. C. J. M. van Veggel, A. G. Pattantyus-Abraham, J. F. Young and M. Raudsepp, *J. Phys. Chem. C,* 2007, **111**, 1086.
37. Q. Dai, Y. Wang, Y. Zhang, X. Li, R. Li, B. Zou, J. T. Seo, Y. Wang, M. Liu and W. W. Yu, *Langmuir,* 2009, **25**, 12320.
38. C. M. Evans, L. Guo, J. J. Peterson, S. Maccagnano-Zacher and T. D. Krauss, *Nano Lett.,* 2008, **8**, 2896.
39. A. Sashchiuk, L. Langof, R. Chaim and E. Lifshitz, *J. Cryst. Growth,* 2002, **240**, 431.
40. A. Sashchiuk, L. Amirav, M. Bashouti, M. Krueger, U. Sivan and E. Lifshitz, *Nano Lett.,* 2004, **4**, 159.
41. K.-S. Cho, D. V. Talapin, W. Gaschler and C. B. Murray, *J. Am. Chem. Soc.,* 2005, **127**, 7140.
42. W.-K. Koh, A. C. Bartnik, F. W. Wise and C. B. Murray, *J. Am. Chem. Soc.,* 2010, **132**, 3909.
43. C. Fang, M. A. van Huis, D. Vanmaekelbergh and H. W. Zandbergen, *ACS Nano,* 2010, **4**, 211.
44. T. Hanrath, D. Veldman, J. J. Choi, C. G. Christova, M. M. Wienk and R. A. J. Janssen, *ACS Appl. Mater. Interfaces,* 2009, **1**, 244.
45. K. L. Hull, J. W. Grebinski, T. H. Kosel and M. Kuno, *Chem. Mater.,* 2005, **17**, 4416.
46. K.-T. Yong, Y. Sahoo, K. R. Choudhury, M. T. Swihart, J. R. Minter and P. N. Prasad, *Nano Lett.,* 2006, **6**, 709.
47. A. J. Houtepen, R. Koole, D. Vanmaekelbergh, J. Meeldijk and S. G. Hickey, *J. Am. Chem. Soc.,* 2006, **128**, 6792.
48. S. Chen, X. Zhang, X. Hou, Q. Zhou and W. Tan, *Cryst. Growth Des.,* 2010, **10**, 1257.
49. J. S. Steckel, B. K. H. Yen, D. C. Oertel and M. G. Bawendi, *J. Am. Chem. Soc.,* 2006, **128**, 13032.
50. J. M. Luther, M. Law, Q. Song, C. L. Perkins, M. C. Beard and A. J. Nozik, *ACS Nano,* 2008, **2**, 271.
51. M. Law, J. M. Luther, Q. Song, B. K. Hughes, C. L. Perkins and A. J. Nozik, *J. Am. Chem. Soc.,* 2008, **130**, 5974.
52. J. Joo, J. M. Pietryga, J. A. McGuire, S.-H. Jeon, D. J. Williams, H.-L. Wang and V. I. Klimov, *J. Am. Chem. Soc.,* 2009, **131**, 10620.
53. W. W. Yu, J. C. Falkner, B. S. Shih and V. L. Colvin, *Chem. Mater.,* 2004, **16**, 3318.
54. M. V. Kovalenko, D. V. Talapin, M. A. Loi, F. Cordella, G. Hesser, M. I. Bodnarchuk and W. Heiss, *Angew. Chem., Int. Ed.,* 2008, **47**, 3029.
55. M. A. Hines and G. D. Scholes, *Adv. Mater.,* 2003, **15**, 1844.
56. A. Lobo, T. Möller, M. Nagel, H. Borchert, S. G. Hickey and H. Weller, *J. Phys. Chem. B,* 2005, **109**, 17422.

57. K. A. Abel, J. Shan, J.-C. Boyer, F. Harris and F. C. J. M. van Veggel, *Chem. Mater.*, 2008, **20**, 3794.
58. L. Bakueva, G. Konstantatos, L. Levina, S. Musikhin and E. H. Sargent, *Appl. Phys. Lett.*, 2004, **84**, 3459.
59. J. Tang, L. Brzozowski, D. A. R. Barkhouse, X. Wang, R. Debnath, R. Wolowiec, E. Palmiano, L. Levina, A. G. Pattantyus-Abraham, D. Jamakosmanovic and E. H. Sargent, *ACS Nano*, 2010, **4**, 869.
60. J. J. Peterson and T. D. Krauss, *Phys. Chem. Chem. Phys.*, 2006, **8**, 3851.
61. J. Joo, H. Bin Na, T. Yu, Y. W. Kim, F. Wu, J. Z. Zhang and T. Hyeon, *J. Am. Chem. Soc.*, 2003, **125**, 11100.
62. L. Cademartiri, J. Bertolotti, R. Sapienza, D. S. Wiersma, G. Von Freymann and G. A. Ozin, *J. Phys. Chem. B*, 2006, **110**, 671.
63. I. Moreels, K. Lambert, D. Smeets, D. De Muynck, T. Nollet, J. C. Martins, F. Vanhaecke, A. Vantomme, C. Delerue, G. Allan and Z. Hens, *ACS Nano*, 2009, **3**, 3023.
64. C. Schliehe, B. H. Juarez, M. Pelletier, S. Jander, D. Greshnykh, M. Nagel, A. Meyer, S. Foerster, A. Kornowski, C. Klinke and H. Weller, *Science*, 2010, **329**, 550.
65. W. Ma, J. M. Luther, H. Zheng, Y. Wu and A. P. Alivisatos, *Nano Lett.*, 2009, **9**, 1699.
66. W. Lu, J. Fang, K. L. Stokes and J. Lin, *J. Am. Chem. Soc.*, 2004, **126**, 11798.
67. T. Mokari, M. Zhang and P. Yang, *J. Am. Chem. Soc.*, 2007, **129**, 9864.
68. N. Ziqubu, K. Ramasamy, P. V. S. R. Rajasekhar, N. Revaprasadu and P. O'Brien, *Chem. Mater.*, 2010, **22**, 3817.
69. M.-K. Lee, T. G. Kim, B.-K. Ju and Y.-M. Sung, *Cryst. Growth Des.*, 2009, **9**, 938.
70. M. A. Caldwell, S. Raoux, R. Y. Wang, H.-S. P. Wong and D. J. Milliron, *J. Mater. Chem.*, 2012, **20**, 1285.
71. M. J. Polking, J. J. Urban, D. J. Milliron, H. Zheng, E. Chan, M. A. Caldwell, S. Raoux, C. F. Kisielowski, J. W. Ager, R. Ramesh and A. P. Alivisatos, *Nano Lett.*, 2011, **11**, 1147.
72. S. G. Hickey, C. Waurisch, B. Rellinghaus and A. Eychmüller, *J. Am. Chem. Soc.*, 2008, **130**, 14978.
73. M. V. Kovalenko, W. Heiss, E. V. Shevchenko, J.-S. Lee, H. Schwinghammer, A. P. Alivisatos and D. V. Talapin, *J. Am. Chem. Soc.*, 2007, **129**, 11354.
74. M. A. Franzman, C. W. Schlenker, M. E. Thompson and R. L. Brutchey, *J. Am. Chem. Soc.*, 2010, **132**, 4060.

CHAPTER 4

The Preparation of Other Chalcogenides and Pnictide Nanomaterials

4.1 Copper-Based Chalcogenides

It could be argued that the organometallic route to quantum dots (QDs) has found the most success with chalcogen-containing semiconducting particles, as the II–VI and IV–VI families of materials are easily prepared yielding exceptionally high-quality QDs. Other families of technically important semiconductors incorporating chalcogens have also been prepared, notably I–III–VI materials, which have potential applications in solar energy, especially CuInSe$_2$, which has a bandgap between 1.04 eV (*ca.* 1190 nm) and 1.10 eV (*ca.* 1125 nm), and an exciton radius of 21.2 nm.[1] The related material CuInS$_2$, with a bandgap of *ca.* 1.53 eV (810 nm) and an excitonic diameter of 8.2 nm,[2,3] has also been prepared. Most of the synthetic pathways described are similar in their approach, apart from the preparation of I–III–VI QDs using single-source precursors, which is covered in Chapter 7.

The ternary compound CuInSe$_2$ has been synthesised in the form of spherical QDs, *ca.* 4 nm in diameter, by the injection of InCl$_3$ and CuCl, dissolved in trioctylphosphine (TOP), in trioctylphosphine oxide (TOPO) at 100 °C, followed by the injection of trioctylphosphine selenide (TOPSe) an hour later at 250 °C and growth for 1 day.[4,5] The particles, isolated by the usual solvent/non-solvent interaction, exhibited a band edge at *ca.* 420 nm and broad emission with a maximum at 440 nm. This emission profile is in stark contrast to other reports of Cu–In–Se materials, which have optical properties predominantly in the red/infrared region, and may have some

RSC Nanoscience & Nanotechnology No. 33
Semiconductor Quantum Dots: Organometallic and Inorganic Synthesis
By Mark Green
Published by the Royal Society of Chemistry, www.rsc.org

origin in the decomposition products of TOPO described later. Tang *et al.* extended this method, using metal acetates and selenium in oleylamine (OAm) at 250 °C to produce a range of ternary compounds such as $CuInSe_2$ (a mix of triangular, spherical and hexagonal, *ca.* 16 nm in size), $CuGaSe_2$ (11 nm plates) and $Cu(InGa)Se_2$ (15 nm triangular particles). The particles displayed diffraction patterns consistent with the target phase, and the stoichiometry of $Cu(InGa)Se_2$ could be changed by altering the amount of reagents. The reaction solvent/capping agent was a key factor in the synthesis; reactions that contained TOP, TOPO, dodecanethiol or stearic acid all failed.[6] Very similar chemistry was used to again prepare QDs of $CuInSe_2$, using the metal chloride salts as precursors, although in contrast to the earlier report, TOP could be used as a precursor.[1] In this case, the addition of the selenium precursor dictated the crystalline phase; with the metal precursors dissolved in OAm, rapid injection of OAm/Se resulted in the formation of sphalerite particles, whereas addition of OAm/Se with the metal precursors followed by gradual heating resulted in chalcopyrite-structured particles. Unusually, hexagonal rings of $CuInSe_2$ were obtained by using TOP/ $InCl_3$, TOP/CuCl and TOPSe as precursors (Figure 4.1). This chemistry was extended to the synthesis of $Cu(In_{1-x}Ga_x)(S_{1-y}Se_y)_2$ nanocrystal inks for photovoltaic applications.[7]

Using a different selenium precursor, Ph_2Se_2, in the OAm system gave another crystalline metastable phase of $CuInSe_2$ with an onset of absorption at *ca.* 1000 nm.[8] Injection of a Ph_2Se_2 solution into a solution of CuCl and $In(CH_3C(O)CH_2C(O)CH_3)_3$ in OAm at 180 °C followed by 3 hours growth resulted in the wurtzite-structured particles of $CuInSe_2$, approximately

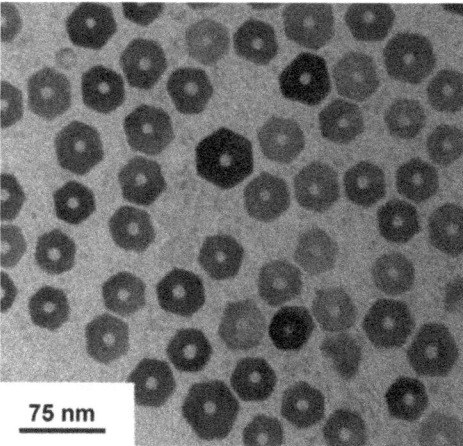

Figure 4.1   Electron microscope images of $CuInSe_2$ rings. Reprinted with permission from Q. Guo, S. J. Kim, M. Kar, W. N. Shafarman, R. W. Birkmire, E. A. Stach, R. Agrawal and H. W. Hillhouse, *Nano Lett.*, 2008, **8**, 2982. Copyright 2008 American Chemical Society.

spherical and 30 nm in diameter. Substitution of OAm for a non-coordinating solvent resulted in the chalcopyrite phase of $CuInSe_2$, as did the use of elemental selenium or selenourea.

A similar reaction was used to grow tetragonal or cubic crystalline $Cu_{1.5}InSe_2$ nanowires using metal nanoparticle catalysts.[9] The stoichiometry of the final product varied, but it was clearly the ternary compound. In this example, $In(O_2CCH_3)_3$, $Cu(O_2CCH_3)$, oleic acid and TOP were mixed and heated to 100 °C, followed by cooling to room temperature and mixing with TOPSe. This precursor solution was injected into a freshly prepared solution of Au/Bi core/shell particles in TOP at 300 °C, resulting in solution–liquid–solid (SLS) catalytic growth of $CuInSe_2$ wires. Changing the order of addition/temperature also yielded solid-state by-products, such as Cu_3Se_2 and Cu_3P depending upon reaction conditions. In this report, the $CuInSe_2$ wires showed an absorption onset at *ca.* 1200 nm.

Most of the above routes to $CuInSe_2$ did not report the optical properties in any depth. A report by Allen and Bawendi[10] did, however, report impressive optical properties for spherical Cu–In–Se QDs with a chalcopyrite structure, after suggesting TOPSe was not ideally suited as a chalcogen source while referring to previous mechanistic studies. Particles of $CuIn_5Se_8$ were obtained by the dispersion of CuI and InI_3 in a mixture of TOP and OAm at between 280 °C and 360 °C, followed by the injection of $Se(SiMe_3)_2$ and growth at 210 °C for a short period of time. The particles clearly exhibited band edge emission towards the red/near infrared end of the visible spectrum, with excitonic features present in the optical spectra and quantum yields up to 25%. Changing the metal precursors from the iodide salt to the chloride salt resulted in a differing stoichiometry, which could be tuned from $CuIn_{1.5}Se_3$ to $CuIn_{2.3}Se_4$ by varying the reaction temperature (injection temperature of 350 °C and growth of temperature of 280 °C for $CuIn_{1.5}Se_3$; injection temperature of 280 °C and growth of temperature of 210 °C for $CuIn_{2.3}Se_4$). Overall, emission from QDs of Cu–In–Se could be tuned between 650 and 975 nm. Replacing CuI with AgI resulted in $AgInSe_2$ QDs which emitted between *ca.* 625 nm and 650 nm, with a maximum quantum yield of 15%.

The related ternary compound system, copper–indium–sulfide, was prepared in a similar manner. Cu-oleate and indium oleate (prepared by the reaction of the metal chlorides and sodium oleate) were mixed in OAm with dodecanethiol, then heated to between 230 °C and 250 °C for up to 1 hour.[11] By varying reaction temperature and time, different sizes and shapes of nanoparticles could be prepared, from *ca.* 30–50 nm 'nanoacorns' with clearly visible regions of Cu_2S and In_2S_3 to 'larva'-shaped particles up to 110 nm in length. X-ray diffraction (XRD) showed the particles were composed of Cu_2S and In_2S_3 rather than the solid solution of copper and indium sulfide; the reaction product could be thought of more as a heterostructure than a true ternary compound. A similar method using a non-coordinating solvent has been used, resulting in $CuInS_2$;[3] a mixture of the metal salts and dodecanethiol in octadecene (ODE) was heated to 240 °C forming a $CuIn(SR)_x$ intermediate, and left for a fixed time, turning a deep red in colour consistent with the formation of $CuInS_2$. The tetragonal phase of the ternary compound was confirmed by XRD and the particles found to be 2.6 nm in diameter, which grew to 3 nm after several hours growth, with the presence of some anisotropic particles. The particles exhibited tuneable absorption edges between 550 nm and 760 nm, with near band edge emission also observed between 600 nm and 750 nm and quantum yields of up to 5%. Dodecanethiol has also been used as a sulfur precursor and capping agent in the preparation of $CuInS_2/ZnS$ core/shell particles with a tetragonal crystalline core, which displayed emission quantum yields of up to 60% in the red region of the visible spectrum, and used in cellular imaging.[12]

Polytypic $CuInS_2$ could also be prepared using metal chlorides as precursors and thiourea as the sulfur source in OAm.[13] The resulting disc-shaped particles were relatively monodispersed and exhibited the wurtzite metastable phase when examined by X-ray powder diffraction; chalcopyrite regions were also observed when the particles were examined by electron microscopy. Other investigations have also uncovered the sensitivity of the crystalline phase to the reaction solvent and, unusually, reagent storage time. Typical metal salts and elemental sulfur were used as precursors with ODE as a solvent, to which were added various surfactants. The reagents were prepared and mixed together, and left to stand for either 3 minutes or 3 hours, prior to heating in an oil bath for up to 5 minutes. Generally, the use of precursor solutions stored for only 3 minutes gave either zinc blende or chalcopyrite structures, the diffraction patterns being too similar to differentiate, whereas the use of precursors stored for longer ultimately gave wurtzite-structured particles.[14]

The use of other sulfur sources as also been reported, with $(SC(CH_3)_3)_2$ being successfully used to prepare wurtzite-structured $CuInS_2$, with $In(CH_3C(O)CH_2C(O)CH_3)_3$ and CuCl as precursors, dodecanethiol and OAm as solvents, an injection temperature of 95 °C and a growth temperature of 180 °C over 33 minutes.[15] The particles were extremely monodispersed, *ca.* 7 nm in diameter with a gradual onset of absorption at around 1000 nm. The presence of dodecanethiol was essential for the production of monodispersed samples, as without it large hexagonal platelets up to 130 nm resulted. The growth kinetics of this reaction were studied and revealed size-focusing and Ostwald ripening.

Quaternary semiconductors of related materials have also been made in QD form using very similar chemistry, reported in several paper published at around the same time; crystalline particles of Cu_2ZnSnS_4 (CZTS) with a tetragonal kesterite phase, *ca.* 10 nm in diameter with an irregular faceted shape were prepared by the thermolysis of $Cu(CH_3C(O)CH_2C(O)CH_3)_2$, $Zn(O_2CCH_3)_2$, $SnCl_2$ and elemental sulfur in OAm at 280 °C for 1 hour.[16] The particles were tin-rich and slightly sulfur-deficient ($Cu_{2.08}Zn_{1.01}Sn_{1.2}S_{3.7}$) and used in photovoltaic devices with a power conversion efficiency of up to 0.23%. In another report, $Cu(CH_3C(O)CH_2C(O)CH_3)_2$, $Zn(CH_3C(O)CH_2C(O)CH_3)_2$, $SnBr_2(CH_3C(O)CH_2C(O)CH_3)_2$ and elemental sulfur were used as precursors in OAm, giving particles 15–25 nm in diameter.[17] The crystal phase of the particles was difficult to determine, due to peak broadening, but was assigned as either a kesterite or stannite structure. The particles were again used in photovoltaic devices with a device efficiency of up to 0.74%. Similarly,[18] $Cu(CH_3C(O)CH_2C(O)CH_3)_2$, $Zn(O_2CCH_3)_2$, and $Sn(O_2CCH_3)_4$ were combined in OAm, and heated to 150 °C. In a separate flask, sulfur was mixed in OAm and sonicated, after which both precursor solutions were injected into TOPO at 300 °C for 45 minutes, giving faceted Cu_2ZnSnS_4 particles *ca.* 13 nm in diameter with a tetragonal crystalline core and an optical band edge of 1.5 eV (*ca.* 830 nm).

The composition of Cu_2ZnSnS_4 has been tuned by altering the ratio of precursors in the compound $(Cu_2Sn)_{x/3}Zn_{1-x}S$, where the material is essentially tuned between ZnS ($x = 0$) and Cu_2ZnSnS_4 ($x = 0.75$), keeping the Cu : Sn ratio at 2 : 1 to maintain charge in the sphalerite phase.[19] The resulting materials maintained the relatively small particle size of *ca.* 3 nm regardless of composition, and the cubic sphalerite structure. The absorption band edge was easily tuned across the entire visible spectrum, with band edges ranging from *ca.* 350 nm for $x = 0$ to *ca.* 1000 nm for $x = 0.75$. The material $(Cu_2Sn)_{0.01}Zn_{0.97}S$ was utilised in simple photovoltaic devices. Likewise, spherical $Cu_2Zn_xSn_ySe_{1+x+2y}$ nanoparticles of a tetragonal crystalline phase, *ca.* 20 nm in diameter have been prepared, using amine complexes of the metal chlorides and TOPSe as precursors in ODE at 295 °C. The identity of the material was confirmed using a number of techniques, including Z-contrast electron microscopy, as XRD alone was not sufficient to identify the phase.[20]

One of the simplest compounds with the most potential when one consider the number of available crystalline phases is the binary copper sulfide system, with numerous naturally occurring forms accessible, several of which are semiconducting with applications in solar energy conversion.[21] Some of the copper(I) sulfides ($Cu_{2-x}S$) are copper deficient and hence give rise to self-doped systems with sufficient free carrier densities to exhibit valence band surface plasmons (when $x > 0$), notably in the near infrared region (Figure 4.2).[22,23] Early studies into the organometallic-related synthesis of copper sulfides used a solventless technique, using copper octanoate and dodecanethiol as the starting materials for a single-source precursor, which was heated at low temperatures to apparently yield Cu_2S nanorods,[24,25] although these were later found to be hexagonal discs which self-assembled face to face.[26]

Other single-source routes using solvents have been reported and are described in Chapter 7. Typical binary routes which are based on high-temperature solvent synthesis have also been reported, using precursors such as $Cu(CH_3C(O)CH_2C(O)CH_3)_2$ and elemental sulfur in OAm at 200 °C for 1 hour, followed by a further hour at 230 °C. The resulting hexagonal β-Cu_2S nanoparticle platelets with a hexagonal crystalline core were isolated using solvent/non-solvent interactions, although spherical particles could also be obtained using a synthesis temperature of 200 °C for 2 hours.[27] A similar

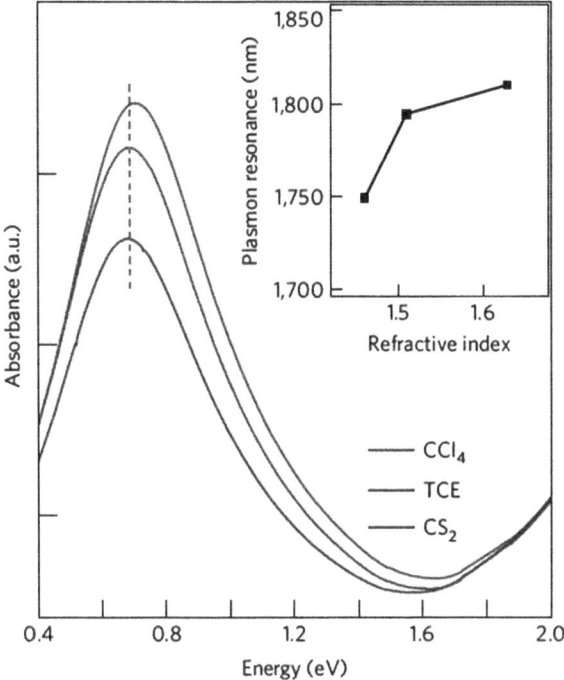

Figure 4.2 Surface plasmon resonance (SPR) in the absorption spectrum of 6 nm diameter $Cu_{2-x}S$ QDs in three different solvents (CCl_4, tetrachlorethylene (TCE), CS_2). Inset shows the shift in SPR with refractive index. Reprinted by permission from Macmillan Publishers Ltd: *Nat. Mater.*, J. M. Luther, P. K. Jain, T. Ewers and A. P. Alivisatos, *Nat. Mater.*, 2011, **10**, 361. Copyright 2011.

route yielded different polymorphs which maintained the disc shape, yielding either CuS or $Cu_{1.8}S$ particles depending on precursor concentration.[28] The use of a thiol to form an *in situ* copper thiolate which was then thermolysed in the presence of TOPO has also been reported, giving monodispersed particles which easily self-assembled on a substrate.[29]

The selenides can also be prepared by similar methods, using hexadecylamine (HDA), TOPSe and $Cu(CH_3C(O)CH_2C(O)CH_3)_2$ at 150 °C for 2 hours, giving $Cu_{2-x}Se$ particles with a cubic crystalline core.[30] The particle size could be tuned between *ca.* 5 and 8 nm by increasing the amount of precursor. Single-crystal nanoplates or nanosheets (depending on precursor ratio) of $Cu_{2-x}Se$ have also been grown using CuCl, 2-ethylhexanoic acid and OAm at 80 °C, which was injected into a paraffin/selenium solution at 250 °C.[31] The resulting 2D materials had absorption spectra in the near infrared region of the spectra.

Interestingly, other selenium sources have been explored, notably the air-stable compound 1,3-dimethylimidazoline-2-selenone, a simply prepared precursor which was dissolved in CH_2Cl_2 and OAm, and injected into $CuCl_2$

and OAm at 175 °C, followed by 10 minutes growth.[32] The resulting $Cu_{2-x}Se$ disc-like nanoparticles were isolated by solvent/non-solvent interactions, and found to have broad absorption spectra covering the majority of the visible region with an excitonic shoulder at 400 nm. It was suggested the resulting heterocyclic carbene was involved with the growth process and also coordinated to the particle surface. In this example, substituting 1,3-dimethylimidazoline-2-selenone with TOPSe resulted in small nanomaterials that could not be observed by electron microscopy. The copper-deficient $Cu_{2-x}Se$ could be prepared without the use of phosphines, using CuCl and ODE/Se at 300 °C, with ODE as solvent and oleylamine—found to be essential as both capping species and reducing agent for the selenium, yielding Se^{2-}.[33] The resulting material included a small amount of large (>100 nm) CuSe hexagonal plates, but the majority product was monodispersed cubooctahedral $Cu_{2-x}Se$ of *ca.* 16 nm diameter. The optical spectra displayed an absorption shoulder at 480 nm, and a broad feature at 1150 nm, assigned as an indirect transition, although later described as a surface plasmon[34] which could be reversibly tuned by varying the copper stoichiometry (oxidation and reduction of the particles).[35]

4.2 Other Chalcogenide-Containing Materials

Investigations into $CuInSe_2$ and related materials lead to the development of simple group III–chalcogenides. Materials such as InSe, a simple layered semiconductor with a hexagonal structure and bulk bandgap of 1.24 eV, were initially prepared by single-source precursors as described in Chapter 7.[36] The preparation of diselenocarbamates precludes the routine preparation of InSe, however, therefore simple routes are desirable. The injection of a solution of $(CH_3)_3In$ in TOP into a solution of TOPO, TOP and Se at 250 °C resulted in InSe QDs *ca.* 3 nm in diameter with a disc-like geometry. The inclusion of a phosphonic acid in the synthesis resulted in few particles nucleating, yielding fewer but larger particles. Using this method, particles as large as 80 nm were obtainable, although these were much larger than the InSe excitonic diameter of 5 nm and only the smaller particles exhibited notable emission. The polarised emission of the smaller (2.9 nm) particles was found to be tuneable, with excitation wavelength between 350 and 450 nm. This emission anisotropy was attributed to the morphology and associated electronic structure, with quantum yields of up to 25% observed in the smaller particles.[37] Park *et al.* explored a one-pot reaction to InSe nanowires, where the reactants, $InCl_3$ and selenium, were heated with OAm to 215 °C and stirred for 5 hours.[38] The method utilised the insolubility of selenium in the long-chain amine below 205 °C, above which the precursor dissolved, becoming a suitable starting material. The nanowires, *ca.* 8 nm thick and 5 μm long, were extremely monodispersed, and were found by XRD to be cubic rather than the inherent hexagonal phase. By altering the amount of selenium, the wires could be made thinner and longer, or thicker and shorter. Rods of In_2S_3 could also be prepared by the injection of

$(CH_3)_3CSSC(CH_3)_3$ into a solution of $In(CH_3C(O)CH_2C(O)CH_3)_3$ and OAm.[39] After growth at 180 °C for 7 hours, nanorods (with a cubic crystalline core) of varying lengths were obtained, although a relatively monodispersed average width of 11.5 nm $\pm$ 1.3 nm was observed. The rods were also found to have a remarkably square edge. The rods had broad, weak emission at *ca.* 445 nm attributed to either trap states or structural defects, slightly Stokes shifted from the absorption edge with an excitonic feature at *ca.* 350 nm. It is worth noting that elemental sulfur could not be used as a precursor in this reaction, although anisotropic In_2S_3 nanotubes were reported elsewhere using $InCl_3$, elemental sulfur and OAm, followed by heating at 50 °C for 30 minutes, and further heating at 220 °C for up to 5 hours.[40] During the reaction, it was found that 13 nm hexagonal nanoplates of β-In_2S_3 formed after 30 minutes. The particles increased in size during prolonged heating, stacking face on, and then converting to nanotubes upon further heating. It was suggested that the crystalline phase of the particles drove the conversion, as full coordination could be achieved by reactions between similar surfaces, reducing the energy of the system as dictated by dangling bonds. The OAm was also suggested to be essential, as it bound to the (111) facet of the crystals, allowing the growth *via* oriented attachment.

Other layered semiconductors, such as GaSe, have been prepared by analogous chemistry.[41,42] GaSe has a direct bandgap of 2.11 eV (although the difference between the direct and indirect transition is only 25 meV), and is composed of weakly bound Se–Ga–Ga–Se sheets. The material is unusual as it display two types of exciton; a 2D exciton associated with the indirect transition, and a 3D exciton, with a Bohr radius of 3.1 nm, associated with the direct transition. The injection of a solution of $(CH_3)_3Ga$ in TOP into TOPO, TOP and TOPSe at 278 °C, followed by growth at *ca.* 270 °C, resulted in the formation of GaSe particles *ca.* 4 nm in diameter. After observation of an excitonic shoulder in the absorption spectrum, the reaction was cooled to room temperature to stop the reaction, a process that took in total *ca.* 2 hours. The particles were isolated by methanol/toluene solvent/non-solvent interactions, although interestingly, if the system was extremely water-free, the particles were retained by methanol. Trace amounts of water resulted in the particles remaining in the non-polar solvent. The TOPO on the resulting GaSe particles could be exchanged for a long-chain amine, which reduced agglomeration. This synthesis is extremely interesting as previous attempts to prepare gallium-containing QDs have proved difficult; this is attributed to the tight binding of gallium precursors to ligands, as described in Chapter 2. The synthesis of monodispersed GaSe particles therefore needed the high temperature to proceed: any higher, and the particles became polydispersed; any lower, and the reaction proceeded too slowly. Analysis of the selected area diffraction patterns obtained from the particles suggested a single tetralayer structure. The emission of the particles was found to be relatively strong, with quantum yields of up to 10%. In a similar manner to InSe particles, the wavelength of the emission was tuneable from *ca.* 425 nm to *ca.* 525 nm by altering the excitation wavelength.

The chemistry described above has also been applied to other metal chalcogenide (S, Se, Te) systems by simply altering the precursors. Nickel sulfide particles, Ni_3S_4, have been prepared by the thermolysis of $NiCl_2$ and elemental sulfur in OAm and TOP at 220 °C for 1 hour. The particles, produced by a two-step reduction and sulfidation mechanism, exhibited irregular shapes with elemental nickel as a by-product.[28] Nickel selenide particles, NiSe, have so far only been made by single-source precursor.[43] One of the earliest applications of organometallic-based chemistry to the synthesis of metal chalcogenide (that was not based on group II materials), was the synthesis of α-MnS, a potentially useful magnetic material, where $MnCl_2$ and $S(SiMe_3)_2$ were thermolysed in TOPO. The particle size varied between 20 and 80 nm depending on precursor concentration.[44] Non-coordinating solvents have also been used;[45] manganese carboxylates and elemental sulfur (molar ratio 1 : 2) were dissolved in ODE, followed by heating at 300 °C for 30 minutes, yielding α-MnS. Isolation by addition of non-solvent resulted in octahedral particles *ca.* 30 nm in diameter. Anisotropic particles (rods and T-shaped) could be obtained when manganese stearate was used as a precursor. The absence of sulfur (or a lower concentration, S : Mn ≤ 0.6) resulted in MnO particles. Similarly, bullet, hexagonal or rod-shaped MnS, with a wurtzite structure (γ-MnS) were obtained when $MnCl_2$ and elemental sulfur were used as precursors in OAm at 280 °C.[46] Large particles of MnSe (several hundred of nanometres) have also been prepared using the metal salt and elemental selenium in octadecylamine (ODA), although few details were provided.[47] Similar experiments resulted in wurtzite-structured MnSe, prepared using $MnCl_2$, oleic acid and selenium in tetraethylene glycol at 235 °C for 1 hour.[48]

Bismuth chalcogenides, notably Bi_2Te_3, are known as excellent thermoelectric materials and these compounds have also been developed due to the theoretical high figures of merit when prepared on the nano scale,[49] although the synthesis of monodispersed sub-10 nm particles is problematic because of the high reactivity of chalcogens with bismuth precursors. Nanowires of Bi_2Te_3 were first prepared by utilising tellurium nanowires (prepared from $TeCl_4$ and TOPO) and $Bi[N(SiMe_3)_2]_3$ or $Bi(C_6H_5)_3$ as precursors. Thermolysis of the precursors in solvents such a polydecene and 1,3-diisopropylbenzene at either 160 or 200 °C for between 2 and 12 hours yielded micrometre-long wires, with amorphous sheaths of either Bi_2Te_3 or an oxide of tellurium.[50] In a simpler fashion, nanowafers of Bi_2Te_3 could be prepared by the reaction between $Bi(CH_3CO_2)_3$ and elemental tellurium in octylamine (OA), which required prolonged refluxing (>24 hours), whereas the analogous reactions with selenium or sulfur required only 2 hours of refluxing, yielding either Bi_2Se_3 nanowafers or Bi_2S_3 nanorods respectively.[51] Interestingly, the bismuth precursor reacted prior to nanoparticle formation to give bismuth oxo-acetate as an intermediate. A similar reaction reported by Malakooti *et al.* described the use of $BiCl_3$ as a precursor, where the intentional inclusion of oxygen yielded BiOCl which was then reduced by the amine to give elemental bismuth.[52] Addition of elemental sulfur in OAm at 130 °C resulted in either monodispersed Bi_2S_3 dots or rods depending on the precursor ratio. Expanding this reaction to use bismuth citrate as a precursor resulted in the formation of ultrathin Bi_2S_3 necklace nanowires, 1.6 nm in diameter, formed of individual nanoparticles separated by amorphous sections or grain boundaries.[53] The crystal structure was found to be similar to the bulk, with 75% of the bismuth atoms found on the surface, available to coordinate to surface ligands.[54] Elemental sulfur was not the only precursor to be used in the synthesis of the sulfides: $S(SiMe_3)_2$ has also been used, along with $Bi(CH_3CO_2)_3$ in oleic acid (forming the oleate) at 170 °C, yielding Bi_2S_3 rods *ca.* 22 nm in length and *ca.* 7 nm wide.[55] In this case, the absorption spectrum was also measured, giving a bandgap close to the bulk value due to the small exciton. Particles with a branched morphology have also been prepared by the reaction of $Bi(OCOC_{17}H_{33})_3$ with sulfur in ODE and oleic acid at either 150 °C or 180 °C, showing evidence of crystal splitting. The structures adopting a multiple sheaf-like morphology, with individual filaments having a diameter of *ca.* 9 nm and an average length of *ca.* 570 nm.[56]

Hexagonal nanoplates of Bi_2Te_3, with edges 200–300 nm long and 15 nm thick, were prepared by the injection of TOPTe into a hot solution of diphenylether, with oleic acid as a capping agent and $Bi(OCOC(C_2H_5)(C_4H_9))_3$ as the bismuth precursor at 150 °C, followed by 30 minutes growth at the same temperature.[57] The shape and thickness was controlled by reaction temperature, with the relatively high temperature necessary for the hexagonal shape. Interestingly, including 5% TOPSe in the reaction in an attempt to dope the material resulted in strings of tellurium with individual Bi_2Te_3 plates incorporated into the rod perpendicularly.

Spherical particle of Bi_2Te_3 with diameters below 10 nm could, however, be obtained by again using a two-step reaction; first preparing bismuth nanoparticles by reducing $Bi(CO_2CH_3)_3$ using OAm (or TOP) in the presence of dodecanethiol, a capping agent. To the metal particles was added TOPTe and a phosphonic acid at 60 °C, followed by stirring for 2–3 days, resulting in the formation of a Bi/Te alloy. The alloy converted to Bi_2Te_3 particles upon heating at 110 °C for 18 hours.[58] The resulting particles exhibited a reduced thermal conductivity and similar electrical conductivity to bulk n-type Bi_2Te_3. By including $Sb(CO_2CH_3)_3$ in the reaction, $Sb_{(2-x)}Bi_xTe_3$ was formed which exhibited an obvious change in particle morphology, with either nanoflakes or agglomerations of nanoflakes (termed nanosheets) resulting, depending on reaction temperature.[59] After ligand removal using a methanolic solution of NH_3, and further processing by plasma sintering, the material showed a 15% increase in the figure of merit.

A very similar method to Bi_2Te_3 and antimony-doped nanomaterials was reported by Zhao and Burda, who used a one-step reaction, where a phenyl ether solution of dodecanethiol and $Bi(CO_2CH_3)_3$ was heated to 100–120 °C, followed by injection of TOPTe and growth for up to 60 minutes.[60] In this case, the resulting Bi_2Te_3 particles were *ca.* 50 nm in diameter and clearly faceted, and surface studies uncovered oxidation of the surface tellurium sites. The surface bismuth sites appeared protected, presumably by the surfactant. This method was also used to prepare $Bi_{0.5}Sb_{1.5}Te_3$ particles of *ca.* 50 nm, by again including $Sb(CO_2CH_3)_3$ in the reaction. These doped materials also exhibited a threefold increase in the power factor.[61] $Sb(CO_2CH_3)_3$ has also been used to prepare antimony chalcogenides, by dissolving the precursor in a long-chain amine followed by mixing with an amine solution/dispersion of the elemental chalcogen, and heating. The resulting particles displayed a variety of structures.[62]

Transition metal-based particles, which often display a layered morphology, include MoS, prepared by the thermolysis of $Mn(CO)_6$ and sulfur in ODE between 270 °C and 330 °C, using TOPO as a capping agent. The resulting particles were less than 5 nm in diameter with a size distribution of up to 15%.[63] Nanoplates of $NbSe_2$ have also been grown from $NbCl_5$ and Se in OAm or dodecylamine (DDA) at 280 °C, although different morphologies were obtained by different synthesis temperatures, giving either lamellar-structured plates or wires.[64]

Possibly the most successful transition metal chalcogenide nanoparticles prepared by the organometallic route are the silver-based materials, such as Ag_2S, which are usually prepared by single-source precursor as described in Chapter 7. An interesting method of preparing Ag_2S has been described by Huxter *et al.*, where $AgNO_3$ and $NH_2CSCSNH_2$ were mixed with HDA and compressed into a pellet in specific layers. The pellet was then purged with nitrogen gas, and heated in an air oven to just above the melting point of HDA for 20 minutes. The resulting black material was then dispersed in acetone, centrifuged, washed, and dispersed in toluene. This solventless process yielded monodispersed particles, *ca.* 10 nm in diameter with a near infrared band edge at *ca.* 1170 nm, consistent with the bulk bandgap of Ag_2S.[65] Other materials which are not accessible by single-source precursor, such as Ag_2Te, have been prepared using AgCl dissolved in TOP, with TOPTe, both of which were injected into OAm at 140 °C, followed by overnight growth and ageing.[66] Similarly, $AgNO_3$ has also been mixed with dodecanethiol in water, then transferred to toluene, followed by injection of TOPTe at 85 °C yielding Ag_2Te particles. Prolonged heating of up to 1 day resulted in large anisotropic nanoparticles up to 15 nm in size, along with smaller particles *ca.* 5 nm in diameter. Further heating resulted in the larger particles stacking together, while the smaller particles formed islands of particles. In all cases, the particles displayed an excitonic peak at *ca.* 1150 nm, significantly shifted from the bulk bandgap of 0.064 eV (*ca.* 19 300 nm).[67] Ternary Ag-based nanorods can also be prepared by similar methods to those described above for copper-based ternary materials. $AgNO_3$, $In(NO_3)_3$ and elemental sulfur were dispersed in ODA at 120 °C, where they were left to decompose giving anisotropic $AgInS_2$ particles with an orthorhombic crystalline core.[68] Using less sulfur resulted in spherical nanoparticle formation, whereas increasing the synthesis temperature to 200 °C resulted in worm-like wires.

4.3 Other Pnictide-Based Materials

The synthesis of the II_3–V_2 family of semiconductors has also been explored using organometallic chemistry. Cadmium phosphide, Cd_3P_2, has a bulk bandgap of 0.5 eV and was initially prepared in nanoparticulate form using aqueous chemistry.[69] The excitonic diameter is suggested to be significantly larger (36.1 nm)[70] than the II–VI analogues, leading to a larger degree of quantum confinement and hence a wider range of available band edges. The initial organometallic routes to Cd_3P_2 were single-source based, as described in Chapter 7, where $[MeCdP^tBu_2]_3$ was thermolysed in TOPO to give Cd_3P_2 QDs with a significantly shifted band edge, with emission reported between *ca.* 525 and 650 nm.[71] The first study using two precursors in a organometallic-type route described the use of $Cd(CH_3CO_2)_2$ and $P(SiMe_3)_3$;[72] in a typical synthesis, the cadmium salt was dispersed in ODE with oleic acid and heated, forming a carboxylate precursor. The solution was then injected with an ODE solution of the phosphorous precursor at 40 °C, which was then gradually heated. The progression of the growth was monitored by absorption spectroscopy, and it was observed that the particles formed a 'magic size'—a stable diameter. Further heating resulted in the disappearance of the excitonic feature in the absorption spectra, suggesting that the particles dissolved back into solution. The magic-sized Cd_3P_2 particles, *ca.* 3 nm in diameter, exhibited narrow (17 nm FWHM) band edge emission at *ca.* 455 nm with a quantum yield of 3%, which rose to 7% upon ageing.

A similar method, using CdO as a precursor and the hot injection of the precursors rather than the slow growth, reported tuneable emission by varying the oleic acid concentration, resulted in particles between 1.6 and over 12 nm in diameter.[73] The particle emission covered the range between 450 nm to beyond 1500 nm, with particles smaller than 7 nm exhibiting band edge emission with quantum yields as high as 30%; particles between 2 and 3 nm in diameter had quantum yields of up to 70%. A similar investigation reported essentially the same reaction with a wider range of capping agents and a variation of precursor ratios.[74] In this case, the use of TOP was found to push the emission out into the near infrared region, but also resulted in a significantly lower quantum yield. This study was also notable for detecting a silicon-containing species on the particle surface, presumed to be a phosphorus precursor decomposition by-product, and for reporting the first XRD pattern of nanosized Cd_3P_2.

Related chemistry has been used to prepare Cd_3As_2 nanoparticles, a material with a bulk inverse bandgap of −0.19 eV and an excitonic diameter of *ca.* 47 nm. In this case, a multiple injection process was utilised, where a solution of $As(SiMe_3)_3$ in TOP was injected into a solution of ODE containing a cadmium carboxylate salt at 175 °C, and left to grow for 20 minutes. At this point, particles emitting at 850 nm were observed. The particles were grown further by multiple injections of the arsenic precursor, confirming previous reports of II_3–V_2 materials that the particles do not grow by the typical Ostwald ripening mechanisms observed in II–VI materials.

The particles could be tuned to emit between 530 and 2000 nm (Figure 4.3), exhibiting quantum yields typically 20–60% with the larger particles having the lower emission values.[75]

A further II_3–V_2 compound, Zn_3P_2, has been prepared by the injection of $(CH_3)_2Zn$ and HP^tBu_2 into either 4-ethylpyridine or TOPO at a range of temperatures, resulting in particles 3–13 nm in diameter, emitting between *ca.* 400 and 600 nm.[76]

Other phosphide-based semiconducting systems have also been explored, including magnetic semiconductors, for which organometallic chemistry appears ideally suited. The various phases of iron phosphide present a wealth of materials exhibiting interesting magnetic and optical characteristics.[77] Although one phase (FeP_2) is known to be a small-bandgap semiconductor,

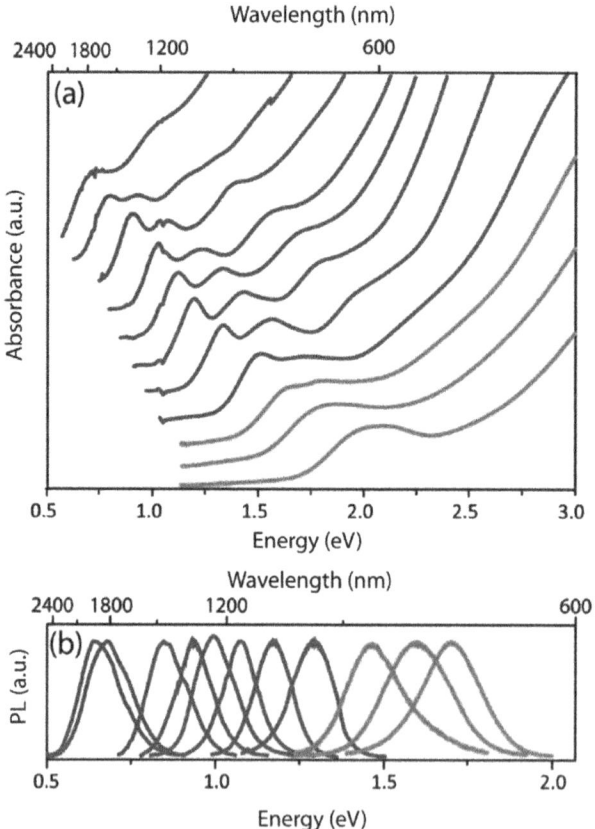

Figure 4.3 Absorption and emission spectra (a and b respectively) of Cd_3As_2 QDs between 2 and 5 nm in diameter.[75] Reprinted with permission from D. K. Harris, P. M. Allen, H.-S. Han, B. J. Walker, J. Lee and M. G. Bawendi, *J. Am. Chem. Soc.*, 2011, **133**, 4676. Copyright 2011 American Chemical Society.

other phases (Fe_2P and Fe_3P) are better known for their magnetic properties. Chemical routes to FeP have been reported *via* the reduction of the phosphate on a surface,[78] although few details have been reported for this method. This material obviously lends itself to the desilylation reaction, similar to the synthetic processes described for III–V materials in Chapter 2. An early report[79] on the synthesis of FeP described the desilylation reaction, where iron(III) acetylacetonate, $Fe(C_5H_7O_2)_3$, was dissolved in a mixture of TOPO, myristic acid and DDA, followed by the injection of $P(SiMe_3)_3$ and the subsequent heating of the reaction to 240–320 °C for 2–3 days. The reaction was then cooled, the materials dispersed in pyridine, and purified by size-selective precipitation using hexane as a non-solvent. The presence of the long-chain amine and the carboxylic acid were found to be essential: their absence resulted in the formation of precipitates due to the weak binding of TOPO. Interestingly, phosphonic acids were found to bind too strongly. The resulting particles, *ca.* 4.6 nm diameter, were determined to be pure-phase FeP, with antiferromagnetic order observed below 124 K.

The desilylation reaction was found to be unsuitable for the synthesis of MnP, a related magnetic material of some interest. However, the expansion of this method to the use of zero-valent carbonyl complexes, such as $Mn_2(CO)_{10}$, $Fe(CO)_5$ and $Co_2(CO)_8$ as cation precursors in a similar reaction resulted in the formation of MnP, FeP and CoP respectively.[80] This route highlights the suitability of the carbonyl complexes as efficient precursors, which can be traced back to the earliest organometallic synthetic routes to compound semiconductors[81] and metal particles.[82] In these reactions, the injection of $P(SiMe_3)_3$ into a hot solution of TOPO, myristic acid and the metal carbonyl complex at 100 °C, followed by heating for 1 day at 220 °C, resulted in the formation of the metal phosphide particles. The particles were isolated into pyridine, and hexane was again used as a non-solvent. The MnP particles prepared by this route were *ca.* 5.1 nm ± 0.48 nm in diameter and exhibited superparamagnetism when below the Curie temperature. The FeP particles (*ca.* 3.1 nm) required a higher synthesis temperature of 270 °C, and CoP particles required a still higher temperature of 320 °C.

The formation of MnP nanoparticles using the carbonyl complex could also be achieved using simple alkylphosphines as a precursor. These are usually used as capping agents, notably TOP, which highlights the transition from inert capping ligand to labile precursor when used at an elevated temperature (>300 °C). TOP has now emerged as a standard synthetic precursor for phosphorus, and has been used to prepare Ni_2P, PtP_2, Rh_2P, Au_2P_3, Pd_5P_2 and PdP_2,[83] Ni_5P_4, Zn_3P_2, Cu_3P, CuP_2, InP and GaP[84] by the reaction of TOP with preformed metal nanoparticles, the formation of which was suggested to occur *via* a diffusion mechanism. The synthesis of FeP from preformed iron metal particles using TOP as a precursor has been explored in some depth by Muthuswamy *et al.* who explored the effect of differing reaction times, precursor concentrations and temperatures on particle formation.[85] It was found that higher reaction temperatures resulted in FeP, with lower temperatures favouring Fe_2P. Short reaction times and excess iron precursor also resulted in Fe_2P rods; long reaction times resulted in spherical FeP with some hollow particles observed, attributed to the Kirkendall effect.[86] The inclusion of a palladium nanoparticle catalyst in the reaction of TOP with preformed iron particles resulted in the synthesis of rods with controllable dimensions, depending on palladium particle size and Fe : Pd ratio.[87] The palladium particles acted as both a decomposition catalyst for the iron particles, and as a catalytic centre for rod growth. The resulting rods contained palladium throughout the length of the structure, indicating that the palladium particles dissolved and reacted with the structure, possibly forming a FePd alloy. Importantly, a closed synthetic system was required for particle formation, and the constant flow of an inert atmosphere was to be avoided. This suggested the phosphorus precursor was PH_3, generated by the elimination of octene. In related work, CoP nanowires were prepared by the thermolysis of $Co(C_5H_7O_2)_2$, with technical-grade TOPO, HDA and tetrade-cylphosphonic acid (TDPA). In this case, the 90% pure TOPO was suggested to be the phosphorus source.[88]

TOP has also been used as a precursor in the preparation of FeP nanorods and wires using $Fe(CO)_5$, because anisotropic particles are of interest due to the potential effect of shape on the magnetic properties of these materials. The precursor was dissolved in TOP forming a TOP–iron carbonyl complex, and injected into TOPO at temperatures above 300 °C. Initially, rods were found to form, which could be grown to micrometre-length wires by multiple injections of precursors.[89] The use of multiple injections was extended to the use of a syringe pump, to ensure a continuous introduction of precursors into the reaction vessel,[90] allowing extra control over wire morphology. The use of a syringe pump was then extended to the preparation of MnP, CoP and Ni_2P nanorods.[91]

Nanowires of FeP have also been prepared using a related carbonyl complex $(\eta^4\text{-}C_6H_8)Fe(CO)_3$, which was thermolysed in TOPO, followed by the addition of more precursor dissolved in TOP at 360 °C.[92] The TOP was suggested to coordinate to the particle surface, and at high temperatures decomposed yielding phosphorus which diffused into the structure and reacted, forming the FeP wire. Nanorods of Fe_2P, 30–260 nm in length, have also been prepared using similar chemistry, with multiple injections of $Fe(CO)_5$ and TOP into a solution of 5 nm Fe_3O_4 seeds, TOP and didode-ceyldimethylammonium bromide at 300 °C. The resulting rods with a Fe_3O_4/Fe_2P core/shell morphology exhibited unusual magnetic properties, notably a variable blocking temperature dictated by the rod length, with a maximum at 60 nm.[93]

The synthesis of anisotropic MnP particles using Mn–carbonyl complexes and TOP as precursors in a syringe pump stressed the importance of the pump in producing particles with a rod morphology. Gregg *et al.* explored this further by preparing anisotropic MnP particles by a simple injection method, negating the need for a syringe pump.[94] In this method, a 50 : 50 (weight) mix of TOPO and TOP was heated at 350 °C for 2 hours prior to injection, to generate the phosphorus precursor. $Mn_2(CO)_{10}$ was dissolved in the high-temperature solvent ODE and heated to *ca.* 70 °C, followed by a single injection of precursor followed by heating for up to 5 hours (or longer if a lower annealing temperature was used). The particles were isolated by size-selective precipitation, into either chloroform or pyridine. Rods isolated after a single hours growth were found to be *ca.* 20 nm long and 5 nm wide, although the aspect ratio could be increased further by multiple injections. Prolonged heating times also increased the anisotropy but also generated cubes and rectangular shaped particles. Interestingly, the shape anisotropy was not found to significantly affect the magnetic properties of MnP.

A related material, MnAs, has been prepared using the same manganese precursor, and C_6H_5AsO as an arsenic precursor.[95] The use of this precursor is somewhat surprising, as the related TOPO molecule has generally been shown to be ineffective as a phosphorus precursor. The reaction can be carried out at both high and low temperatures, yielding particles with different crystalline structures. In a typical high-temperature reaction, both precursors were dissolved in ODE and injected into TOPO at 330 °C, followed by 18 hours growth and isolation by solvent/non-solvent interactions. The resulting particles exhibited the β-MnAs structure, usually unstable at room temperature. Alternatively, slow heating of the precursors up to 200 °C then increasing the reaction temperature to 250 °C resulted in α-MnAs (hot injection at 250 °C did not yield an isolatable product). Both sets of particles were *ca.* 25 nm in diameter, ferromagnetic and consisted of a crystalline core and an amorphous shell, clearly visible by electron microscopy.

The synthesis of FeP does not use TOP or $P(SiMe_3)_3$ exclusively as phosphorus precursors. Single-source precursors, as detailed in Chapter 7, have also been used in the preparation of the related Fe_2P.[96] In this case, no separate pnictide precursor was required, as the compound $H_2Fe_3(CO)_9P^tBu$ possessed both anionic and cationic constituents. The precursor was heated in trioctyl amine (TOA) and oleic acid to *ca.* 315 °C followed by 20 minutes further heating. The precursor underwent a rearrangement to $[Fe_3(CO)_9P^tBu]^{2-}$ and decomposed at *ca.* 140 °C, forming Fe_2P particles, which were precipitated with ethanol after cooling. The particles exhibited varying morphologies, depending on experimental conditions, forming rods, bundles, split rods, X-shaped and T-shaped structures.

A significant number of reports describe the synthesis of the nickel phosphide system. The rapid injection of bis(1,5-cyclooctadiene)nickel and TOP in TOPO at 345 °C followed by 24 hours stirring resulted in monodispersed Ni_2P particles, *ca.* 11 nm in diameter, which could be isolated using pyridine and hexane as solvent and non-solvent respectively.[97] Unlike other reports, the formation of spherical particles was attributed to the rapid injection of precursors, resulting in the conversion of all reactive monomer into particles, leaving none for anisotropic particle growth. Several related reports also describe the synthesis of hollow NiP particles, attributable to the Kirkendall effect.[86,98] In a typical example,[99] $Ni(CH_3COCHCOCH_3)_2$, OAm, TOP and ODE were mixed together and heated to 320 °C over a 15 minute period, and left for 1 hour, followed by cooling and precipitation of the resulting solid using acetone. The resulting hollow Ni_2P particles were 5–15 nm in diameter, with a wall thickness of 2–3 nm. Further work revealed that nickel particles formed initially, which then reacted with the phosphorus, forming an Ni_2P layer through which the remaining nickel diffused, reacting with the available reagents, forming the hollow particle. Tributylphosphine (TBP) was also found to work as a precursor, and CoP hollow spheres were prepared in a similar manner. A detailed study reported the P : Ni ratio was a key factor in determining the product outcome using essentially the same reaction.[100] A ratio of 1 : 3 resulted in nickel particles forming below 240 °C, converting to the hollow mixed-phase $Ni_2P/Ni_{12}P_5$ particles at 300 °C and above. Reactions with a ratio of 1 : 6 were found to generate a mixture of hollow and solid mixed-phase particles, whereas reactions with a larger ratio (>1 : 9), resulted in a Ni–TOP complex, and ultimately in the direct synthesis of the mixed phase which was amorphous below 240 °C and crystalline above 300 °C. Again TBP was found to be an effective precursor as was triphenyl-phosphine. The use of TOPO instead of TOP did not generate the phosphide, but did improve the solubility of the product when compared to the use of ODE. Triphenylphosphine was again used in a similar study, resulting in the formation of Ni/Ni_2P core/shell particles by preparing nickel particles at 200 °C followed by the reaction with the surface triphenylphosphine ligand at 280 °C, giving the core/shell structure.[101] The reduced temperature can be used to explain the differing reaction product. Solid Ni_2P particles have also been made by the reaction of either nickel particles or $Ni(C_8H_{12})_2$ with P_4.[102,103]

In this chapter, we have shown how the chemistry described previously can be applied to the synthesis of more unusual materials. This demonstrates how general the chemistry is and how most materials should be accessible using the synthetic pathways described.

References

1. Q. Guo, S. J. Kim, M. Kar, W. N. Shafarman, R. W. Birkmire, E. A. Stach, R. Agrawal and H. W. Hillhouse, *Nano Lett.,* 2008, **8**, 2982.
2. M. V. Yakushev, A. V. Mudryi, I. V. Victorov, J. Krustok and E. Mellikov, *Appl. Phys. Lett.,* 2006, **88**, 011922.
3. H. Zhong, Y. Zhou, M. Fe, Y. He, J. Ye, C. He, C. Yang and Y. Li, *Chem. Mater.,* 2008, **20**, 6434.
4. M. A. Malik, P. O'Brien and N. Revaprasadu, *Adv. Mater.,* 1999, **11**, 1441.
5. M. A. Malik, P. O'Brien, N. Revaprasadu and G. Wakefield, *Mater. Res. Soc. Symp. Proc.,* 1999, **536**, 371.
6. J. Tang, S. Hinds, S. O. Kelley and E. H. Sargent, *Chem. Mater.,* 2008, **20**, 6906.
7. Q. Guo, G. M. Ford, H. W. Hillhouse and R. Agrawal, *Nano Lett.,* 2009, **9**, 3060.
8. M. E. Norako and R. L. Brutchey, *Chem. Mater.,* 2010, **22**, 1613.

 9. A. J. Wooten, D. J. Werder, D. J. Williams, J. L. Casson and J. A. Hollingsworth, *J. Am. Chem. Soc.,* 2009, **131**, 16177.
10. P. M. Allen and M. G. Bawendi, *J. Am. Chem. Soc.,* 2008, **130**, 9240.
11. S.-H. Choi, E.-G. Kim and T. Hyeon, *J. Am. Chem. Soc.,* 2006, **128**, 2520.
12. L. Li, T. Jean Daou, I. Texier, T. T. K. Chi, N. Q. Liem and P. Reiss, *Chem. Mater.,* 2009, **21**, 2422.
13. B. Koo, R. N. Patel and B. A. Korgel, *Chem. Mater.,* 2009, **21**, 1962.
14. K. Nose, Y. Soma, T. Omata and S. Otsuka-Yao-Matsuo, *Chem. Mater.,* 2009, **21**, 2607.
15. M. E. Norako, M. A. Franzman and R. L. Brutchey, *Chem. Mater.,* 2009, **21**, 4299.
16. C. Steinhagen, M. G. Panthani, V. Akhaven, B. Goodfellow, B. Koo and B. A. Korgel, *J. Am. Chem. Soc.,* 2009, **131**, 12554.
17. Q. Guo, H. W. Hillhouse and R. Agrawal, *J. Am. Chem. Soc.,* 2009, **131**, 11672.
18. S. C. Riha, B. A. Parkinson and A. L. Prieto, *J. Am. Chem. Soc.,* 2009, **131**, 12054.
19. P. Dai, X. Shen, Z. Lin, Z. Feng, H. Xu and J. Zhan, *Chem. Commun.,* 2010, **46**, 5749.
20. A. Shavel, J. Arbiol and A. Cabot, *J. Am. Chem. Soc.,* 2010, **132**, 4514.
21. Z. Yang, C.-Y. Chen, C.-W. Liu, C.-L. Li and H.-T. Chang, *Adv. Energy Mater.,* 2011, **1**, 259.
22. J. M. Luther, P. K. Jain, T. Ewers and A. P. Alivisatos, *Nat. Mater.,* 2011, **10**, 361.
23. Y. Zhao, H. Pan, Y. Lou, X. Qiu, J. Zhu and C. Burda, *J. Am. Chem. Soc.,* 2009, **131**, 4253.
24. T. H. Larsen, M. Sigman, A. Ghezelbash, R. C. Doty and B. A. Korgel, *J. Am. Chem. Soc.,* 2003, **125**, 5638.
25. L. Chen, Y.-B. Chen and L.-M. Wu, *J. Am. Chem. Soc.,* 2004, **126**, 16334.
26. M. Sigman, A. Ghezelbash, T. Hanrath, A. E. Saunders, F. Lee and B. A. Korgel, *J. Am. Chem. Soc.,* 2003, **125**, 16050.
27. H.-T. Zhang, G. Wu and X.-H. Chen, *Langmuir,* 2005, **21**, 4281.
28. A. Ghezelbash and B. A. Korgel, *Langmuir,* 2005, **21**, 9451.
29. Y. Wang, Y. Hu, Q. Zhang, J. Ge, Z. Lu, Y. Hou and Y. Yin, *Inorg. Chem.,* 2010, **49**, 6601.
30. Y. Hu, M. Afzaal, M. A. Malik and P. O'Brien, *J. Cryst. Growth,* 2006, **297**, 61.
31. Z. Deng, M. Mansuripur and A. J. Muscat, *J. Mater Chem.,* 2009, **19**, 6201.
32. J. Choi, N. Kang, H. Y. Yang, H. J. Kim and S. U. Son, *Chem. Mater.,* 2010, **22**, 3586.
33. S. Deka, A. Genovese, Y. Zhang, K. Miszta, G. Bertoni, R. Krahne, C. Giannini and L. Manna, *J. Am. Chem. Soc.,* 2010, **132**, 8912.
34. F. Scotognella, G. Della Valle, A. R. S. Kandada, D. Dorfs, M. Zavelani-Rossi, M. Conforti, K. Miszta, A. Comin, K. Korobchevskaya, G. Lanzani, L. Manna and F. Tassone, *Nano Lett.,* 2011, **11**, 4711.

35. D. Dorfs, T. Härtling, K. Miszta, N. C. Bigall, M. R. Kim, A. Genovese, A. Falqui, M. Povia and L. Manna, *J. Am. Chem. Soc.*, 2011, **133**, 11175.
36. N. Revaprasadu, M. A. Malik, J. Carstens and P. O'Brien, *J. Mater. Chem.*, 1999, **9**, 2885.
37. S. Yang and D. F. Kelley, *J. Phys. Chem. B*, 2005, **109**, 12701.
38. K. H. Park, K. Jang, S. Kim, H. J. Kim and S. U. Son, *J. Am. Chem. Soc.*, 2006, **128**, 14780.
39. M. A. Franzman and R. L. Brutchey, *Chem. Mater.*, 2009, **21**, 1790.
40. Y. H. Kim, J. H. Lee, D.-W. Shin, S. M. Park, J. S. Moon, J. G. Nam and J.-B. Yoo, *J. Chem. Soc., Chem. Commun.*, 2010, **46**, 2292.
41. V. Chikan and D. F. Kelley, *Nano Lett.*, 2002, **2**, 141.
42. H. Tu, S. Yang, V. Chikan and D. F. Kelley, *J. Phys. Chem. B*, 2004, **108**, 4701.
43. W. Manwwprakorn, C. Q. Nguyen, M. A. Malik, P. O'Brien and J. Raftery, *Dalton Trans.*, 2009, 2103.
44. S. Kan, I. Felner and U. Banin, *Isr. J. Chem.*, 2001, **41**, 55.
45. A. Puglisi, S. Mondini, S. Cenedese, A. M. Ferretti, N. Santo and A. Ponti, *Chem. Mater.*, 2010, **22**, 2804.
46. J. Joo, H. B. Na, T. Yu, J. H. Yu, Y. W. Kim, F. Wu, J. Z. Zhang and T. Hyeon, *J. Am. Chem. Soc.*, 2003, **125**, 11100.
47. D.-S. Wang, W. Zheng, C.-H. Hao, Q. Peng and Y.-D. Li, *Chem.–Eur. J.*, 2009, **15**, 1870.
48. I. T. Sines, R. Misra, P. Schiffer and R. E. Schaak, *Angew. Chem., Int. Ed.*, 2010, **49**, 4638.
49. Y.-M. Lin, X. Sun and M. S. Dresselhaus, *Phys. Rev. B: Condens. Matter Mater. Phys.*, 2000, **62**, 4610.
50. H. Yu, P. C. Gibbons and W. E. Buhro, *J. Mater. Chem.*, 2004, **14**, 595.
51. P. Christian and P. O'Brien, *J. Mater. Chem.*, 2005, **15**, 3021.
52. R. Malakooti, L. Cademartiri, Y. Akçakir, S. Petrov, A. Migliori and G. A. Ozin, *Adv. Mater.*, 2006, **18**, 2189.
53. L. Cademartiri, R. Malakooti, P. G. O'Brien, A. Migliori, S. Petrov, N. P. Kherani and G. A. Ozin, *Angew. Chem., Int. Ed.*, 2008, **47**, 3814.
54. J. W. Thomson, L. Cademartiri, M. MacDonald, S. Petrov, G. Calestani, P. Zhang and G. A. Ozin, *J. Am. Chem. Soc.*, 2010, **132**, 9058.
55. G. Konstantatos, L. Levina, J. Tang and E. H. Sargent, *Nano Lett.*, 2008, **8**, 4002.
56. J. Tang and A. P. Alivisatos, *Nano Lett.*, 2006, **6**, 2701.
57. W. Lu, Y. Ding, Y. Chen, Z. L. Wang and J. Fang, *J. Am. Chem. Soc.*, 2005, **127**, 10112.
58. M. Scheele, N. Oeschler, K. Meier, A. Kornowski, C. Klinke and H. Weller, *Adv. Funct. Mater.*, 2009, **19**, 3476.
59. M. Scheele, N. Oeschler, I. Veremchuk, K.-G. Reinsberg, A.-M. Kreuziger, A. Kornowski, J. Broekaert, C. Klinke and H. Weller, *ACS Nano*, 2010, **4**, 4283.
60. Y. Zhao and C. Burda, *ACS Appl. Mater. Interfaces*, 2009, **1**, 1259.

61. Y. Zhao, J. S. Dyck, B. M. Hernandez and C. Burda, *J. Am. Chem. Soc.,* 2010, **132**, 4982.
62. P. Christian and P. O'Brien, *J. Mater. Chem.,* 2005, **15**, 4949.
63. H. Yu, Y. Liu and S. L. Brock, *Inorg. Chem.,* 2008, **47**, 1428.
64. P. Sekar, E. C. Greyson, J. E. Barton and T. W. Odom, *J. Am. Chem. Soc.,* 2005, **127**, 2054.
65. V. M. Huxter, T. Mirkovic, P. Sreekumari Nair and G. D. Scholes, *Adv. Mater.,* 2008, **20**, 2439.
66. D.-K. Ko, J. J. Urban and C. B. Murray, *Nano Lett.,* 2010, **10**, 1842.
67. J. J. Urban, D. V. Talapin, E. V. Shevchenko, C. R. Kagan and C. B. Murray, *Nat. Mater.,* 2007, **6**, 115.
68. D. Weng, W. Zheng, C. Hao, Q. Peng and Y. Li, *Chem. Commun.,* 2008, 2556.
69. A. Kornowski, R. Eichberger, M. Giersig, H. Weller and A. Eychmüller, *J. Phys. Chem.,* 1996, **100**, 12467.
70. M. Green and P. O'Brien, *J. Mater. Chem.,* 1999, **9**, 243.
71. M. Green and P. O'Brien, *Adv. Mater.,* 1998, **10**, 527.
72. R. Wang, C. I. Ratcliffe, X. Wu, O. Voznyy, Y. Tao and K. Yu, *J. Phys. Chem. C,* 2009, **113**, 17979.
73. R. Xie, J. Zhang, F. Zhao, W. Yang and X. Peng, *Chem. Mater.,* 2010, **22**, 3820.
74. S. Miao, S. G. Hickey, B. Rellinghaus, C. Waurisch and A. Eychmüller, *J. Am. Chem. Soc.,* 2010, **132**, 5613.
75. D. K. Harris, P. M. Allen, H.-S. Han, B. J. Walker, J. Lee and M. G. Bawendi, *J. Am. Chem. Soc.,* 2011, **133**, 4676.
76. M. Green and P. O'Brien, *Chem. Mater.,* 2001, **13**, 4500.
77. S. L. Brock, S. C. Perera and K. L. Stamm, *Chem.–Eur. J.,* 2004, **10**, 3364.
78. K. L. Stamm, J. C. Garno, G.-Y. Liu and S. L. Brock, *J. Am. Chem. Soc.,* 2003, **125**, 4038.
79. S. C. Perera, P. S. Fodor, G. M. Tsoi, L. E. Wenger and S. L. Brock, *Chem. Mater.,* 2003, **15**, 4034.
80. S. C. Perara, G. Tsoi, L. E. Wenger and S. L. Brock, *J. Am. Chem. Soc.,* 2003, **125**, 13960.
81. S. M. Stuczynski, Y. Kwon and M. L. Steigerwald, *J. Organomet. Chem.,* 1993, **449**, 167.
82. C. B. Murray, S. Sun, H. Doyle and T. Betley, *MRS Bull.,* 2001, **26**, 985.
83. A. E. Henkes, Y. Vasquez and R. E. Schaak, *J. Am. Chem. Soc.,* 2007, **129**, 1896.
84. A. E. Henkes and R. E. Schaak, *Chem. Mater.,* 2007, **19**, 4234.
85. E. Muthuswamy, P. R. Kharel, G. Lawes and S. L. Brock, *ACS Nano,* 2009, **3**, 2383.
86. Y. Yin, R. M. Rioux, C. K. Erdonmez, S. Hughes, G. A. Somorjai and A. P. Alivisatos, *Science,* 2004, **304**, 711.
87. H. Kim, Y. Chae, D. H. Lee, M. Kim, J. Huh, Y. Kim, H. Kim, H. J. Kim, S. O. Kim, H. Baik, K. Choi, J. S. Kim, G.-R. Yi and K. Lee, *Angew. Chem., Int. Ed.,* 2010, **49**, 5712.

88. Y. Li, M. A. Malik and P. O'Brien, *J. Am. Chem. Soc.*, 2005, **127**, 16020.
89. C. Qian, F. Kim, L. Ma, F. Tsui, P. Yang and J. Liu, *J. Am. Chem. Soc.*, 2004, **126**, 1195.
90. J. Park, B. Koo, Y. Hwang, C. Bae, K. An, J.-G. Park, H. M. Park and T. Hyeon, *Angew. Chem., Int. Ed.*, 2004, **43**, 2282.
91. J. Park, B. Koo, K. Y. Yoon, Y. Hwang, M. Kang, J.-G. Park and T. Hyeon, *J. Am. Chem. Soc.*, 2005, **127**, 8433.
92. J.-H. Chen, M.-F. Tai and K.-M. Chi, *J. Mater. Chem.*, 2004, **14**, 296.
93. C.-T. Lo and P.-Y. Kuo, *J. Phys. Chem. C*, 2010, **114**, 4808.
94. K. A. Gregg, S. C. Perera, G. Lawes, S. Shinozaki and S. L. Brock, *Chem. Mater.*, 2006, **18**, 879.
95. K. Senevirathne, R. Tackett, P. R. Kharel, G. Lawes, K. Somaskandan and S. L. Brock, *ACS Nano*, 2009, **3**, 1129.
96. A. T. Kelly, I. Rusakova, T. Ould-Ely, C. Hofman, A. Lüttge and K. H. Whitmire, *Nano Lett.*, 2007, **7**, 2920.
97. K. Senevirathne, A. W. Burns, M. E. Bussel and S. L. Brock, *Adv. Funct. Mater.*, 2007, **17**, 3933.
98. I. Zafiropoulou, K. Papagelis, N. Boukos, A. Siokou, D. Niarchos and V. Tzitzios, *J. Phys. Chem. C*, 2010, **114**, 7582.
99. R.-K. Chiang and R.-T. Chiang, *Inorg. Chem.*, 2007, **46**, 369.
100. J. Wang, A. C. Johnston-Peck and J. B. Tracey, *Chem. Mater.*, 2009, **21**, 4462.
101. X. Zheng, S. Yuan, Z. Tian, S. Yin, J. He, K. Liu and L. Liu, *Chem. Mater.*, 2009, **21**, 4839.
102. S. Carenco, I. Resa, X. Le Goff, P. Le Floch and N. Mézailles, *Chem. Commun.*, 2008, 2568.
103. S. Carenco, M. Demange, J. Shi, C. Boissière, C. Sanchez, P. Le Floch and N. Mézailles, *Chem. Commun.*, 2010, **46**, 5578.

CHAPTER 5

The Synthesis of Core/Shell Quantum Dots

5.1 Early Studies on Core/Shell Materials

Surface passivation is a key element in a robust nanoparticle system. As the number of surface atoms is high (50% of particles are at the surface in a 3 nm diameter spherical particle) the preparation of a 'clean' surface with as few defects, dangling bonds or charge carrier trapping sites as possible is desirable. Early studies relied on polymers and surfactants to passivate the surface, although these often desorbed leaving unprotected particles.

A further inorganic layer is ideal for numerous reasons; once deposited, it is unlikely to desorb and may provide a more consistent, complete surface passivation. Surfactant cone angles no longer matter, the particles are more resistant to processing, and, by a judicious choice of material, quantum dot (QD) heterojunctions can be formed where the band energy offsets of the two materials can result in materials with novel and interesting optical properties.[1,2]

The two main factors to consider when choosing a shell material are the band energy levels and lattice mismatch. The lattice constants need to be different enough to avoid alloy formation, yet close enough to allow even shell growth. It is worth noting that although some core/shell systems have large lattice mismatches (*ca.* 15%), homogenous shells can form easily. The dynamics of shell growth on a small particle are significantly different from multilayer growth in thin film, and structures with large mismatches that would not form in flat systems may form in nanoparticle systems because of the short facets and reconstructed surfaces.

RSC Nanoscience & Nanotechnology No. 33
Semiconductor Quantum Dots: Organometallic and Inorganic Synthesis
By Mark Green
© Mark Green 2014
Published by the Royal Society of Chemistry, www.rsc.org

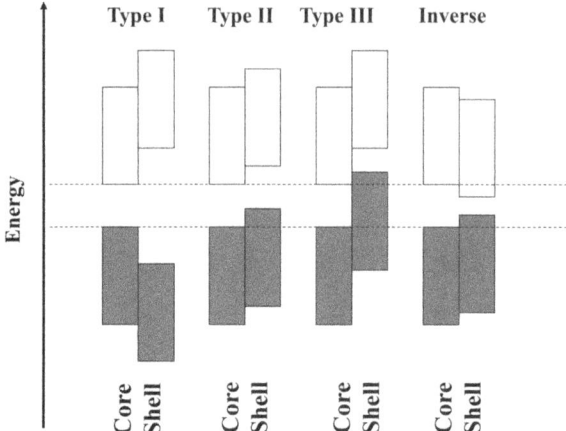

Figure 5.1   Band edge alignments and mismatches for the differing classes of core/shell materials. Grey boxes represent the valence band, white boxes the conduction band.

The energy level mismatch is the key parameter. Four types of core/shell semiconductor particles can be described, depending on the alignment of the energy levels (Figure 5.1). It must be noted that the values taken for these materials are usually based on bulk systems, and hence do not take into account the widening bandgap in a semiconductor nanoparticle; whether a shell of semiconductor a few monolayers thick can have a band structure at all is debatable. The bulk energy diagrams are, however, a good indication of what might be expected and some key publications describe relevant energy levels and mismatches.[3-9]

The most common are termed *type I*, where the energy levels of the core are narrower than those of the shell. By choosing a semiconducting shell with a wider bandgap than that of the core, photoluminescence should be enhanced (although not shifted), because the charge carriers are restricted from the shell by the energetic potential barriers, forcing charge recombination in the core species. The shell protects the emitting core from environmental factors by passivating surface defects and hindering oxidation. The other structures are classified as *type II*, where the energy levels are staggered; *type III*, where the energy levels are completely misaligned or broken; and *inverse*, effectively the opposite of type I structures.

For type II structures, electron and holes are spatially separated between the core and the shell, and emission is from the recombination at an effectively smaller indirect bandgap, making the emission red-shifted. As a result of these dynamics, type II particles have long excited-state lifetimes, which may find applications in photovoltaics. There are, to date, no examples of type III core/shell nanoparticles prepared by organometallic precursors in solution. Inverse structures can exhibit either type I or type II behaviour, depending on core size and shell thickness.

There are also instances where metal or metal oxide shells can be deposited on a semiconducting core, giving bifunctional core/shell materials, and in these cases the optics are dependent on the particular system.[10] For example, the deposition of gold on to a CdSe surface resulted in the quenching of the semiconductor emission. However, deposition of gold on to a CdTe surface resulted in enhanced emission.[11] In these cases, the absorption of the metal must be taken into account, and in the case of gold, silver and copper (which are free-electron metals), the surface plasmon resonance of the nanosized metal must be considered. Not all semiconductor/metal or metal oxide particles form core/shell particles, and these will be discussed later.

Initial work into what can be described as the earliest core/shell systems was reported in 1984, where CdS islands were grown on ZnS.[12] This was followed by numerous notable investigations, including $CdS/Cd(OH)_2$,[13] and the CdS/HgS series,[14] which also included the first multilayer structure CdS/HgS/CdS,[15-19] which could be considered the most studied and technically advanced core/shell system before the advent of organometallic chemical routes. The genesis of organometallic core/shell systems is covered in Chapter 1, with CdSe/ZnS[20] and CdSe/ZnSe[21] systems being prepared by a pseudo-organometallic route, using a mixture of organometallic and inorganic chemistry.

5.2 Organometallic Routes to Core/Shell Nanoparticles

The synthesis of CdSe nanoparticles capped with trioctylphosphine oxide (TOPO) initiated numerous studies into size quantisation effects; the relative ease of preparation, the manipulation of the high-quality material obtained and the superior optical properties make QDs prepared by the organometallic route an ideal method of studying low-dimensional systems. The particles, capped with TOPO, were crystalline, monodispersed and exhibited band edge luminescence. However, the surfactant monolayer that passivates surface defects, ensuring band edge emission while protecting the particles from agglomeration, is not ideal as the surfactant does not completely cap the surface and often desorbs. Gradual oxidation of TOPO-capped CdSe particles over *ca.* 24 hours was found to result in the formation of SeO_2, which desorbed off the particle surface over a 96 hour period.[22] The resulting surface vacancies provided non-radiative recombination pathways for the charge carriers and induced deep trap emission from the particle, giving optical properties similar to poorly passivated particles. It is common to see nanoparticles capped with just organic surfactants referred to as 'naked' or 'bare' dots. The majority of 'bare' particles prepared with a monolayers of surfactant eventually display deep trap emission, usually manifest as a low-energy shoulder or feature in emission spectra. Different capping agents result in differing degrees of protection, with amine-capped particles exhibiting excellent stability when compared to the phosphine oxide analogues, attributable to the linear shape and hence efficient packing of the molecule. Organic monolayers, however, provide only minimal protection against oxidation and a more complete passivating agent is usually required. The overcapping of bare particles prepared by organometallic precursors with a further inorganic phase can be traced back to 1994, where work by Danek *et al.* examined capping CdSe particles with ZnSe and ZnS.[23–26] Although most of these process were based on metal–organic chemical vapour deposition (MOCVD), where bare CdSe particles were incorporated into a ZnS matrix, one report described the reaction of CdSe with TOPSe and Et_2Zn prior to deposition by an electrospray process. This can be seen as the first paper to examine core/shell structures of these particular materials.[25]

5.3 Type I Materials

5.3.1 CdSe/ZnS

Hines and Guyot-Sionnest reported the first complete solution-based core/shell semiconductor system (CdSe/ZnS, type I) prepared by organometallic precursors in a single-pot reaction.[27] Typically, the core CdSe particles were prepared as described in Chapter 1, by the injection of a stock solution consisting of Me_2Cd, trioctyl phosphine (TOP) and TOPSe in TOPO at 350 °C. The reagents were then allowed to cool to 300 °C, whereupon Me_2Zn and $S(SiMe_3)_2$ in TOP were then injected in five portions over a period of about a minute, to avoid the nucleation and growth of separate ZnS particles. The molar ratio of Cd/Se : Zn/S precursors was 1 : 4, consistent with earlier attempts to grow core/shell particles where the shell thickness was comparable to that of the core diameter.[12] The particles were found to be composed of *ca.* 3 nm diameter CdSe cores capped with a ZnS shell (6 ± 3 Å thick), as determined by transmission electron microscopy (TEM) comparisons of particle size before and after addition of the shell precursor. A surfactant layer on the ZnS shell ensured solubility in non-polar hydrocarbons in a similar manner to the bare dots.

Dabbousi *et al.* reported a similar, more detailed preparation[28] using the same precursors, but notably isolated the core particles and narrowed the size distribution by selective precipitation before redispersing the particles in TOPO/TOP. Once the size of the particles had been determined by either small-angle X-ray spectroscopy or electron microscopy, the amount of precursor required to form a shell of the desired thickness was then calculated. The core particle mixture was heated to the required optimum temperature (140 °C for 2.3 nm diameter dots, to 220 °C for 5.5 nm diameter dots) where the ZnS precursors were added dropwise over 10 minutes. After addition, the mixture was left for several hours at 90 °C before isolation. A similar procedure was used to prepare CdSe/CdS core/shell particles, replacing Et_2Zn with Me_2Cd.

The slow rate of addition, low concentration of precursor solution and specific addition temperature were found to be critical, ensuring the ZnS grew heterogeneously on to the CdSe seeds instead of nucleating and forming separate ZnS particles. Although this could not be completely avoided, size-selective precipitation on the final product ensured that all ZnS particles were removed.

Absorption spectroscopy of the CdSe core and the CdSe/ZnS core/shell particles displayed only a slight red shift in the band edge after addition of 1–2 monolayers of ZnS (a monolayers was defined as a shell 3.1 Å thick, the distance between planes in the (002) direction in bulk, hexagonal ZnS[29]), attributed to a leakage of the exciton into the ZnS shell. Emission spectroscopy showed essentially the same position of emission after the addition of the shell with an increased intensity, relating to an increase in quantum yields from a maximum of 15% for the bare dots to 50% for the core/shell particles. The emission wavelengths accessible were between 470 nm and 620 nm. It was noted that for the smallest bare dot size, emission was normally dominated by trapping states giving broad white emission. Upon deposition of the ZnS shell, intense blue emission was observed. (This synthesis of blue-emitting CdSe/ZnS materials was later extended, using so-called green precursors to prepare core CdSe particles 1.6–2.2 nm in size, with wavelengths as short at 475 nm reported with quantum yields of up to 50%.[30])

An investigation into the optimum shell thickness probed a range of particles with shells varying from 0.65 to 5.3 monolayers thick. Emission increased upon shell addition up to 1.3 monolayers, then declined as a result of the incoherent growth of the thicker shell due to lattice strain (CdSe/ZnS particles have a 12% lattice mismatch) leading to grain boundaries and dislocations resulting in non-radiative exciton recombination in the shell. Similar effects have also been reported for CdSe with a CdS shell.[31]

Modelling of the electronic structure of CdSe/ZnS gave radial probability functions (first excitonic transition) for holes and electrons in bare, CdS- and ZnS-capped CdSe particles, as shown in Figure 5.2. In bare dots, the wave function of the electron was suggested to spread over the entire particle and tunnel very slightly into the capping agent. In ZnS-capped particles, the electron wave function was suggested to tunnel into the shell and the probability of shell penetration for the hole wave function was low. Notably, the increased delocalisation of the electron narrowed the energy gap in the particle, although the maximum narrowing observed was only 36 meV, depending upon particle size and shell thickness.

In CdS-capped particles, the barriers height for the electron was found to be only 0.2 eV, less than the energy of the electron; therefore the wave function extended deeply into the shell. The hole had a barrier of 0.55 eV, so the wave function had a low probability of shell penetration, which was still smaller than the hole barrier for the ZnS shell and therefore extended further. The red shift in the absorption feature was therefore larger, with a maximum of *ca.* 390 meV.

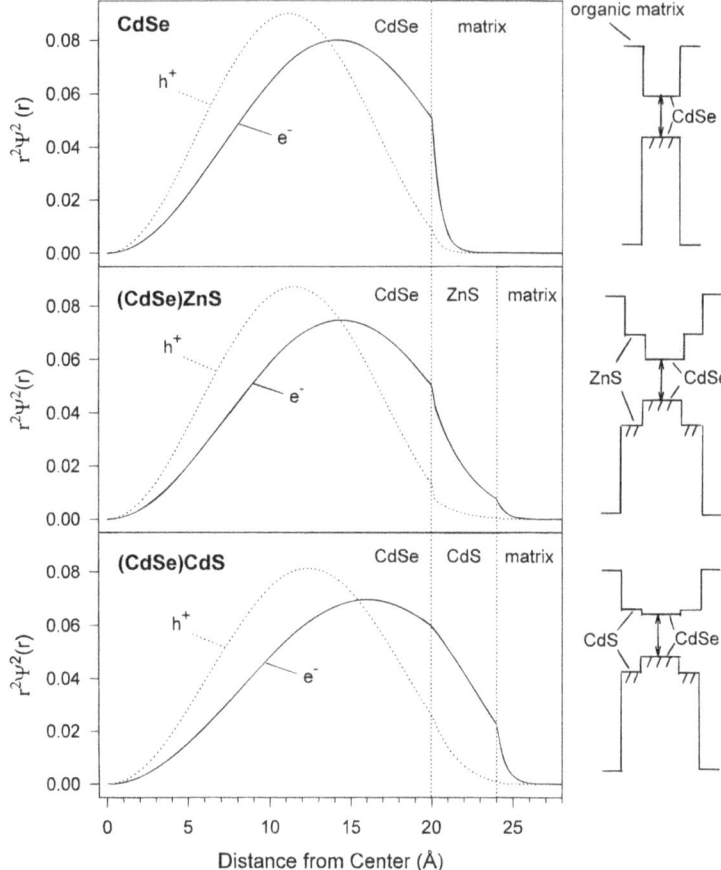

Figure 5.2 Radial probability functions for hole and electron in (a) 2 nm diameter CdSe particle; (b) 2 nm CdSe with 4 Å ZnS shell; (c) 2 nm CdSe with 4 Å CdS shell. Right-hand side shows associated band offsets. Reprinted with permission from B. O. Dabbousi, J. Rodriguez-Viejo, F. V. Mikulec, J. R. Heine, H. Mattoussi, R. Ober, K. F. Jensen and M. G. Bawendi, *J. Phys. Chem. B*, 1997, **101**, 9463. Copyright 1997 American Chemical Society.

Particles with shells of 1.3 monolayers were also found to be significantly protected against surface oxidation, as determined by X-ray photoelectron spectroscopy (XPS), whereas shells of only 0.65 monolayers still showed evidence of surface oxide generation. Electron microscopy of the particles showed continuous lattice fringes throughout the samples, with no interfaces between the core and the shell, indicating clean epitaxial growth. Slightly patchy growth with preference along the (001) axis was also observed, resulting in particles with a slight increase in the aspect ratio. Wide angle X-ray scattering patterns revealed diffraction patterns increasingly consistent with the hexagonal phase of ZnS upon addition of thicker shells, confirming the epitaxial growth of an ordered crystalline phase of ZnS.

A similar report on the preparation of CdSe/ZnS core/shell particles described the effect on the incorporation of long-chain amines into the TOPO surfactant mixture.[32] A maximum quantum yield of 60% was reported with a shell 1.6 monolayers thick. Interestingly, the growth dynamics of CdSe capped with hexadecylamine (HDA) and TOPO were found to be significantly different from CdSe particles grown in TOPO. With the amine present, fast focusing of the size distribution was observed (as opposed to defocusing, Ostwald ripening) upon prolonged heating, therefore the narrowest size distribution coincided with the largest particle size at a specified growth temperature. It was also observed that large amounts of amine in the surfactant mix resulted in precipitation of nanoparticles above 200 °C.

Core/shell particles of CdSe/ZnS are now standard materials when robust, luminescent and stable particles are required. Numerous studies have been carried out on various aspects of photophysics utilising these materials. Their inherent stability makes them ideal candidates for various applications, and although elemental sulfur or $S(SiMe_3)_2$ are the usual precursors for the shell chalcogen deposition, other precursors such as P_2S_5[33] and $C_6H_{11}NCS$[34] have also been used.

Core/shell particles of rods have also been grown.[35] Using the chemistry described in Chapter 1, the core rods were prepared using hexylphosphonic acid (HPA) and multiple injections of precursors. A large aliquot of rod sample was isolated directly without separation. This was followed by the addition of HDA (critical for high quantum yields) and subsequent heating at 120 °C for 20 minutes allowed the amine to exchange with TOPO. The temperature was then stabilised at 190 °C and the standard ZnS precursors added dropwise. If the rods were isolated by non-solvent interactions prior to shell growth, as is usual, the morphology of the rods changed when redissolved in surfactant and the shell precursors added. By avoiding isolation, the precursors for rod growth were present during shell growth, maintaining the anisotropic structure. The optimum ratio of Zn : S precursor was found to be 1 : 2, with deviation resulting in lower quantum yields.

In a standard experiment, a bare rod (22 × 4 nm) exhibited a quantum yield of 3%. Addition of 1.3 monolayers, the optimum shell thickness for spherical particles, resulted in an increase in quantum yields to 18%. However, further deposition increased the quantum yield up to 28% (1.7 monolayers). Increasing the shell thickness even further decreased the quantum yield, again attributed to the lattice mismatch, which was accommodated in thin shells but resulted in defect-prone thicker shells. The maximum quantum yield in this system (40%) was observed in rods 11 × 3 nm, with a shell 2.3 monolayers thick. Photostability experiments showed that rods without a shell displayed a rapid reduction in quantum yield and precipitated irreversibly from solution upon intense illumination, while core/shell rods remained stable. Core/shell rods with a low quantum yield exhibited an increase in quantum yield after irradiation, attributed to photochemical annealing. Interestingly, energy-dispersive X-ray analysis spectroscopy (EDS) indicated the growth of a CdS buffer layer between the CdSe and ZnS shell.

The Alivisatos group prepared graded structures of CdSe/CdS/ZnS 'core/buffer/shell' rods using metal alkyls and standard precursors.[36] This work was initiated after the observation that attempts to grow a ZnS layer on CdSe rods often resulted in uneven structures with very thin (on average less than one monolayer) shells. This was overcome by adding a small amount (8 : 1 Zn : Cd ratio) of cadmium precursor to the shell stock solution, resulting in the formation of the buffer layer. What is important is that the layers were *not* prepared sequentially; the buffer layer formed when the ZnS precursors were added to the preformed CdSe. The segregation of growth could be due to numerous factors, including the lower solubility of CdS *versus* ZnS in TOPO, or the tendency for zinc atoms to form more stable complexes with TOPO than cadmium atoms do.

CdS has a lower lattice mismatch than ZnS, and this is a key factor when growing shells on CdSe nanorods, which have a relatively 'flatter' surface than spherical dots and hence strain becomes an issue. The introduction of a buffer layer resulted in even, epitaxial growth of the shell, with no evidence of separate CdS or ZnS particles observed. Shell ups to 6.5 monolayers thick (on rods of *ca.* 30 nm × 7 nm) could be grown, with Zn : Cd ratio in the shell being 2 : 1 in a 2 monolayer shell, 4 : 1 in a 4.5 monolayer shell, and 4.5 : 1 in a 6.5 monolayer shell. The shell thickness was controlled by the amount of precursor added, until a certain shell thickness was reached and the rods developed 'tails'. The addition of a ZnS shell improved the surface regularity of the structures, overcoming the stacking faults in long rods described in Chapter 1. Interestingly, as the shell thickness increased, the reflections from the X-ray diffraction (XRD) patterns shifted position towards a larger 2θ value not consistent with ZnS, CdS or CdSe phases, which was attributed to lattice compression of the core particle.

Upon the deposition of the shell, the absorption and emission spectra red-shifted slightly as the exciton tunnelled through the relatively low energy barriers into the shell as described above, confirming the CdS layer. The quantum yields of the core particles alone was less than 1%, while particles with a thin shell (2 monolayers) displayed quantum yields of up to 4%, and particles with a medium shell (4.5 monolayers) displayed quantum yields of 1%, consistent with strain-induced defects of thicker layers. Under constant laser illumination for 20 hours, the bare rods either exhibited no change in quantum yield or a decrease, whereas the particles with shells photo-annealed to exhibit quantum yields of up to 16% (4.5 monolayers). This implied that a photochemical process induced an internal structural reorganisation, as thermal annealing did not result in the same increase in emission. The importance of the interface and reducing lattice strain in preparing highly luminescent structures has been explored, and the highly graded CdS/ZnS shell/shell particles have been suggested to be the optimum material for such structures by supplying uniform coverage.[37]

Interestingly, rods of CdSe have also been used to make alloyed rods of ZnCdSe by adding ZnSe precursors to CdSe rods, followed by a prolonged annealing step.[38] The emission quantum yield of these rods was up to 10%, significantly higher than that of simple CdSe rods.

Spherical CdSe particles have also been capped with a ternary shell of CdZnS using an amendment on the usual precursors.[39] In this case, purified TOPO, TOP, HDA and HPA were degassed, and injected with a mixture of $Cd(CH_3COCHCOCH_3)_2$, TOPSe and hexadecanediol in TOP at 360 °C. The particles were left to grow at 280 °C, annealed overnight at 80 °C and then isolated. The size distribution was narrowed by size-selective precipitation, followed by redispersion in TOPO with HPA. Dropwise addition of a TOP solution of Et_2Zn, Me_2Cd (4 : 1 ratio) and a threefold excess of $S(SiMe_3)_2$ at 145 °C followed by overnight annealing resulted in core/shell particles with a quantum yield maximum of 46%.

5.3.2   CdSe/CdS

Nanoparticles of CdSe can also be prepared with a shell of just CdS.[40] The energy gap is aligned to confine charge carriers suggesting a type I structure, and the lattice mismatch (only 3.9%) is sufficient to prevent alloying while allowing epitaxial shell growth. In the seminal report of CdSe/CdS, the core particles were grown by the injection of Me_2Cd and TBPSe (with a slight excess of cadmium precursor) into TOPO at elevated temperatures, followed by growth to the required size. Isolation by non-solvent interactions was followed by dissolution in pyridine, exchanging the surface cap for the more labile ligand. After prolonged refluxing under nitrogen, the shell stock solution of Me_2Cd and $S(SiMe_3)_2$ (1 : 0.4 Cd : S ratio) was added dropwise at 100 °C. After growth, DDA was added until the particles precipitated, exchanging the capping ligand. The particles were then soluble in chlorinated solvent, but not pyridine. Growth of discrete CdS particles was also observed, and avoided by changing the Cd : S ratio to 1 : 2.1, whereupon only shell growth was observed. This unusual method of growth was developed after the failure to grow CdS shells on CdSe cores in TOPO at 200 °C: a more labile surface ligand was required. The low Cd : S ratio used in shell growth was explained due to the cadmium-rich precursor solution for the core particles, which may have given particles with a metal-rich surface. The particles were analysed by the usual methods, notably XPS which was used to probe shell thickness using the mean free path of the emitted photoelectron. High-resolution electron microscopy also revealed particles with a reduced image contrast on the edges, consistent with a material with fewer electrons per unit cell than the core.

The absorption spectra were found to be shifted, which was explained by the perturbation of the core energy levels by the shell, as previously described for CdSe/ZnS core/shell particles. This is neatly explained in Figure 5.3a and b, where the new energy levels emerged when the core/shell particle formed. The molecular-like energy levels, a consequence of quantum confinement,

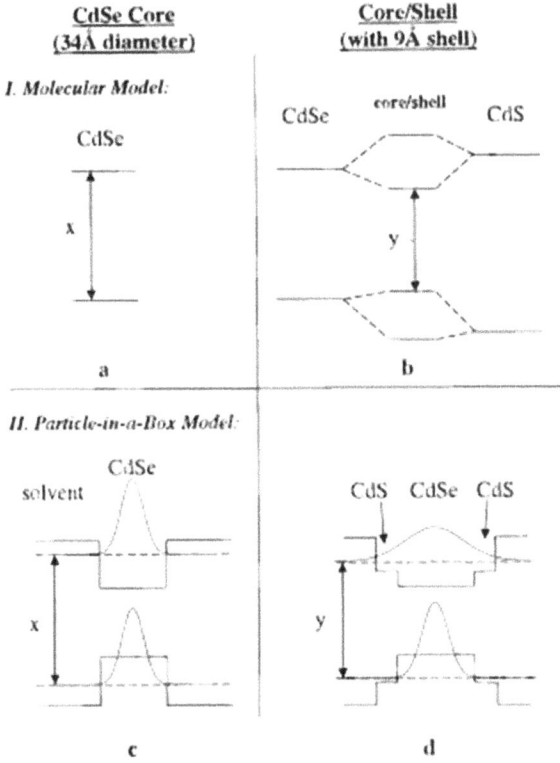

Figure 5.3   Molecular models (a and b) and 'particle in a box' models (c and d) of core and core/shell electronic structures, showing the narrowing of the bandgap upon addition of a shell. Reprinted with permission from X. Peng, M. C. Schlamp, A. V. Kadanavich and A. P. Alivisatos, *J. Am. Chem. Soc.*, 1997, **119**, 7019. Copyright 1997 American Chemical Society.

mixed to form a new HOMO–LUMO gap. A more detailed explanation was provided by the 'particle in a box' model (Figure 5.3c and d) using the potentials (solid lines) and the actual hole and electron energy levels (dashed lines). In the example shown, the potentials were estimated from bulk electron affinities and ionisation potentials, showing a type I heterostructure. However, calculations where the conduction band potential offset between CdSe and CdS was altered to give a gap of -0.3 eV (rather than $+0.3$ eV), giving a type II, structure demonstrated the hole and electron wave functions were insensitive to small changes in energetic barriers at the interface, suggesting that the difference between type I and type II heterojunctions in this case may be negligible.

The slight decrease in bandgap is accompanied by an increase in quantum yield, as high as 84% for 3 nm diameter core CdSe particles with 1.8 monolayers of CdS. There are reports of core/shell particles with quantum yields of almost 100%, although no specific details were provided. Typically, shells up to 3 monolayers thick were grown on core between 2.3 and 3.9 nm in diameter. After 2 monolayers of shell growth, the quantum yield was found to drop, consistent with other core/shell particles. The ability to grow thicker shells of CdS on CdSe while still increasing the emission when compared to the CdSe/ZnS system can be attributed directly to the smaller lattice mismatch, possibly providing a heterojunction with increased stability towards environmental factors such as oxidation.

Altering the ratio of Cd : S in the precursor solution for shell growth resulted in the formation of core/shell particles with unusual morphologies.[41] Core particles of CdSe were prepared using the precursors described immediately above, with the inclusion of HDA. The shells were grown by the slow addition of a TOP solution of Me_2Cd and $S(SiMe_3)_2$ to the crude CdSe solution at 130 °C followed by growth at 90 °C. When the ratio of Cd : S was between 1 : 1 and 1 : 1.6, spherical core/shell particles were formed. When an excess of sulfur precursor was used (ratio 1 : 3–1 : 5), an anisotropic shell grew along the *c*-axis giving rods with an aspect ratio of up to 5. The width of initial rods was the same as the core particle diameter (3–6 nm); however, when a growth temperature of 140–180 °C was used, the width of the rod also increased. Two main points were highlighted for the anisotropic shell growth; the lattice mismatch along the (001) *c*-axis was larger than along the perpendicular (100) plane (*ca.* 4.2% compared to 3.8%) and CdS growth on the (100) CdSe face proceeded with more strain than on the (001) or (00$\bar{1}$) face. Also, cadmium atoms on the (100) and (001) had only one dangling bond, whereas cadmium atoms on the (00$\bar{1}$) facet had three, making it more active. The increased amount of sulfur precursor resulted in preferential growth along the (00$\bar{1}$) facet on the cadmium sites. The growth of isotropic shells at higher temperature was explained by the temperature-dependent transition from the low-temperature reaction kinetic-limited growth, to the higher-temperature diffusion-limited growth. The particles prepared exhibited large extinction coefficients of 10^7 cm^{-1} M^{-1}, large Stokes shifts and quantum yields greater than 70% for rods with aspect ratios of 1.6–2, the largest recorded for anisotropic core/shell particles at that time. An increase in the quantum yields was also observed for rods with the same diameter as the core particle, suggesting there was no CdS shell growth on the (100) facet of the CdSe core. This implied a surface reconstruction or annealing effect during the shell growth process, and that the passivation of the dangling bonds on the (001) and (00$\bar{1}$) facets were essential for high emission efficiencies. The CdSe-seeded CdS-rod core/shell systems could also be tuned to exhibit type II behaviour, as discussed later, by controlling the morphology of the structures by using a significantly smaller (<2.8 nm diameter) CdSe core.[42]

Despite the high quantum yields obtained for spherical CdSe/CdS particles, the reaction required the isolation of the cores, and a one-pot synthesis is preferred utilising simple precursors. A 'green' one-pot route to CdSe/CdS has been described based on the organometallic synthesis.[43] In this case,

CdSe core particles were grown using $Cd(CH_3CO_2)_2$, TOPO, HDA, a phosphonic acid and TOPSe. Unlike previous reactions, the cadmium precursor was injected to the hot surfactants/selenium solution, resulting in particle formation and the conversion of only 20% of the $Cd(CH_3CO_2)_2$, compared to almost 100% conversion of Me_2Cd under similar conditions. The excess cadmium precursor was then used to form an epitaxial CdS shell by the addition of H_2S gas into a closed reaction flask (above, not into the reaction mixture) at 140 °C. The particles prepared exhibited narrow emission (FWHM 27–35 nm) in the green to red regions of the spectrum, with quantum yields up to 85%, while CdSe core with *ca.* 3.5 monolayers of CdS exhibited similar stability to the analogous CdSe/ZnS system.

The control over the deposition of the shell material is a key factor as described above. The growth of too thick a shell resulted in the quenching of the emission due to strain-induced defects, therefore a method of epitaxially growing individual monolayers controllably was of immense importance. An interesting development is the use of successive ion layer adsorption and reaction (SILAR),[44,45] a solution analogue of atomic layer epitaxy (ALE), where cationic and anionic species were deposited sequentially, essentially half a monolayer at a time. Since both species are not present at the same time, separate nucleation is avoided.

This method was initially used to prepare CdSe/CdS particles.[46] In this case, the core particles of CdSe, *ca.* 3.5 nm in diameter were prepared using the green method described by Peng (see Chapter 1), utilising CdO/stearate and TBPSe as precursors in a non-coordinating solvent. The particles were purified and isolated into hexane, removing excess capping agents, and the concentration of nanoparticles calculated using Beer's law. The isolation of core nanoparticles as a powder was avoided as this resulted in substandard core/shell materials. The amount required to add a monolayer was calculated and the cadmium precursor then added, followed by the sulfur precursor; this process was repeated to add a specific number of monolayers. When the required number of monolayers had been grown, the reaction was quenched by removal from the heat source. Contrary to other methods of core/shell preparation, a high temperature (240 °C) was required for discrete shell growth, avoiding the formation of CdS particles. The absorption spectra were consistent with shell growth, and the emission intensity increased as expected. Confirmation of a core/shell structure was obtained using XPS measurements. Emission quantum yields were found to increase with the increasing number of shells to a maximum of *ca.* 40%, with no drop-off after 5 monolayers had been deposited, in contrast to CdSe/ZnS. What is notable is that the core particles capped with amines exhibited quantum yields of up to 80%, which actually dropped upon shell addition. Core particles capped with just TOPO had low quantum yields, which increased to a similar level to that of CdSe/CdS particles prepared from amine-capped cores upon addition of the shells. The emission width remained narrow throughout the experiments. The emission spectra of particles with thick shells (five monolayers) displayed a blue-shifted shoulder, the origins of which were not clear. The particles were found to photo-brighten upon annealing, purification and photo-oxidation. The quantum yield was found to dip after the injection of each shell component, after which the emission then increased. This was attributed to the rapid growth of the shell causing structural disorder. This same method has been used to prepare structures with a maximum quantum yield of 92% when 5 monolayers of CdS had been deposited.[47] The SILAR technique has been used to prepare CdSe/CdS in an in-depth study examining reaction conditions and optical properties, allowing a calibration plot to be realised, allowing the accurate determination of shell thickness without the need for high-resolution electron microscopy.[48] The study also highlighted that charge carriers can be found at the surface, despite shell thicknesses of up to 2.2 nm.

The use of the SILAR method has been extended to the preparation of CdSe particles with a giant CdS shell, designed to isolate the core wave function from the surface.[49] Core CdSe particles with a diameter of 3–4 nm had 18–19 monolayers of CdS deposited sequentially. The shell materials dominated the absorption spectra, while the emission profiles were significantly shifted from the original band edge, although they were not attributed to the shell materials. As expected, the optical properties of the materials were insensitive to ligand exchange, with quantum yields of up to 40%. The synthesis also allowed for the preparation of CdSe/11CdS/6Cd$_x$Zn$_y$S/2ZnS shelled materials, with a graded alloy Cd$_x$Zn$_y$S shell, which had quantum yields of *ca.* 10%. Interestingly, the materials exhibited significantly reduced optical blinking, and long multiexciton lifetimes, consistent with suppressed Auger recombination.[50]

The growth of CdSe/CdS core/shell particles can also be achieved by a two-phase method in an autoclave as a reaction vessel, using similar precursors. In this case, Cd(CO$_2$(CH$_2$)$_{12}$CH$_3$)$_2$ and oleic acid were dissolved in toluene and heated until dissolved, followed by the addition of an aqueous solution of SeC(NH$_2$)$_2$. After the reaction at 180 °C for *ca.* 20 minutes the reaction was cooled and the nanocrystals precipitated with ethanol. The particles were then dissolved in toluene, and fresh batches of cadmium precursor and oleic acid were added. Further autoclaving was affected until the cadmium precursor had dissolved, followed by addition of an aqueous solution of thiourea. The reagents were then heated at 140 °C for 4 hours, and then allowed to cool and isolated. The material showed band edge emission with quantum yields of 60–80%, with no evidence of deep trap emission. The resulting emission was between *ca.* 450 and 550 nm. The reaction was notable for the small size of the core particles used; between 1.2 and 1.5 nm core with up to 5 monolayers of shell deposited.[51]

5.3.3 Other Type I Core/Shell Materials Based on Group II Chalcogenides

The preparation of materials that emit effectively in the blue region of the spectrum is desirable for nanoparticle chemists. The most studied system, CdSe, would require particles below 2 nm in diameter to emit at such a wavelength. In this case, surface traps dominate the photoluminescence process and the resulting emission is often predominantly white in colour. The absorption cross-section of particles *ca.* 2 nm in diameter is also reduced, making them less than ideal for incorporation into devices. The overcapping of small particles with the relevant shell (such as ZnS) is difficult to achieve while maintaining a narrow size distribution and high quantum yields. Zinc chalcogenides, as described in Chapter 1, are potential blue emitters, although they are relatively unexplored and currently emit just beyond the visible spectrum at *ca.* 440 nm. Cadmium sulfide, with a bulk band edge at 2.42 eV (512 nm) and excitonic diameter of just 3.9 nm, is an ideal material for blue emission; however, photoluminescence from bare CdS nanoparticles is often dominated by trapping states giving a broad emission profile resulting in a loss of colour purity. The deposition of a suitable semiconducting shell on the CdS particles removed the trapping states and significantly improved the emission.[52] The core particles were prepared using the established green organometallic-based routes; using $Cd(CH_3CO_2)_2$ with a phosphonic acid to prepare the cadmium precursor, sulfur in oleylamine (OAm) as both the chalcogen source and capping agent source, and OAm and TOP as capping agents into which the precursors were injected. The core particles were *ca.* 5 nm in diameter, the largest of them exhibiting an excitonic feature at 448 nm and predominantly near band edge luminescence with lower energy emission evident, consistent with surface traps. The quantum yields were estimated at up to 6%. The particles were exhaustively purified by precipitation to remove excess precursors and capping agents before shell deposition was attempted.

Deposition of a wide-bandgap material on the particle surface resulted in significant increase in the emission quantum yield and a decrease in trap emission. ZnS was chosen due to its wide bandgap of 3.7 eV and low lattice mismatch of only 8%. Also, maintaining the same anion at the interface resulted in a larger offset in the conduction band.[53] The shell was deposited using Et_2Zn and $S(SiMe_3)_2$ in a TOP solution. The addition of the shell precursors resulted in the gradual growth of the shell, which could be observed visually by the change in emission from a weak violet (from the trapping states) to a strong blue band edge emission. The quantum yield maximum of about 30% was achieved after the addition of *ca.* 3 monolayers. The emission profile of the band edge emission broadened slightly upon addition of the shell solution, from a FWHM of *ca.* 19 nm to *ca.* 26 nm, and a slight red shift in emission position was also observed, both attributed to the leakage of the charge carriers into the shell. Notably, the presence of the amine appeared essential for efficient shell growth, as, without it, evidence of trap emission was still observed after the addition of the shell.

What is evident about the majority of synthetic pathways to core/shell materials is the continuing evolution of the core particle chemistry, whereas the chemistry of shell deposition has, to date, barely changed from the initial reports. The reasons for this are not clear. One notable advance is the preparation of CdSe/ZnSe particles, where green organometallic-based routes have been used for the deposition of the ZnSe shell.[54] ZnSe is an ideal shell material for CdSe dots; the bandgap (2.72 eV) is wider than that of CdSe (1.74 eV) giving a type I structure, the lattice mismatch is low (6.3%) and the same anion is maintained. The core particles were prepared as described in Chapter 1, by the thermolysis of a cadmium carboxylate and TOPSe in TOPO, TOP and a long-chain amine at 250 °C. The particles were isolated and redispersed in a hexane solution of TOPO and long-chain amine, then heated to 190 °C. A syringe pump then injected a toluene solution consisting of an equimolar amount of $Zn(CO_2(CH_2)_{16}CH_3)_2$ and TOPSe within 1 hour, at a temperature of 190–200 °C, which was followed by annealing for up to 1.5 hours. The quantum yield increased up to a maximum of 85%, although the number of monolayers this related to was not reported. The absorption spectra red-shifted slightly as the exciton leaked into the shell, although the width (FWHM) of the emission profile remained unchanged at *ca.* 28 nm. The CdSe/ZnSe particles were stable when phase-transferred to water and exhibited no drop-off in emission intensity, an essential property of particles to be used in biological imaging.

An unusual route to core/shell particles has been reported by Schreder *et al.* who described the synthesis of CdTe/CdS structures starting from CdS particles, *i.e.*, with the shell material first.[55] In this route, the initial CdS particles were prepared by complexing $CdCl_2$ to TBP in $CHCl_3$, followed by the addition of $S(SiMe_3)_2$ and eventual refluxing. Rigorous purification resulted in a yellow powder of CdS nanoparticles, estimated to be *ca.* 3 nm in diameter as determined from XRD measurements. The particles were then dissolved in CH_2Cl_2, to which was added $CdCl_2$ and $Te(SiMe_3)_2$, with a Cd : Te ratio of 2 : 1. It is worth noting that no excess stabilising ligand was used. The reaction started immediately and took up to 30 hours at room temperature to complete. The cause of this seemingly reverse reaction was the rapid anion exchange, with the tellurium ions expelling and replacing the sulfur ions, which then in turn reacted with the excess cadmium chloride in solution to yield the shell. The final material was estimated to be *ca.* 3 nm in diameter as determined by the position of the excitonic peak in the absorption spectra, which supported the ion exchange hypothesis. The material structure was confirmed by XRD, although reflections from the as-prepared particles were weak as a result of the low temperature of synthesis. Annealing of spin-coated films confirmed the zinc blende structure. Emission quantum yields increase from virtually zero to 10% after the reaction and formation of the core/shell structure, with a shift in the absorption edge observed during the conversion from the CdS core to the CdTe core. Addition of excess sulfur precursor to the final product resulted in the further red shift in the emission spectrum from 550 nm to *ca.* 650 nm. This highlights the complex growth mechanism, and the shift was attributed to the increased shell straining the core.

One interesting point regarding the growth of tellurium-containing nanoparticles is that aqueous-based routes utilising thiols as capping agents appear to be more effective than organometallic routes, attributed to the thiol blocking trapping states. Wuister has nicely highlighted this, where

CdTe particles prepared by organometallic chemistry were phase-transferred to water using long-chain thiols.[56] Nanoparticles of CdTe made by this method are expected to have all fluorescence quenched in a matter of hours, and phase transfer is usually accompanied by further loss of emission. In this case, where the thiol was used to phase-transfer, the emission was enhanced and no evidence of trapping was observed. To take advantage of the suitability of the CdTe/thiol system in water, a hybrid approach to core/shell particles has been developed in which core CdTe (and CdHgTe) particles were prepared by the aqueous methods.[57] The particles were then phase-transferred from the aqueous phase to the organic phase using dodecanethiol. Once in the organic phase (toluene) the particles were capped with a ZnS shell using $S(SiMe_3)_2$ and dialkylzinc precursors using TOPO and TBP as surfactants. Once phase-transferred to the organic phase, the quantum yields dropped, although this increased to the initial value (and above) upon shell deposition. After shell growth, the absorption and emission spectra were found to have significantly red-shifted. The shift was too large to be explained by particle growth or exciton leakage into the shell. Although the exact cause of the red shift is not clear, it increased upon increasing shell thickness, so was possibly related to lattice strain.

Other core materials have also been investigated. As highlighted in Chapter 1, zinc chalcogenides are potentially ideal blue emitters, but their air sensitivity restricts potential applications. The deposition of a suitable cap should therefore make the particles more stable and useful in devices. ZnSe nanoparticles capped with HDA, synthesised as described earlier, have been capped with a ZnS shell using dialkylzinc and $S(SiMe_3)_2$, giving strong emission at *ca.* 3 eV (*ca.* 415 nm) with quantum yields of around 15%, even months after synthesis.[58] Chen *et al.* have reported a more in-depth study, where ZnSe particles were capped with either ZnS or ZnSeS.[59] The core particles, made from ZnO and lauric acid in HDA, were 2–6 nm in diameter, emitting at 400–440 nm with quantum yields of up to 10%. Interestingly, ZnO was found only to fully dissolve with over 3 molar equivalents of lauric acid, attributed to the decomposition of the carboxylic acid at high temperatures or the slow formation of the zinc–acid complex. The ZnS shell was deposited using ZnO/lauric acid and trioctylphospine sulfide (TOPS). After deposition of 1.8 monolayers, the emission quantum yield was found to increase to 32%. To prepare the ZnSeS shell, TOPS was injected into the reaction flask while the core was still growing. After deposition of 1.6 monolayers of the alloy shell, the emission was found to increase to 26%. In related reports, the optical properties of ZnSe/ZnS have been examined although few experimental details were provided.[60,61]

The desire for cadmium-free materials (notably in bioimaging) has also led to other materials being examined, notably doped materials. Since transition metals have been successfully used as dopants in numerous nanomaterials, a natural progression is the use of doping chemistry in core/shell synthesis, resulting in unique structures that can be considered hybrid doped or core/shell. Doped structures can be grown using techniques referred to as 'nucleation doping' or 'growth doping' and extended to pseudo-core/shell material.[62] In a seminal example of growth doping, ZnSe core particles were prepared as described in Chapter 1 and the growth arrested by reducing the growth temperature. The dopant ions (Cu) were then added to produce the small doped structure, with an emergence of the doped emission being observed at *ca.* 525 nm. Further growth of ZnSe resulted in the structural embedding of the dopant ions well within the particle structure, avoiding the usual problem with doping; ensuring the ions remain in the interior of the QD. As recombination occurred between a ZnSe generated electron and a hole from the d-orbital of the Cu^{2+} ions, the resulting emission could be red-shifted with changing particle size due to the overall change in ZnSe bandgap. In this example, no core/shell structure was prepared. However, in nucleation doping, the dopant precursor (*e.g.* manganese carboxylates) and host precursor (ZnSe) were added together in the initial injection, resulting in the immediate nucleation of both species. By then tailoring the growth conditions, only the host material (ZnSe) grew, resulting in essentially a core/shell species with a MnSe/ZnSe structure. The optical properties were consistent with manganese-doped ZnSe, with emission at *ca.* 580 nm from the $^4T_1 \rightarrow {}^6A_1$ transition in Mn^{2+} and no emission attributable to ZnSe; these materials were therefore referred to as doped nanostructures despite having, essentially, a core/shell structure. In the case of MnSe/ZnSe, pure MnSe cores of up to 4 nm were observed, with an overall zinc blende crystalline structure detected. The emission quantum yields of MnSe/ZnSe were up to 30%, with the emission tuneable between 575 nm and 595 nm with increasing shell thickness. In this case, the shift was attributable to the crystal field splitting of the dopant ion becoming smaller. Further work on this system highlighted the importance of the diffused interface between the core MnSe particles and the ZnSe shell, and the use of small MnSe clusters.[63] Related to this is the preparation of ZnSe/ZnMnS/ZnS core/shell/shell materials, using standard techniques.[64] The particles consisted of 2.5 nm core ZnSe particles, with a further 1.5–2 nm shell. This material was again developed to overcome issues with doping simple QDs, and emission from the Mn^{2+} state was found to be poor until the final shell layer was deposited. Design of the structure took into account several factors; the preference for Mn^{2+} to interact with a ZnS matrix rather than a ZnSe one, and the increased efficiency of impurity emitters when they are located within a shell or on a nanoparticle edge. The absorption edge red-shifted with the deposition of an increasing number of shells, while the emission from the Mn^{2+} state reached an average quantum yield of 25%. Quantum yields of up to 65% have also been reported when a thick ZnS shell was deposited, in conjunction with specific ligand chemistry and a specific ratio of precursors.[65] Similar results have been achieved by doping copper on to the surface of small (*ca.* 2.9 nm diameter) ZnSe particles, followed by further shell growth, yielding particles with emission quantum yields of up to 20%.[66] Larger particles (>3.6 nm) were found to be too unreactive to adsorb the dopant ions onto the surface.

In a related example, MnS/ZnS was also prepared. In this example, $Mn(CO_2(CH_2)_{16}CH_3)_2$ was dissolved in octadecene (ODE) and heated to 270 °C with octadecylamine (ODA), into which was injected an excess of sulfur in hot ODE. This was then immediately cooled to obtain small clusters of MnS.[67] The reaction mixture was then heated to 250 °C, followed by addition of $Zn(COOC_{17}H_{35})_2$, forming an initial overcoating of ZnS, which was then repeated to obtain a thicker shell. The deposition of the initial shell resulted in a diffusion layer of ZnS : Mn, thus effectively forming a MnS/ZnS : Mn/ZnS heterostructure. The core particles did not appear to emit, and exhibited an excitonic feature below 300 nm. Addition of the shell resulted in an absorption feature at *ca.* 315 nm due to the ZnS, with orange emission at *ca.* 600 nm, the intensity of which could be controlled by the ratio of Zn : Mn. The orange emission, with a maximum quantum yield of *ca.* 35% at 6 monolayers, consistent with manganese in a ZnS host, was again attributed to the $^4T_1 \rightarrow {}^6A_1$ transition in Mn^{2+}. Slight emission was also observed at *ca.* 500 nm, from the trap states of ZnS.

Another interesting report describes the use of a $Zn_{1-x}Mn_xS$ shell, up to 6 monolayers thick, to passivate CdSe, which introduced the paramagnetic species into the shell material.[68] CdSe particles were prepared by a typical green synthesis as described in Chapter 1. The shell was then deposited onto purified core particles using Et_2Zn and Me_2Mn as precursors, which was introduced into the reaction vessel containing the core particles, TOPO and HDA at 170 °C, while H_2S was simultaneously introduced, and the resulting reaction was maintained for at 170 °C for 2 hours. The material was then phase-transferred to water using an amphiphilic polymer described in Chapter 6. The final particles were purified to remove surface manganese species, and electron spin resonance (ESR) confirmed that manganese was incorporated into the ZnS shell. The amount of manganese deposited in the shell was dependent on the amount of precursor used and the shell thickness, and was controlled between 2 Mn^{2+} ions per particle (for 1.5 monolayers of shell) and 52 ions per particles in the thicker-shelled species. The optical properties of the $CdSe/Zn_{1-x}Mn_xS$ closely resembled typical CdSe/ZnS particles, with quantum yields of 30–60%, although the emission quantum yield was found to reduce when large amounts of manganese were incorporated into the shell, attributed to the manganese accumulating at the CdSe/ZnS interface. No evidence was found for Mn^{2+} emission. The particles were successfully used in optical cell imaging and in simple magnetic resonance imaging (MRI) experiments.

Core/Shell Alloys

Alloyed $CdTe_xSe_{1-x}/CdS$ core/shell particles have been specifically designed for use in biological imaging, where the emission wavelength could be tuned between 600 and 850 nm, ideal for imaging applications.[69] Upon precursor addition, it was observed that the tellurium precursor (TOPTe) reacted faster than the selenium precursor (TOPSe), thus effecting the final composition of the materials. The wavelength of emission could be tuned by varying the selenium and tellurium composition, the particles with the higher selenium content emitting further towards the red end of the spectrum, and, to a lesser degree by varying the particle size which was dictated by the reaction time. The particles could not be capped by ZnS (10% success rate in capping), attributed to the lattice mismatch but could successfully be capped by CdS which surprisingly reduced the emission quantum yield. The reduced emission did however recover in most cases, attributed to photoannealing of the particles giving quantum yields above 30%.

The use of alloyed materials was extended to the preparation of the related CdTeSe/CdZnS QDs, utilising graded CdTeSe alloyed core particles. In this case, attempts to grow either a CdS or a ZnS shell failed, and a method of depositing a CdZnS shell using metal carboxylates as precursors with trioctylamine (TOA) was reported. The TOA was found to be essential, resulting in the formation of a high-quality zinc blende shell on the zinc blende core particles. The resulting particles emitted at 780–800 nm, with emission quantum yields of 50%, which dropped to 30% upon phase transfer to water using phospholipids.[70]

Alloy-based core/shell materials have also been prepared that emit in the blue region of the visible spectrum, in a simple one-pot reaction. In a typical reaction CdO and an excess of $Zn(CO_2CH_3)_2$ and oleic acid were dissolved and dried in ODE, followed by the injection of a solution of sulfur in octadecene (ODE/S) at 300 °C.[71] This was followed shortly after by the injection of tributylphosphine (TBPS), overcapping the particles with ZnS, yielding $Cd_{1-x}Zn_xS$/ZnS, QDs, *ca.* 9 nm in diameter, with a wurtzite crystalline structure. The particles exhibited an absorption band edge at *ca.* 450 nm, with band edge emission also observed with quantum yields of 42–81%. The emission of the core/shells was reported to slightly blue-shift after shell deposition, attributed to the intradiffusion of the zinc atoms from the shell into the core. Notably, a related material, CdZnSe/ZnSe, exhibited continuous non-blinking photoluminescence, a significant advance in the preparation of materials with controllable, stable optical properties.[72]

Alternative Core/Shell Structures

An interesting alternative shelling material, AsS, has been reported and used to passivate CdSe particles, giving CdSe/AsS QDs.[73] In this example, the $C_4H_9NH_2$–As–S cluster was coordinated to the CdSe surface by gentle heating in isopropanol, ultimately resulting in the formation of the AsS shell, followed by addition of long-chain amines or TOP to help passivate the core/shell particle. The shelling technique could be applied to a wide range of CdSe sizes, increasing the quantum yield from *ca.* 5% to up to 50%, resulting in strong emission across the visible spectrum. The particle exhibited upconversion and were used in imaging HeLa cells, despite having arsenic in the shell material.

CdSe has also been used as the shelling material on magnetic particles such as NiPt,[74] CoPt,[75] Co[76] and FePt[77] using standard precursors. In the case of CoPt/CdSe, quantum yields of emission from the shell was found to be 3–5%, the relatively low value attributed to the core species. Shell thickness increased with growth time, and hence the shell emission also red-shifted across the entire visible spectrum. Similarly, for Co/CdSe, a maximum quantum yield of 3% was reported, whereas for FePt/CdSe, the quantum emission was found to be up to 9% at *ca.* 500 nm. Other semiconducting species, such as PbS and PbSe, have also been deposited on FePt, although in some cases the structure was better described as FePt *inside* a cubic PbS particle (Figure 5.4).[78] These methods, however, used materials with a large lattice mismatch and the resulting shells varied in quality. In an interesting example to counter this, semiconductor shells were grown on metal core using a non-epitaxial mechanism.[79,80] In a typical synthesis, a strong acid silver layer was first deposited on a gold core, followed by a chalcogen (E) giving an amorphous AgE shell, which was then cation-exchanged using TBP, giving the semiconductor shell. The shells were of a high quality, and various structures and materials could be grown, such as Au/CdS/CdSe.

In an unusual example of core/shell particles, plasmonic fluorescent QDs have been prepared where a precise space was engineered between the core and the shell—important in this case as plasmonic materials such as gold

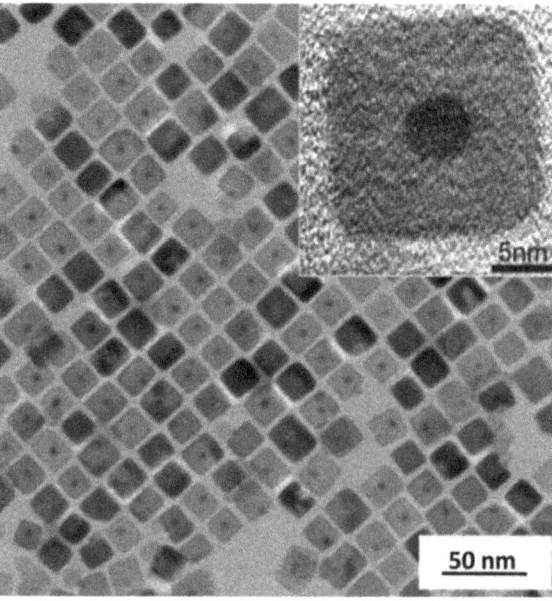

Figure 5.4 Electron microscope images of FePt/PbS core/shell particles. Reprinted with permission from J.-S. Lee, M. I. Bodnarchuk, E. V. Shevchenko and D. V. Talapin, *J. Am. Chem. Soc.*, 2010, **132**, 6382. Copyright 2010 American Chemical Society.

quench QD emission.[81] In this example, water-soluble CdSe/ZnS QDs were coated with a layer of poly-L-histidine, which is capable of immobilising Au^{3+} ions at high densities. The poly(ethylene glycol) (PEG) chains on the lipid, on to which the poly-L-histidine was deposited, dictated the degree of separation. Addition of a reducing agent, NH_2OH, induced the formation of the 2–3 nm thick gold shell on the histidine layer, and addition of a thiolated polyethylene glycol (PEG) was added to ensure overall colloidal stability. The absorption profile of the QD was hidden by the plasmon resonance of the gold layer at *ca.* 580 nm, while the emission from the QDs was found to be reduced from *ca.* 75% to *ca.* 18%, attributed to the addition of the Au^{3+} ions. The emission was also found to reduce with increasing shell thickness, being significantly quenched when the shell thickness reached 5 nm. The QD–gold separation could be increased by introducing polyelectrolyte monolayers between the two inorganic layers, which hindered the diffusion of Au^{3+} to the core that was responsible for the emission quenching. By using this spacing technique, the emission quantum yield was increased to up to 39% by the addition of two bilayers.

5.4   Type II Materials

Type II core/shell structures are of interest due to the unusual mechanism of charge recombination; the band energy levels are offset such that one charge carrier is predominantly confined to the core while the other is confined to the shell. By correctly choosing materials with a notable band energy mismatch, an effectively smaller recombination gap is produced, which results in emission wavelengths that cannot easily be obtained by the single-core II–VI materials. Infrared-emitting type II core/shell dots have been successfully used in oncology studies, as the wavelength, once shifted to the near infrared region, is compatible with tissue imaging.[82] The resulting optical properties are usually characterised by a significant red shift in the emission profile with a prolonged recombination lifetime, while retaining roughly the same absorption position (although a loss of excitonic features is common). One must be careful not to interpret all red shifts in core/shell QD emission as type II, as type I structures may exhibit red-shifted emission (although usually much smaller) due to leakage of charge carriers into shell materials that possess a low potential barrier. The optical properties of these materials can also be manipulated by the injection of electrons, resulting in significant shifts in the emission profiles due to spectral switching.[83]

  The first type II core/shell materials to be prepared by solution-based organometallic-type routes (CdTe/CdSe and CdSe/ZnTe) were synthesised using the metal alkyl-based methods described in Chapter 1; the band energy diagrams are shown in Figure 5.5.[84] As a result of recombination from an effectively indirect structure with a smaller bandgap, the emission was shifted into the near infrared region. The emission could be tuned further by controlling the size of the core particle and the shell thickness. Increasing the particle size and the shell thickness pushed the emission further towards

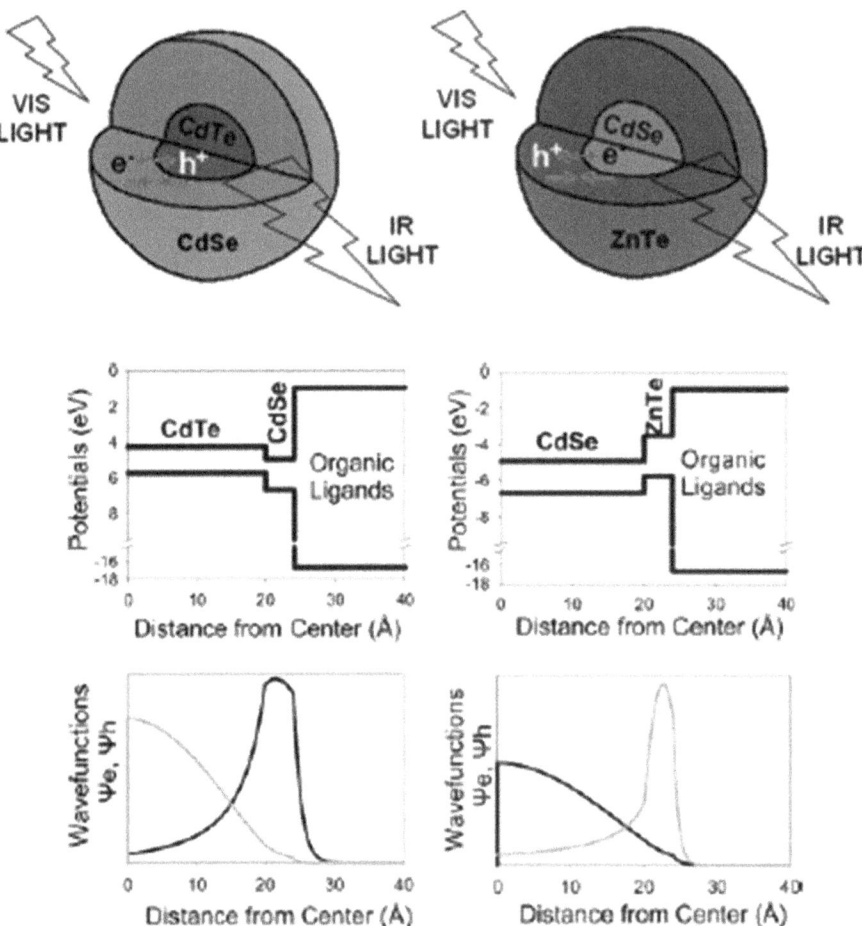

Figure 5.5 Cartoon showing charge recombination and emission in type II dots, potential diagrams and radial wave function distributions for charge carriers in CdTe/CdSe and CdSe/ZnTe core/shell QDs. Reprinted with permission from S. Kim, B. Fisher, H.-J. Eisler and M. G. Bawendi, *J. Am. Chem. Soc.*, 2003, **125**, 11466. Copyright 2003 American Chemical Society.

the red end of the spectrum. This means the emission can be tuned not only by controlling the particle size, but also by carefully choosing a specific core/shell structure and its dimensions. The emission radiative lifetime was also found to be significantly longer than the simple core material, with quantum yields of less than 4% observed. Quantum yields of up to 20% have been obtained when a further ZnTe layer was deposited on CdTe/CdSe particles. The low quantum yield and the unusual recombination mechanism suggested the normal non-radiative pathways (such as surface states) may be a factor to consider when examining type II materials.

The synthesis of CdTe/CdSe has been improved by using CdO as a precursor rather than the metal alkyls, and using TOP rather than TOPO, building on the earlier results described in Chapter 2 which highlighted the preferred surfactant system for CdTe.[85] Sequential addition of the selenide precursor to the CdTe solution allowed the core/shell system to be prepared, with particles growing from *ca.* 3 nm up to 7 nm upon shell addition. Using this system, the initial core CdTe particles were prepared with quantum yields of up to 31%. Upon addition of the CdSe shell, the emission shifted from *ca.* 660 nm to *ca.* 715 nm over a growth period of 4 hours at 200 °C. During this period, the quantum yield dropped immediately from 31% to below 5%, but recovered over the growth period to almost 40%, then dropped slightly to below 30% upon further shell growth. Also noteworthy were the change in the emission FWHM which narrowed then widened upon prolonged growth, and the gradual overall increase in the Stokes shift. The highest quantum yield obtained (38%) was attributed to CdTe with one monolayer of CdSe on the surface. The use of a SILAR-type technique in the preparation of CdTe/CdSe in more traditional solvents such as long-chain amines and TOPO resulted in particles with quantum yields of up to 82%. Initially, CdSe 'tips' were observed on the CdTe particles, growing into pyramids and ultimately multipods. The remarkably high quantum yields were attributed to a low defect density due to the slow epitaxial growth and the low degree of strain at the interface due to formation of the rod-like particles.[86] The shape tuning of CdTe/CdSe heterostructures has also been achieved by preparing CdTe cores by standard chemistry, followed by the addition of a cadmium precursor to the particles with a mixture of phosphonic acids and the dropwise addition of TOPSe at a set temperature.[87] The defining parameter was determined to be the size of the CdTe seed, where small (<4 nm) seeds incurred unidirectional growth of the CdSe component, while larger (*ca.* 4.3 nm) seeds resulted in Y-shaped particles, and seeds larger than 5 nm resulted in the formation of tetrapod heterostructures. Surprisingly, growth of CdSe on 3.5 nm seeds resulted in a significant red shift in the emission profile of up to *ca.* 200 nm, consistent with type II behaviour, while maintaining impressive emission quantum yields of up to *ca.* 39%.

Type II CdSe/CdTe have also been prepared using the SILAR method, using standard precursors in a non-coordinating solvent.[88] The use of the standard SILAR technique resulted in peanut-shaped particles, whereas the use of thermal cycling—the repeated low injection temperature of the shell precursor followed by a high growth temperature—resulted in spherical particles. The resulting particles had significantly higher quantum yields (up to 60%) than other similar type II materials, suggested to be due to the presence of CdTe as a shell layer where the heavy hole was better confined to the interface junction by the electrostatic attractions from the CdSe core, and the improved shell deposition afforded by the SILAR shell deposition technique. Further work on the structure of typical CdTe/CdSe particles confirmed the material was indeed a core/shell structure, although CdSe/CdTe particles prepared in a similar manner were actually a single-phase structure (solid solution) with a uniform distribution of CdSe and CdTe, more akin to an alloy than a heterostructure.[89]

Related to this is the preparation of CdTe/CdSe bar bells, where CdTe tips are grown on the end of short CdSe rods. These materials are of course not strictly core/shell materials, and although type II charge carrier separation was observed, no emission was detected.[90] Similar results were reported by Kumar *et al.*, who also reported the growth of CdTe particles on the tips of CdSe rods when the tellurium precursor was added slowly with an excess of tellurium relative to selenium.[91] The same group also reported that rapid addition of the tellurium precursor to the CdSe rods resulted in the deposition of a CdTe layer along both the lateral and *c*-axis, forming a type II quantum rod heterostructure. The heterostructures exhibited a clear extra excitonic feature in the near infrared region which could be further shifted towards the red with further growth, consistent with formation of the heterostructure. Interestingly, two distinct emission features were observed: weak band edge emission from the CdTe exciton, and a stronger red-shifted charge transfer emission originating from recombination of the CdSe electron and CdTe hole, with quantum yields of 5–10%, with trap states playing a key role in the population of radiative states.[92] Similar chemistry was employed to prepare curved CdSe/CdTe structures, using CdSe rods as seeds. Addition of standard precursors for CdTe resulted in the rods growing slightly longer and 'fattening' at the ends, while the centre of the rods increased by only a few monolayers. The rods also exhibited a significant curvature, attributed to a deflection in the lattice planes due to deposition of CdTe on the tips and the sides of the rod.[93]

It is worth noting that in related work, Shieh *et al.* reported CdSe/CdTe/CdSe heterostructured rods, again with almost totally quenched emission, yet reported CdTe/CdS/CdTe rods with enhanced emission with quantum yields up to a maximum of 20% depending upon reaction conditions.[94] Xi *et al.* have also reported the synthesis of CdTe/CdSe multiblock heterostructures in a one-pot reaction.[95] In this case, CdTe rods were prepared by the green route at 330 °C,[96] using TOPTe as a precursor. After the rods were prepared by multiple injections, TOPSe was introduced by two further separate injections, finally followed by more TOPTe. Electron microscopy and powder diffraction confirmed the growth of the CdSe block epitaxially on CdTe, followed by a further CdTe block. The structures varied in morphology, with some materials exhibiting branching. The addition of the secondary precursors resulted in the quenching of the rod emission.

Using similar chemistry, core/shell particles of CdSe/ZnTe 4–6 nm in diameter have also been made, although quantum yields were found to be in the region of 10^{-3}%,[97] while particles of CdTe/ZnSe have also been prepared in ODE, which had quantum yields of up to 24%. These particles were of especial interest as the material was reported to exhibit a novel carrier distribution along the material junction, and as such exhibited a blue shift in emission due to the recombination of charge carriers at the interfacial alloyed layer.[98] CdTe/CdS and CdTe/CdSe have also been prepared in the non-coordinating solvent ODE, using either sulfur or selenium in ODE as chalcogen precursors in shell deposition.[99] An interesting method of preparing CdTe/CdSe has been reported where CdTe particles were initially made, followed by the addition of TOPSe and either Me$_2$Zn or Me$_3$Al which acted as reducing agents for the selenium.[100] The metal alkyls formed adduct compounds, shielding the selenium and making it more reactive with the cadmium monomer retained in solution. The resulting rectangular particles of CdTe/CdSe had quantum yields of up to 25% and appeared more photostable than the core CdTe.

It is worth noting that when examining the band structure offset for CdTe and ZnSe in bulk materials, one would expect type I behaviour from core/shell structures prepared from these materials; however, type II optical characteristics are routinely observed. An investigation into the deposition of a range of compressive II–VI shells on a small, soft CdTe core highlighted significant lattice strain that increased the CdTe bandgap due to the applied compressive force, which simultaneously depressed the conduction energy band of the shell material (typically ZnSe) due to tensile strain, yielding new bandgap offsets consistent with type II materials (termed strain-induced type II behaviour).[101] By choosing the correct shell materials, and by tuning the number of shell monolayers and hence the degree of strain, the emission could be tuned between 500 and 1050 nm with a maximum quantum yield of 60%. It is quite possible that the majority of type II structures mentioned above based on CdTe cores do in fact exhibit strain-induced optical behaviour.

Other type II structures include ZnSe/CdS, which again were produced using simple precursors, such as zinc carboxylates, TOPSe, cadmium oleate and sulfur in a non-coordinating solvent.[102] In this case, the ZnSe core particles were prepared as outlined in Chapter 1, isolated and purified up to four times using solvent/non-solvent cycles, an essential step if high quantum yields were to be realised. The shell was then deposited using the SILAR technique. These particles exhibited emission at 480 nm, which was shifted to 610 nm (using one specific core size) when more than three monolayers of shell were deposited, consistent with the spatial separation of charge carriers. The resulting large Stokes shift was attributed to the large offset of the band edges. The emission had a maximum quantum yield of 18%, which could be slightly increased upon particle purification. An interesting related structure in which ZnSe QDs were embedded in a CdSe rod has been reported.[103] In this example, preformed ZnSe particles were mixed with CdS precursors and injected into a mixture of TOPO and phosphonic acids, resulting in rods *ca.* 47 nm long with the ZnSe particles (of *ca.* 3.7 nm diameter) incorporated into part of the structure. The emission of the ZnSe particles at *ca.* 400 nm was significantly red-shifted to *ca.* 600 nm after 8 minutes rod growth, consistent with type II behaviour, with a remarkably high maximum quantum yield of 45%.

5.4.1   Inverted Core/Shell Structures

Formation of a core/shell material where the shell is a narrower bandgap material than the core should, in theory, result in photo-generated charge carriers residing in the shell, as this would present the lowest energy state. A report by the Klimov group, who investigated ZnSe/CdSe particles, their recombination lifetimes and the theoretical spatial distribution of charge carriers in particles of varying shell thickness using an effective mass approximation, found that the particle could be tuned between type I and type II structures by varying the shell dimensions.[104]

Taking a fixed core (ZnTe) with a radius of 1.5 nm, the calculated spatial distribution of the density function for the charge carriers within a range of shell thickness (CdSe) is shown in Figure 5.6b, where three distinct regions could be seen. For shells less than 1.1 nm thick, the electron and holes were delocalised over the entire structure while the density maximum was found to be in the ZnTe core, despite the fact that bulk energy diagrams would suggest carrier localisation in the shell. This behaviour is type I, where the charge carriers were confined to the core. With a shell thickness of between 1.1 and 1.6 nm, the carriers were separated between the core and the shell, which equated to type II behaviour. With shells thicker than 1.6 nm, the

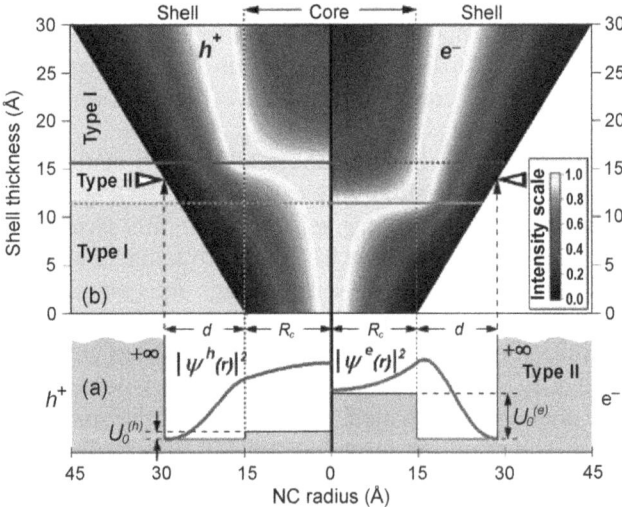

Figure 5.6   (a) Diagram of the conduction band (right-hand side) and valence band (left-hand side), showing radial distribution for a 1.5 nm core ZnTe particle with a 1.4 nm thick CdSe shell (type II system, with carriers delocalised over the entire particle). (b) Plot of radial distribution for electron and hole for a 1.5 nm ZnTe particle with a CdSe shell of varying thickness, showing the three distinct regimes. Reprinted with permission from L. P. Balet, S. A. Ivanov, A. Piryatinski, M. Achermann and V. I. Klimov, *Nano Lett.*, 2004, **4**, 1485. Copyright 2004 American Chemical Society.

energy levels were aligned such that carrier localisation was in the shell (type I behaviour—it is worth noting that this is still considered type I even though the confinement is in the shell, not the core as found in CdSe/ZnS particles). This highlighted that the charge carriers can be restricted to certain parts of the nanoparticle and hence the recombination tuned by simply altering the shell thickness. ZnTe-based core/shell materials have been reported with CdS, CdSe and CdTe shells, prepared by a one-pot reaction.[105] In this case, the ZnTe cores were prepared using TOPTe and Me$_2$Zn in TOP as precursors, which were then injected into an ODE solution of ODA at 280 °C, followed by cooling to 240 °C in readiness for shell growth. Shell precursors (CdO in ODE and oleic acid, TOPSe, TOPTe or ODE/S) were then injected sequentially using the SILAR technique to controllably grow the required number of monolayers. The resulting type II structures exhibited tuneable emission, in the case of ZnTe/CdSe, from *ca.* 550 nm (0.5 monolayers of CdSe on a 4.5 nm ZnTe core) to *ca.* 900 nm (4.5 monolayers of CdSe on a 7.6 nm ZnTe core) with emission quantum yields of up to 20%. When OAm was used as a ligand during shell growth, the spherical ZnTe/CdSe particles eventually grew into a pyramidal structure and ultimately into tetrapods.[106] The ZnTe/CdSe tetrapods exhibited a feature in the absorption spectra at 650 nm, consistent with the presence of CdSe, but showed no evidence of emission.

Type II structures with a ZnTe/ZnSe structure have also been prepared.[107] In this case, the core particles were made by the thermolysis of Et$_2$Zn and TOPTe at 270 °C in ODE, using HDA as a capping agent. Isolation and purification were carried out before the core particles were redispersed in fresh HDA/ODE, and Et$_2$Zn and TOPSe were added dropwise over 6 hours at 200–250 °C. A further ZnS shell could be added without purification, by slowly adding Et$_2$Zn and S(SiMe$_3$)$_2$ at 220 °C over 2 hours followed by 1 hour stirring. The bare ZnTe particles were not luminescent, but were emissive between 500 nm and 580 nm when the ZnSe shell was deposited. The emission, tuneable by the shell thickness, had a maximum quantum yield of 6%, which could be improved to 12% with the deposition of a further ZnS shell.

Related materials (ZnSe/CdSe) have been prepared using metal alkyl-based routes as described in Chapter 1. The ZnSe cores were prepared by the thermolysis of Et$_2$Zn and TOPSe in long-chain amines at *ca.* 300 °C. Shell growth was achieved by addition of a required amount of Me$_2$Cd and TOPSe to give the required shell thickness. An annealing stage of up to 2 days was required to achieve quantum yields of up to 80%. Emission could be tuned between 430 and 600 nm, with FWHM of 20–40 nm. These particles were then used in light amplification studies.[108]

Core/shell ZnSe/CdSe particles have also been prepared using CdO as a precursor and ODE as a solvent for the shell deposition, using 2.8 nm ZnSe cores, giving materials with similar optical properties to those described above.[109] In this case, the diffraction patterns of the ZnSe particles suggested a cubic core, which changed to a hexagonal pattern with the deposition of 6 monolayers of CdSe shell. The emission and absorption spectra provided evidence of the core/shell structure, displaying band edge emission with no evidence of surface trapping states. The quantum yields of the core particles were reported as 40%, which rose to 85% with the deposition of one monolayer. This value was maintained until 3 monolayers were deposited, dropping to 25% with the deposition of 6 monolayers. This drop was attributed to weaker confinement effects as the particle increased in size. In this case, the charge carriers were mainly delocalised in the CdSe core. This is an excellent example of how the emission wavelength can be varied in core/shell structures, with photoluminescence being tuned through the entire visible region of the spectrum by growth conditions. This range of colours cannot be achieved by either the CdSe or ZnSe materials on their own. Also, the emission obtained appeared to show no evidence of trapping states. In comparison, emission from CdSe/ZnS nanoparticles cannot be tuned, with the emission of the CdSe remaining constant.

5.4.2   Multiple-Shell Structures

The core/shell CdSe/ZnS system with a 'buffer layer' of CdS was described earlier and found to produce better-quality nanorods, as a result of the decreased lattice strain. This strain resulted in uneven particle formation and defects at the core/shell boundary resulting in the reduction of quantum yields. The incorporation of a buffer layer has been developed further, leading to the preparation of double shell particles that overcame the strain between the interface of core/shell systems, notably CdSe/ZnS. The CdSe/ZnS system is ideal in many respects, when compared to CdSe/ZnSe, for example. The barriers for charge carrier confinement are better in the ZnS shell, where the hole has an energetic offset of 0.6 eV as compared to only 0.1 eV for the analogous ZnSe shell. The lattice mismatch is, however, better for the ZnSe shell with a mismatch of only 6.3%. CdSe/ZnSe/ZnS nanoparticles have been reported, building on the synthesis of CdSe/ZnSe core/shell particles described earlier including the addition of the ZnS shell using $Zn(CO_2(CH_2)_{16}CH_3)_2$ and either $S(SiMe_3)_2$ or $CH_3C(S)NH_2$. The ZnS shell was deposited at 200 °C with no evidence of separate ZnS nucleation observed.[110] Thick shells, 5 monolayers of both the ZnSe and ZnS, were deposited, and shell growth was evidenced by the bathochromic shift of the emission spectra as the exciton leaked into the shell. The quantum yields were found to be 14.8% for the CdSe/ZnS system, 17.6% for the analogous CdSe/ZnSe system, and 25.3% for CdSe/ZnSe/ZnS.[111] CdSe/ZnSe/ZnS nanoparticles have also been prepared for optical data storage applications; in this case, the core particles of CdSe were synthesised using CdO, oleic acid and TOPO in ODE as described by Peng (see Chapter 1). The shell of ZnSe, 1 nm thick, was deposited using Et_2Zn and TOPSe and the ZnS shell, 1.5 nm thick, was deposited using Et_2Zn and $S(SiMe_3)_2$.[112] Few other details were given in this case. CdSe/CdS/ZnS particles have also been grown using octanethiol as the sulfur source.[113] As octanethiol is significantly more stable than silylated sulfur sources, it can be used at higher deposition temperatures which allowed the formation of an alloy layer rather than a distinct core/shell structure.[114]

Talapin *et al.* have carried out a detailed study into the preparation of CdSe/CdS/ZnS and CdSe/ZnSe/ZnS nanorods and spherical nanoparticles.[115] In the case of spherical CdSe/CdS/ZnS particles, the CdSe/CdS constituent of the material was prepared using the green chemical route described earlier. The core/shell particles were then isolated and a ZnS shell deposited by addition of Et_2Zn and $S(SiMe_3)_2$ to a TOPO solution of the preformed core/shell particles. CdSe/CdS/ZnS rods were made in a similar fashion, using CdSe/CdS rods prepared as described above, with a spherical CdSe core and elongated CdS shell.[116] CdSe/ZnSe/ZnS particles were prepared by preparing CdSe particles capped with TOPO, TOP and HDA, with shells deposited using metal alkyls, TOPSe and $S(SiMe_3)_2$. The materials showed strong emission (quantum yields of 70–85%) at *ca.* 550 nm, with the expected red shift upon deposition of a ZnSe shell on the CdSe core, consistent with exciton leakage due to the relatively small potential barrier. Again, ZnS shell thicknesses beyond 2 monolayers resulted in a decrease in quantum yield. Prior to this work, CdSe/ZnS was considered the most photochemically stable core/shell system, surpassing CdSe/CdS and CdSe/ZnSe for resistance to environmental influences. Examination of comparable CdSe/ZnS and CdSe/ZnSe/ZnS (shell thickness of 1.5/3 monolayers for ZnSe/ZnS) particles under intense laser illumination showed that the double shell particles were the most stable.

The preparation of core/shell/shell materials and the gradual easing of the lattice strain was advanced further using SILAR deposition to prepare a three-shell system, $CdSe/CdS/Zn_{0.5}Cd_{0.5}S/ZnS$.[117,118] The nature of the deposition allowed careful, exact shell growth. The CdSe core particles were grown from CdO in TOPO and ODA as described by Peng (see Chapter 1), followed by redispersion in an ODE solution of ODA. The shells were grown (using the metal oxides and sulfur in ODE) by sequential addition of either ion in solution to build the structure half a monolayer at a time. Over a 3 hour period, a structure was grown that consisted of CdSe cores with 2 monolayers of CdS, 3.5 monolayers of $Zn_{0.5}Cd_{0.5}S$ and 2 monolayers of ZnS, capped with the amine. The particles were found to be highly spherical in shape, in contrast to CdSe/ZnS particles made by metal alkyl precursors. The gradual addition of the shells could be tracked through the emission and absorption spectra, as the exciton leaked into the extended shell structure, with the final material exhibiting a quantum yield of 80–90%. The gradual overall red shift was accompanied by a final blue shift after the addition of 1.5 monolayers of the ZnS shell, attributed in this case to the formation of a $Zn_xCd_{1-x}S$ alloy shell. Phase transfer to water resulted in the quantum yield dropping to almost half its original value. Photo-oxidation studies of the three-shell material showed enhanced stability, although the material gradually oxidised over 120 hours of illumination, with the quantum yield dropping to approximately half its original value, although this was found to be slightly better than CdSe/ZnS material made using metal alkyls as described above.

The deposition of an extra shell has also been made on type II QD core/shell structures.[119] As the charge carriers are not necessarily confined to the core in type II materials, extra passivation with a wide-bandgap material further protects the structure, preventing charge carriers from reaching surface defects and trapping sites. In the case of CdSe/ZnTe, the energy levels highlighted the charge carrier distribution was spread across the particle, with holes residing in the shell. If holes migrated to surface defects, emission would be quenched to some degree. Particles of CdSe/ZnTe (5 nm in diameter) synthesised as described earlier using metal oxides as precursors have been prepared, isolated, and then dispersed in TOPO and ODA. Deposition of a 0.4 nm thick ZnS shell was achieved by thermolysis of $Zn(CO_2(CH_2)_{16}CH_3)_2$ and TBPS at 190 °C for up to 1.5 hours, giving a core/shell/shell material with the electronic structure as shown in Figure 5.7. The wide bandgap of ZnS dictated the electrons and holes were restricted to the CdSe/ZnTe structure. The particles were isolated as normal, and could be phase-transferred to water using dihydrolipoic acid (DHLA, see Chapter 6). Upon addition of the ZnS shell, a slight red shift was observed as the exciton leaked into the ZnS layer. This ruled out formation of an alloy, which would have resulted in a slight blue shift. The emission quantum yield for the CdSe/ZnTe particles was estimated at 0.004%, which increased to 0.12% upon addition

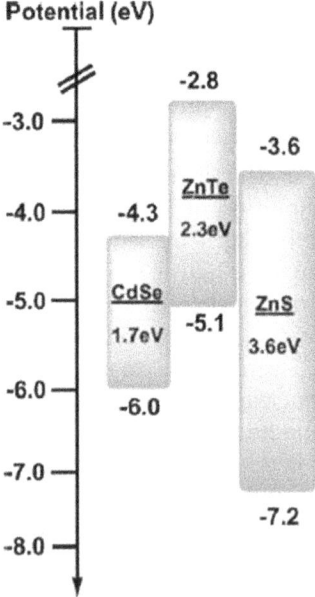

Figure 5.7 Diagram depicting the energy levels for CdSe/ZnTe/ZnS core/shell/shell particles. Reprinted with permission from C.-T. Cheng, C.-Y. Chen, C.-W. Lai, W.-H. Liu, S.-C. Pu, P.-T. Chou, Y.-H. Chou and H.-T. Chiu, *J. Mater. Chem.*, 2005, **15**, 3409. Copyright 2005 The Royal Society of Chemistry.

of the shell. Phase transfer resulted in a slight drop to 0.10% and a slight emission shift of *ca.* 15 nm. Other similar structures with a type II/type I alignment have been prepared such as CdSe/CdTe/ZnSe.[120] The core CdSe/ CdTe was prepared as described earlier using the SILAR-thermal cycling technique[88] and simply expanded to include the deposition of a final ZnSe shell using similar precursors. Using this method, spherical and peanut-shaped CdTe/CdSe particles were capped with the extra shell. The particles were found to emit in the near-infra red as described earlier, and the addition of the ZnSe shell slightly red-shifted and broadened the emission further, attributed to a compressed CdSe lattice. As expected, the quantum yields improved from *ca.* 20% to between 50% and 60% upon ZnSe shell growth and efficient phase transfer to water was possible.

Likewise, CdS/ZnSe/ZnS core/shell/shell particles have also been prepared using standard green precursor chemistry.[121] Core CdS particles and ZnSe shell have staggered bandgaps as described earlier, and band edge emission (at *ca.* 400–475 nm) for CdS was significantly red-shifted by over 100 nm by the deposition of a ZnSe shell. The addition of a ZnSe shell increased the emission quantum yield from (a maximum of) 33% to a maximum of 56%, and the addition of a final ZnS shell increased the quantum yield slightly to up to 60% while red-shifting the emission even further.

An interesting related advance in the formation of type II structures is the development of a cascade core/shell/shell system, where all three energy bands are offset.[122] In the case of CdSe/CdTe/ZnTe, excitation resulted in the electron residing in the CdSe core, the hole was confined in the ZnTe shell, and recombination occurred across the CdTe layer. Core particles of CdSe, 3.6–5.7 nm in diameter, were prepared using CdO as a precursor, with CdTe shells (*ca.* 1.7 nm thick) and a final shell of ZnTe (*ca.* 1.3 nm thick) deposited using $CdCl_2$, $Zn(CO_2(CH_2)_{16}CH_3)_2$, TBPSe and TBPTe in TOPO and HDA at *ca.* 200 °C. The absorption spectra remained *ca.* at 600 nm but lost the excitonic peak upon shell formation. The core particles exhibited narrow band edge emission at *ca.* 550 nm, which shifted to a broad emission at *ca.* 1000 nm upon CdTe shell deposition, and *ca.* 1500 nm upon deposition of the ZnTe shell. The emission quantum yield dropped from 28%, to 0.12% upon addition of the CdTe shell, and to 1.1×10^{-5}% when the final ZnTe shell was added.

Multiple-shell systems present a unique opportunity to build particles with a more specific confinement; core/shell particles can be tuned to have charge carriers in the core, the shell or across the whole particle. Multiple-shell particles can be prepared where the charge carriers are restricted to the middle layer, *i.e.* the first shell. By using a wide-bandgap core particle (template), carefully depositing a narrow-bandgap shell, followed by a final wide-bandgap layer, a solution analogue of a quantum well can be produced. In materials prepared in this manner, quantum confinement effects should only manifest themselves along the radial dimensions. Although early studies investigated such systems, such as CdS/HgS/CdS,[13–17] the quality of the nanoparticles produced was lower than that of particles produced by organometallic-based precursors.

To explore such systems, particles of CdS/CdSe/CdS core/shell/shell materials were grown[123] which emitted between 520 and 650 nm with emission quantum yields of over 40%. CdS core particle templates 2.5–5 nm in diameter were prepared using CdO and sulfur in ODE as described in Chapter 1. Up to 7 monolayers of CdSe were grown on the template epitaxially using the SILAR technique described earlier, with CdO/oleic acid in ODE and selenium in ODE being added sequentially at 180 °C to grow the shell to the required thickness. The final CdS shell was added using the SILAR technique, having replaced selenium with sulfur and fatty acids. Up to 5 monolayers of the final shell, followed by annealing at 230–240 °C and purification resulted in particles with the highest quantum yields, of about 40%. The emission and absorption spectra resembled those of CdSe QDs, although there was less detail in the high-energy region of the absorption spectra. Interestingly, the extinction coefficient at the first excitonic peak showed a linear relationship with shell thickness revealing a simple way to probe the exact dimensions of the particle. The emission position red-shifted with increasing CdSe monolayers, as expected from size quantisation effects. An investigation into emission lifetimes suggested the existence of free two-dimensional excitons which decayed radiatively when the CdSe layer was 1–3 monolayers thick; 4 monolayers resulted in an increase in non-radiative recombination due to defect formation in the thicker layer.[124]

ZnS/CdSe/ZnS structures are again typical of such materials, and have been prepared by the growth of ZnS core particles using $(Et_3NH)_4[Zn_{10}S_4(SPh)_{16}]$ (discussed in Chapter 7) in HDA at 300 °C, followed by particle isolation.[125] Addition of the ZnS cores to a fresh batch of HDA at 100 °C, followed by addition of $(Et_3NH)_4[Cd_{10}Se_4(SPh)_{16}]$, slowly resulted in the deposition of a CdSe shell. During shell addition, monitoring of the emission highlighted a gradual red shift, and growth was stopped when the desired wavelength was reached. The final shell was then deposited on freshly isolated and purified ZnS/CdSe particles in HDA by adding Et_2Zn and $S(SiMe_3)_2$/TOP dropwise at 180 °C until the maximum emission was observed *in situ*, usually at 1 monolayer of ZnS. The emission from the resulting particles could be tuned between 450 nm and 650 nm, with quantum yields of up to 60%. XPS highlighted that the separate components of the core/shell/shell systems were not distinct phases with sharp interfaces, but the core region existed as a ZnS/CdSe graded alloy, although the final shell layer had a sharp interface.

A multiple-shell system has also been developed containing two quantum-confined materials in one single particle; the core (zero-dimensional confinement), followed by a wide-bandgap buffer layer, and a final shell of narrow-bandgap material exhibiting two-dimensional (2D) confinement. Double quantum well systems, such as CdS/HgS/CdS/HgS/CdS have previously been investigated using wet colloidal chemistry;[126,127] however, emission from the two separate wells was hard to distinguish, possibly due to coupling. A similar system, CdSe/ZnS/CdSe, has been prepared by an organometallic-based route that showed two emissive regions.[128,129] CdSe cores were prepared by green chemical methods as outlined by Peng, followed by the growth of 1–5 monolayers of ZnS deposited by the SILAR technique. The final shell, of either CdSe or CdS, was again grown by SILAR. All precursors were metal oxides/oleic acid, and elemental chalcogens used with ODE and ODA. Temperatures for growth and annealing varied between 190 °C and 240 °C. High-resolution TEM showed particles with a spherical shape and narrow size distribution, highlighting the effectiveness of shell deposition by SILAR. Optical spectra are shown in Figure 5.8, where the two emission bands are clearly visible. The total emission quantum yield was measured at 30%, and each peak could be tuned, by altering the core size and/or the shell thickness. The intensity of emission could also be tuned by varying the length of time the particles were annealed, with the 2D emission increasing upon prolonged heating. Coupling between the two emission centres was observed, with no emission being detected from the 2D layers when only one monolayer of ZnS was used, and core emission shifted towards the blue end of the spectrum with the increase in thickness of the buffer layer. Förster energy transfer was also observed in the system, especially when the outer layer of CdSe was 5 monolayers thick.

Few multiple-shell systems have been reported with a CdTe core, although CdTe/CdSe/ZnS particles have been prepared using typical green chemistry as described in Chapter 1 for the CdTe cores, a modified SILAR route for the middle CdSe shell, and a single-source precursor (Chapter 7) for the final ZnS shell.[130] The resulting structures exhibited emission which could be tuned between 550 and 850 nm, with a maximum quantum yield of 94% for CdTe with 1 monolayer of CdSe, which dropped significantly upon further layer depositions to *ca.* 46%.

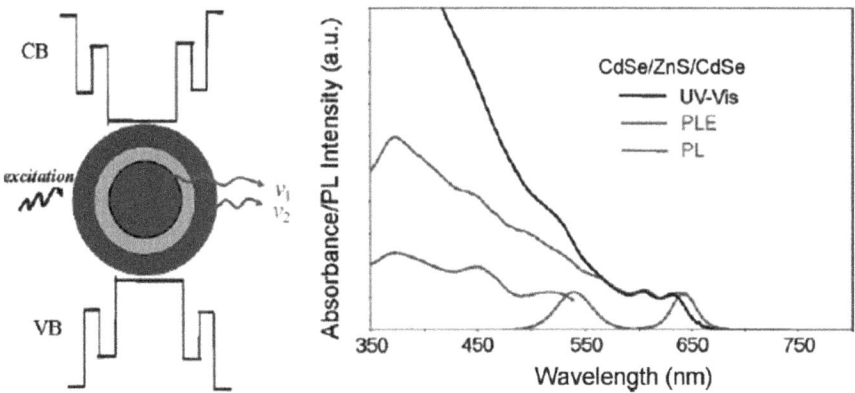

Figure 5.8 (Left) Diagram showing band structure of CdSe/ZnS/CdSe. (Right) Absorption, emission and excitation spectra of both emission peaks for CdSe/ZnS/CdSe. Reprinted with permission from D. Battaglia, B. Blackman and X. Peng, *J. Am. Chem. Soc.*, 2005, **127**, 10889. Copyright 2005 American Chemical Society.

Core/shell systems are not restricted to semiconducting materials of the same family. For example, CdTe cores, prepared by the standard green route, have been layered with a shell of InP, which was grown using standard chemistry used to prepare III–V nanomaterials.[131] The energy gap configuration suggested an inverse structure, with InP possessing a narrower bandgap than CdTe. This resulted in the red shift of the emission wavelength while the excitonic feature in the absorption spectra disappeared, not unlike the optical characteristics in type II core/shell materials. Quantum yields of the core/shell structures improved from 20% to 70% upon shell deposition, which dropped to 40% with increasing shell thickness. At this point, attempts to phase-transfer the material to water resulted in significant quenching of the emission, and attempts to grow a ZnSe shell on the CdTe/ InP failed. A ZnS shell was successfully deposited on the core/shell system using Et_2Zn and S as precursors in ODE at 160 °C. Notably, the use of $S(SiMe_3)_2$ as a precursor resulted in the complete quenching of the emission. Addition of the final shell did not affect the emission quantum yield or wavelength, and yielded material stable enough to be successfully phase-transferred into water and utilised in imaging experiments.

5.5   Core/Shell Structures Based on III–V Materials

So far, we have discussed core/shell particles with the core based on the II–VI family of materials. However, other materials have been explored as cores, notably the III–V family of QDs. Simple unshelled III–V particles such as InP have notably low quantum yields, often well below 1%, which limits their applications. The first attempts to deposit wide-bandgap semiconductor shells on III–V materials were reported by the Banin group who described the synthesis of InAs/CdSe and InAs/InP[132] which might be considered unusual, as InAs is a less popular material than, for example, InP. Shells were deposited by dispersing 1.7 nm diameter InAs cores (prepared, purified and the size distribution narrowed as described in Chapter 2, with a clear excitonic shoulder at 990 nm) in TOPO, followed by stabilisation at 260 °C. This high temperature often leads to Ostwald ripening when depositing shells on II–VI materials, resulting in a wide size distribution, although this was not observed when using InAs cores. Standard precursor solutions in TOP (either $InCl_3$ and $P(SiMe_3)_3$, or Me_2Cd and TOPSe) were added dropwise at this temperature, and the growth monitored by absorption spectroscopy. Isolation of the particles was achieved by standard solvent/non-solvent interactions. The addition of either the InP or CdSe shell had a distinct effect on the absorption spectra, red-shifting the excitonic peak by almost 100 nm. The resulting emission from InAs/CdSe particles was found to be band edge, with quantum yields of 18% (1.2 monolayers), up from 1% for the naked cores. Emission from InAs particles was quenched upon InP shell deposition, due to the electrons residing in the shell, as confirmed by the increase in emission when oxidised, as opposed to InAs/CdSe, which quenched upon surface damage. XRD confirmed epitaxial shell growth with a shift in, and a narrowing of, the reflections, and electron microscopy confirmed their spherical shape.

A larger study by the same group explored the use of GaAs, ZnS and ZnSe as shell materials on InAs cores and discussed their suitability with regards to lattice mismatch, band offset (Figure 5.9) and chemical compatibility.[133] Attempts to grow a GaAs shell on a InAs core were limited in their success, with a maximum of no more than two monolayers of shell. This was attributed to the strong gallium–TOP interactions, which is also a limiting factor in the preparation of gallium-containing QDs, such as GaAs. The report also highlighted the unsuitability of TOPO as a capping agent for CdSe shell growth, as this induced CdSe particle nucleation, unlike the use of TOP. Increasing the Se : Cd ratio was also found to reduce nucleation, although this affected the solubility of the final material. The use of TOP for ZnSe and ZnS shell growth was found to be unsuitable, as the resulting particles were poorly soluble after processing with CH_3OH. A mixture of TOP and TOPO was found to reduce nucleation while making the particles soluble.

The optical properties of InAs capped with ZnS and ZnSe differed from those capped with InP or CdSe, as no shift in the band edge was observed, due to the larger band offset which was larger than the confinement energy of the charge carriers. The emission quantum yields were up to 20% for ZnSe-capped InAs, with a maximum of 8% for InAs/ZnS with a shell of 1.2–1.8 monolayers, after which further shell growth reduced the quantum yields due to charge carriers being trapped at the interface. One should also consider that ZnS prefers the wurtzite structure, which might cause

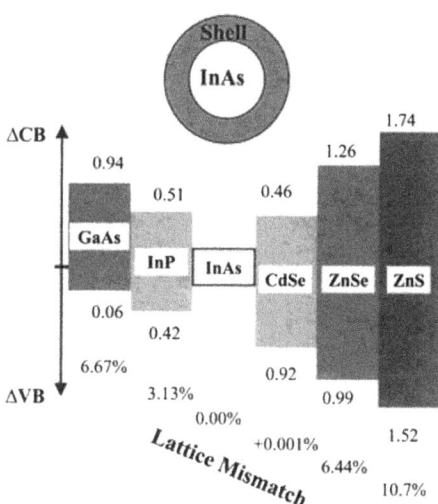

Figure 5.9 A summary of lattice mismatches and band offsets for InAs cores and a range of potential shell materials. Reprinted with permission from Y.-W. Cao and U. Banin, *J. Am. Chem. Soc.*, 2000, **122**, 9692. Copyright 2000 American Chemical Society.

enhanced strain when grown on a cubic core. Electron microscopy, XRD and XPS all confirmed the core/shell structure, particularly important with InAs/CdSe which have identical lattice constants and might therefore have formed alloys. These materials (InAs/ZnSe) have been used to probe the electron structure of QDs, notably in imaging the artificial atom-like states.[134,135]

Although InAs QDs were the first III–V particles to be shelled, other groups have explored this system. InAs/CdSe core/shells have also been grown in non-coordinating solvents. The shell was grown using the SILAR technique at 190 °C with cadmium oxide and TOPSe, the resulting particles again exhibiting a red shift in the optical properties.[136] Impressively, the emission quantum yields were reportedly as high as 90%, and emission could be tuned from *ca.* 800 nm to *ca.* 1400 nm. The use of non-coordinating solvents has also been explored in the synthesis of InAs/ZnSe, the shell being grown *in situ* without isolation of the core particles, using Me_2Zn and TOPSe as precursors[137] (the use of $S(SiMe_3)_2$ in place of TOPSe resulted in InAs/ZnS particles[138]). This could be amended to allow the growth of InAs/ZnCdSe by the addition of a small amount of Me_2Cd during shell synthesis, which extended the emission wavelength of the particles further into the red. The resulting particles were notable by possessing a clear excitonic peak in the absorption spectra past 800 nm, which became more pronounced on shell deposition, with quantum yields of up 10% in hexane which dropped only slightly upon phase transfer into water. The QDs also possessed an extremely small hydrodynamic diameter of below 10 nm when phase-transferred using variants of DHLA, and hence were extremely useful in sentinel lymph node imaging. The lattice constant for InAs (6.058 Å) is almost identical to that of CdSe (6.05 Å) which inspired the same research group to explore other alloy shell systems based on the successful CdSe/shell materials described earlier, with the aim of preparing material exhibiting near infrared emission between 700 and 900 nm, ideal for biological imaging.[139] The growth of a low lattice mismatched CdS shell pushed the emission beyond this window, in a similar manner to CdSe shells described above, but materials with the high lattice mismatched ZnS shell were found to be difficult to phase-transfer, hence the alloy system was deemed ideal. The core particles were grown again in ODE, using TOP as a capping agent, giving extremely small InAs cores (1.4 nm diameter), with an emission wavelength of *ca.* 700 nm. The growth of the shell was achieved with the addition of OAm to provide surface passivation prior to precursor injection, followed by dropwise addition of Me_2Cd, Me_2Zn and $S(SiMe_3)_2$ over several hours at 170 °C, which was stopped once the ideal emission wavelength (800 nm) had been reached. The resulting shell was 2.5 monolayers of $Zn_{0.7}Cd_{0.3}S$, producing particles with an overall diameter of 2.9 nm and an emission quantum yield of 35–50%. Once the surface ligands were exchanged for a water-soluble polymer, the hydrodynamic diameter was less than 10 nm, and the water-soluble core/shell QDs were found to be useful in cell imaging.

A ZnS shell has also been deposited on InAs particles that were grown using $In(OCOCH_3)_3$ and AsH_3 in ODE, giving InAs/ZnS particles that had a quantum yield of *ca.* 10%.[140] The shell was synthesised using zinc stearate and dodecanethiol (DDT) with an injection temperature of 130 °C and a growth temperature of 240 °C. Replacing DDT with TOPSe resulted in a ZnSe shell, giving InAs/ZnSe particles with quantum yields of up to 15%. The InAs/ZnSe particles actually increased in brightness over 30 days in ambient conditions.

Multiple shells can also be grown on an InAs core, introducing a buffer layer to minimise the interfacial strain of a mismatched shell which ultimately leads to a higher quantum yield.[141] A core/shell/shell structure of InAs/CdSe/ZnSe has been reported, prepared using the SILAR technique. The InAs cores (slightly larger than most InAs cores at 3.8 nm diameter) were prepared as outlined in Chapter 2, using TOP as a capping agent. Following redispersion of the particles in toluene with ODE and ODA, $Cd(CO_2(CH_2)_7(CH=CH)(CH_2)_7CH_3)_2$ and Se/ODE were added sequentially at 15 minute intervals at 260 °C, followed by Me_2Zn in toluene/ODE and the selenium precursor to deposit the final shell. After addition of the first cadmium layer, the emission increased notably from 1%, and continued to increase as further shells were added up to a maximum quantum yield of 50%. After the growth of the CdSe shell, a red shift in emission was observed from *ca.* 1050 nm to 1150 nm due to the small potential barrier, although this returned to the original position after the deposition of the ZnSe shell. Notably, the optical spectra appeared largely unchanged after the final deposition, suggesting the particles retained their monodispersed character. This impressive increase in emission quantum yield could not be achieved by simply depositing CdSe or ZnSe shells alone, and a combination of the two was found to be essential. The emission could be tuned by altering the core size, with 1.9 nm cores giving emission with a 70% quantum yield at 885 nm, while the larger core of 6.3 nm diameter gave emission at 1425 nm, although the final quantum yields were only 2.5%.

Multiple-shell particles based on III–V materials have also been prepared using alloyed cores of $InAs_xP_{1-x}$, with stoichiometries from $InAs_{0.33}P_{0.66}$ to $InAs_{0.82}P_{0.18}$, designed for applications in biological imaging.[142] The cores were prepared as outlined in Chapter 2 by the simultaneous addition of $As(SiMe_3)_3$ and $P(SiMe_3)_3$ into an ODE solution of $In(CO_2(CH_2)_7(CH=CH)(CH_2)_7CH_3)_3$ at 270 °C, followed by growth for 1 hour, giving graded structures with a phosphorus-rich core due to the faster decomposition of the phosphine. An InP shell was then deposited by cooling the reaction mixture to 140 °C, followed by addition of $In(CH_3CO_2)_3$ and $P(SiMe_3)_3$ with growth at 180 °C for 1 hour. A further ZnSe shell could be added by increasing the reaction temperature to 200 °C, followed by the dropwise addition of Me_2Zn and TOPSe over a further hour, after which the reaction was cooled to room temperature and precipitated with ethanol. Emission from the cores could be varied from 614 nm (pure InP), through 652 nm ($InAs_{0.33}P_{0.66}$), 699 nm ($InAs_{0.66}P_{0.33}$), 738 nm ($InAs_{0.82}P_{0.18}$), to 755 nm (pure InAs) with quantum yields of about 2%. Addition of an InP shell to $InAs_{0.82}P_{0.18}$ resulted in red shifts in the emission profile of up to over 60 nm, due to exciton leakage into the shell (not from the formation of a type II structure) with quantum yields increasing to up to 6%. Addition of a further shell to aid phase transfer into water was essential, although the deposition of a ZnS shell failed, possibly due to the lattice mismatch. A shell of ZnSe was successfully grown using Me_2Zn and TOPSe, which resulted in a further red shift of up 15 nm, with a final quantum yield of 3.5% and particles 5.78 nm in diameter. An interesting observation has been reported by Mokari *et al.*, where addition of a gold precursor to InAs QDs resulted in diffusion of gold into the QD, resulting in a crystalline gold core with an amorphous InAs or In_2O_3 shell.[143] Impressively, the absorption spectrum of InAs was maintained although the excitonic fine features were lost.

Although InAs QDs were the first core materials to be used in the synthesis of heterostructures based on III–V materials, their uses are primarily limited to infrared applications, such as tissue imaging where they have been extremely successful. The more common III–V material is InP which, in theory, has a wider spectral range, notably across the visible region. The problem with InP QDs is the inherently low-emission quantum yield, which rarely reaches 5% without surface etching procedures with HF, as described in Chapter 2. The material is therefore an ideal candidate for shell deposition, as the poor emissive properties demonstrably improved upon surface treatment.

Mićić *et al.* were the first to explore the possibility of improving the optical properties of InP QDs by depositing a shell layer,[144] shortly after the first reports of InAs-based core/shell particles were described by Cao *et al.* With reference to bulk systems where $Zn_{0.475}Cd_{0.525}Se_2$ is an ideal lattice match for InP, a shell of $ZnCdSe_2$ was deposited on core particles of InP. The QDs of InP with average sizes between 2.5 and 4.5 nm in diameter were synthesised as described in Chapter 2, then precipitated and dispersed in pyridine (in a similar style to the first report of CdSe particles being prepared for a CdS shell).[40] The shells were deposited in pyridine at 100 °C by the dropwise addition of Me_2Zn, Me_2Cd and TBPSe at a ratio of 1 : 1 : 4 to ensure complete formation of the shell, followed by addition of OAm to the solution. The formation of small $ZnCdSe_2$ particles also occurred, although these were easily removed by size-selective precipitation due to their poor solubility in pyridine. Shells of up to 5 nm thick were added to InP cores, with particles up to 20 nm in total diameter being reported which still exhibited excellent colloidal stability due to OAm on the particle surface.

A red shift was observed in the absorption spectra of $InP/ZnCdSe_2$ after shell deposition, consistent with the exciton leakage, which became more pronounced with the addition of further monolayers. Band edge emission with quantum yields of 5–10% were observed upon shell growth on 2.2 nm and 4.2 nm cores, which exhibited little or no emission prior to deposition. The emission could be tuned from *ca.* 600 nm for a 3 nm InP QD with a 1.5 nm shell, to *ca.* 700 nm for the same core with a 5 nm shell. Interestingly, InP QDs etched with HF to initially improve the emission could not be effectively passivated with the inorganic shell; this was suggested to be due to the etchant blocking the surface and hindering effective growth.

The prototypical shelling material (ZnS) has also been grown on InP dots, giving materials with significantly improved emission.[145] In the first report of the preparation of InP/ZnS, Et_2Zn and $S(SiMe_3)_2$ were added to a TOP solution of InP cores (in equimolar amounts to In and P), then injected into hot TOP at 200 °C, followed by quickly raising the temperature to 260 °C to induce precursor decomposition. The reaction was then cooled to 100 °C for 1 hour, then allowed to cool to room temperature, whereupon the reagents were diluted with toluene, filtered, and left standing under nitrogen for several days, slowly allowing up to 2 monolayers of ZnS to grow. The particles could then be isolated by the usual solvent/non-solvent interactions, although rapid agglomeration was observed. The absorption spectra of the core/shell particles lost some definition although appeared to shift very little, possibly due to the thin shell. The emission quantum yield improves from negligible emission up to a maximum of 23% after 3 weeks. Electron microscopy suggested that the shell formation was not complete and that the increase in emission might be improved upon by better shelling techniques, although XPS confirmed the core/shell structure.[146] Langof *et al.* reported a similar route, where Et_2Zn and $S(SiMe_3)_2$ were added to TOPO-capped InP in pyridine at 100 °C, with the shell growth complete after the addition of the precursors.[147] The final InP/ZnS QDs were notably anisotropic, especially apparent with the larger particles. Interestingly, the particles still exhibited band edge emission along with band edge recombination at low temperatures.

The obvious problem with this method of preparing core/shell particles is the use of dangerous precursor and the prolonged length of time required to obtain luminescent materials. Xie *et al.*, who reported a simple method for producing high-quality InP QDs using OA as a capping agent, also reported a simple shelling method to prepare InP/ZnS, using ODE solutions of sulfur and $Zn(CO_2(CH_2)_{16}CH_3)_2$ using thermal cycling.[148] In this method, the precursors were added separately, 10 minutes apart at relatively low temperatures, followed by periods of heating at high (220 °C) temperatures. This was repeated to build up the shell with differing amounts of precursor. Using this method with the high-quality core particles, strong emission (up to 40% quantum yield) could be achieved, ranging from the blue end of the visible spectrum to the near infrared. The use of simple precursors was also explored by Bharali *et al.* in a synthetic procedure that resulted in luminescent InP/ZnS QDs in 6–8 hours.[149] The core particles of InP were prepared using standard precursors and capped with myristate groups, and although they were not isolated from the growth solution prior to shell growth, they were cooled to room temperature and centrifuged to remove waste side products. The supernatant was then added to a reaction vessel which contained the correct amount of $Zn(CO_2CH_3)_2$ and sulfur to produce a 0.7 nm shell of ZnS. After the addition of a solvent (ODE) and degassing at 80 °C, the reaction was then heated to 140 °C for 1.5 hours, after which the product was cooled to room temperature, the volatiles removed *in vacuo*, and the particles isolated by centrifugation. The core/shell particles had an inorganic diameter of between 4.9 and 12.6 nm, up notably from the naked particles which had diameters between 4.0 and 5.7 nm. The absorption spectra of both naked InP particles and the InP/ZnS particles were very similar with an extremely small shift. The emission of the core particles was weak and predominantly trap based, while the core/shell particles exhibited band edge emission with quantum yields of 15%. The particles could be simply phase-transferred using mercaptoacetic acid, and then conjugated to folic acid. These particles were then used in the imaging of folate-positive receptor cells, and can be considered one of the first reports of the use of III–V QDs in bioimaging. Further work by the same group used sulfur dissolved in OAm and $Zn(CO_2(C_2H_5)CH(CH_2)_3CH_3)_2$ in ODE as shell precursors, which were added dropwise at 210 °C to the core particles and the temperature maintained for 45 minutes, followed by direct dissolution of the product in toluene.[150] A similar method to prepare InP/ZnS has been reported by Narayanaswamy *et al.*[151,152] who explored the high temperature decrease in emission and spectral broadening and the link to coupling the emissive state with acoustic phonons.

An effective and elegant method of overcoming the prolonged and time-consuming synthesis is the one-pot reaction to InP/ZnS described by Xu *et al.*[153] In this reaction, InP are prepared using ODE and methyl myristate, $CH_3(CH_2)_{12}COOCH_3$, as solvents with the usual precursors. The obvious difference is the inclusion in the reaction of zinc undecylenate, $Zn(CO_2(CH_2)_8CHCH_2)_2$, which replaced surface indium ions and blocked dangling bonds on the resulting InP particles, massively increasing the quantum yields from under 5% to up to 30%. Once the InP nanoparticles had been formed, which required just 20 minutes, the reaction flask was cooled to room temperature, whereupon a single-source precursor, $Zn(S_2CNEt_2)_2$, was added and reheated to 240 °C for a further 20 minutes. Shell growth could be also be obtained using further $Zn(CO_2(CH_2)_8CHCH_2)_2$ and $C_6H_{11}NCS$ as precursors. By tuning the reaction conditions, tuneable emission between 480 and 750 nm with quantum yields of up to 60% could be achieved. These materials have also been capped with SiO_2 shells and used as converter materials in down-conversion white-emitting LEDs.[154]

A similar approach has been described by Li and Reiss, in which all precursors $(In(CO_2(CH_2)_{12}CH_3)_2$, $Zn(CO_2(CH_2)_{16}CH_3)_2$, $P(SiMe_3)_3$, and $CH_3(CH_2)_{11}SH)$ were mixed in ODE and rapidly heated to 300 °C under argon, leading to the *in situ* formation of what was initially suggested to be InP/ZnS particles, due to the difference in precursor reactivity.[155] Particle growth started with InP core formation followed by shell growth as the thiol precursors decomposed at 230 °C, although indium incorporation into the particles reportedly continued even during the shell growth step. The optical properties red-shifted from a band edge of 450 nm, with clear excitonic features clearly observed after 3 minutes, which broadened over the duration of the reaction, with narrow emission shifting from *ca.* 500–600 nm.

The brightness of the emission was dependent on emission wavelength, with quantum yields reaching up to 70% for particles that emitted at *ca.* 540 nm. The zinc and sulfur precursors were also found to affect growth rate, with lower concentration of precursor resulting in larger particles, but reducing the emission stability. Further photoelectron spectroscopy studies have since shown that the structures grown from such a one-pot reaction are actually an InPZnS alloy with a thin ZnS shell,[156] and further work on the role of zinc carboxylates suggests that inclusion of such materials during QD growth leads to the formation of a luminescent InZnP alloy rather than InP particles passivated with the zinc species.[157] A related route by Ryu *et al.* explored the role of the zinc precursor in the preparation of InP/ZnS, and also used a long-chain thiol as a sulfur precursor.[158] The group reported a notable increase in emission and red shift in wavelength when $Zn(CO_2CH_3)_2$ was added to a solution of InP dots and heated at 230 °C for 5 hours, reportedly forming a zinc carboxylate on the surface. Later addition of the thiol increased the emission quantum yield up to 38%. It is still unclear whether the addition of the $Zn(CO_2CH_3)_2$ actually formed a surface carboxylate species, or, referring to results reported by Reiss, formed an alloy. The increase in emission intensity of InP dots by the addition of a zinc carboxylate was also reported by Li *et al.*[159] The emission was then increased to a final quantum yield of 22% by further addition of zinc ethylxanthate, a single-source precursor for ZnS deposition. Whether the addition of the zinc stearate also resulted in alloy formation was not reported.

Other similar families of core/shell QDs have been reported, such as InGaP/ZnS materials, which displayed band edge emission at *ca.* 675 nm, although this was a commercially available product (at the time) and no experimental details were provided.[160] Emission at this wavelength is of interest to biologists, and these alloyed core materials have been utilised in deep tissue labelling experiments where the emission can be detected even through a mouse skull.[161] $GaInP_2$ shells of 0.5–2 nm have been deposited on InP cores, using $GaCl_3$, $InCl_3$ and $P(SiMe_3)_3$ (in the ratio of 2 : 2 : 1), added slowly to the core particles immediately after growth, at 260 °C followed by heating for 16 hours using TOPO and hexylamine as capping agents.[162] Few optical properties were reported. InP shells were also grown on MnP cores using $InCl_3$ and $P(SiMe_3)_3$, dissolved in TOP, which were added to the core solution of MnP.[163] The resulting MnP/InP core/shell structures maintained the magnetic properties of the MnP core, but did not show any optical characteristics consistent with the InP shell.

5.6 Core/Shell Structures Based on IV–VI Materials

Lead chalcogenide QDs are predominantly infrared-emitting materials, with excellent optical and electronic properties which have been exploited in numerous applications. Although the particles have superior attributes and have few organic dye equivalents, they are extremely sensitive to oxidation. The optical properties have reportedly exhibited blue shifts, emission photobrightening and photodarkening, and even total decomposition under prolonged UV exposure, as outlined in Chapter 3. This extreme air sensitivity makes them ideal candidates for the deposition of a further inorganic shell.

The use of PbS as a shell on PbSe QDs is an excellent choice. The lattice constants are similar (PbSe = 6.12 Å, PbS = 5.93 Å) while the bandgap has been predicted to be slightly offset[164] in what normally might be assumed to be a type II alignment, although theoretical investigations into PbSe/PbS QD heterostructures have predicted the behaviour to be consistent with a type I offset at the usual particle sizes because the lowest occupied molecular orbital remains confined.[165]

One of the earliest reports on the synthesis of PbSe described the reaction between $Pb(O_2CCH(C_2H_5)C_4H_9)_2$ and TBPSe at room temperature, giving QDs between 2 nm and 5 nm diameter. The addition of a solution of TBPS resulted in the formation of a PbS shell through anion exchange, which could be controlled at 1–4 monolayers in thickness.[166] This resulted in the red-shifting of the absorption edge assigned to electronic mixing of the PbSe and PbS conduction states, and increasing the band edge emission of PbSe while reducing deep trap emission, due to increased surface passivation. The addition of more than 2 monolayers of a shell narrowed the emission profile, although no quantum yield measurements were given.

The same group extended the above study by developing a traditional two-stage synthesis of PbSe/PbS core/shell heterostructures rather than using anion exchange.[167] PbSe core particles, 3–9 nm in diameter with an 8% standard deviation, were synthesised as described by Murray *et al.*, by the injection of lead oleate and TOPSe into a hot diphenyl ether. The particles were isolated, purified, then redissolved in TOP. The particles were then mixed with TOPS, then lead oleate, and injected into hot diphenylether at 180 °C, then grown at 120 °C for 15 min, depositing 1–3 monolayers of PbS. The emission was band edge, with a maximum quantum yield of 40%. Interestingly, core/shell particles of $PbSe/PbSe_xS_{1-x}$ were also reported,[167] this time by a single injection route, in an amendment to the synthetic procedure towards PbSe. In this synthesis, a mixture of selenium and sulfur was dissolved in TOP, added to the lead precursor and injected into the hot solvent. In this manner, the PbSe nucleated primarily followed by growth of the $PbSe_xS_{1-x}$ shell, the stoichiometry of which could be controlled by the ratio of precursor. The dots formed were monodispersed in nature, exhibiting emission quantum yields of up to 55%. A further study explored the optical properties of these materials in more depth.[168] The blue shift in the absorption spectra associated with oxidation of PbSe was found to be significantly smaller in PbSe/PbS particles, and almost entirely absent in structures with more than 3 monolayers of PbS. Notably, the absorption edge and emission profile both red shifted significantly after the addition of a single monolayer and the emission profile again narrowed. The red shift increased upon further shell addition. Unusually, once 3 monolayers of PbS had been added, the emission profile exhibited an anti-Stokes shift relative to the excitonic peak. Similarly, an anti-Stokes shift in emission was observed in

PbSe particles with an alloyed PbSe/PbSe$_x$S$_{1-x}$ shell. In this case, the shift became more pronounced with increasing sulfur content, becoming extremely predominant at PbSe/PbSe$_{0.5}$S$_{0.5}$. The origin of this anti-Stokes emission is unknown, although it has been attributed to the smaller particles to the blue side of the excitonic peak having a larger oscillator strength.

Further explorations into PbSe/PbS structures suggested that the addition of the PbS shell did not in fact improve stability, unlike the II–VI analogue heterostructures.[169] The core particles were prepared using a standard route to oleic acid-capped PbSe, after which the particles were isolated and purified. The particles were then redissolved in ODE with a small amount of oleic acid. The lead precursor was then added, the solution heated to 130 °C, which was followed by the addition of S(SiMe$_3$)$_2$ in TOP. Emission profiles of the resulting particles displayed two emission peaks, consistent with the presence of separate PbS particles, due to the highly reactive S(SiMe$_3$)$_2$, and a red-shifted emission peak due to the PbSe/PbS. The use of TOPS as a sulfur precursor avoided this side reaction. Unfortunately, the PbSe/PbS particles did not show enhanced emission, unlike CdSe/ZnS for example, and readily blue-shifted upon storage in ambient conditions, like the core materials, attributed to incomplete shell formation.

Talapin *et al.* also explored the synthesis of PbSe/PbS nanostructures.[170] Using methods similar to those described above, it was reported that the sulfur precursor made a notable difference when growing a PbS shell on quasi-spherical PbSe particles. The use of TOPS as a sulfur precursor resulted in even shell growth, yielding approximately spherical particles of PbSe/PbS. The use of the more reactive sulfur in ODE at lower temperatures, however, resulted in cubic particles. The growth of PbS shell on PbSe wires was also reported, with an interesting difference in the growth mode. The growth of PbS shells on PbSe occurred *via* layer-by-layer deposition, but the growth of PbS on PbSe wires occurred by a Stranski–Krastanov mechanism, yielding strain-grown pyramids along the wire. Notably, pre-synthesised particles of PbS also attached to preformed PbSe wires *via* oriented attachment at 120 °C. Similarly, gentle heating of HAuCl$_4$ in a solution of preformed PbSe nanowires with a surfactant resulted in the growth of gold islands at relatively regular intervals along the wire. The growth of gold on PbS has also been observed previously by others.[171] The growth of a PbTe shell on PbSe wires has also been achieved using a similar method and utilising TOPTe as a precursor.[172] The growth of a PbTe shell required a slightly higher growth temperature of 190 °C—unsurprising, as TOPTe has previously been noted to be relatively unreactive.[173]

The capping of PbSe with II–VI materials has also been undertaken. Materials such as CdSe are excellent candidates for capping due to the low lattice mismatch (*ca.* 1%), relatively better stability to ambient conditions, and the much wider bandgap which theoretically should lead to better charge carrier confinement. The usual method of depositing a II–VI shell by the slow introduction of precursors was found to be unsuccessful, so a different approach was taken,[174] similar to the ion exchange method of preparing new nanomaterials.[175] In this synthesis, preformed PbSe particles were reacted with cadmium oleate at 100 °C, resulting in cation exchange on the surface, yielding PbSe/CdSe particles with a CdSe shell of up to 1.5 nm thick irrespective of reagent concentration and reaction conditions. XRD confirmed the growth of a core/shell species rather than an alloyed material. The emission was found to shift to shorter wavelength as the emitting core effectively became smaller, while the quantum yield of the smaller particles remained high (*ca.* 70%), and the smaller core/shell particles had significantly improved emission of up to 17%. The deposition of the shell improved the particle stability, from a time scale measured in days, to materials that still exhibited excellent properties months later. Once a CdSe shell had been deposited, the epitaxial growth of a further ZnS shell was attempted onto the pre-existing CdSe layer, using the typical precursors such as $S(SiMe_3)_2$ and Me_2Zn in the presence of TOP, giving PbSe/CdSe/ZnS particles. Although successful, the additional layer actually reduced emission efficiency. A similar method was used to grow PbSe/CdSe/CdS particles, using cadmium oleate and sulfur dissolved in ODE at a growth temperature of 240 °C.[176] It was found that the reduction in emission quantum yield while depositing the final shell was actually due to annealing at the core/shell interface due to heating, and the observed blue shift was due to alloy formation, which was reversed upon CdS deposition as the electron wavefunction delocalised into the new shell. The emission quantum yield was found to stabilise at 10%. These particles are notable for their ultra-long recombination life time of 80 µS due to the reduced electron/hole overlap, almost two orders of magnitude longer than the core particles. The use of a lower deposition temperature (170 °C) for the CdS shell resulted in an unusual tetrapod morphology, with CdS arms on spherical PbSe/CdSe particles.

Core/shell particles of PbS/CdS could also be grown using ion exchange, and these particles could be phase-transferred to aqueous solution using an amphiphilic polymer while maintaining an impressive emission quantum yield of *ca.* 30% in the infrared region.[177] The cation exchange method was extended to the preparation of PbTe/CdTe particles[178] as both semiconducting materials exist in the rock salt phase, with almost no lattice mismatch. Extensive high-resolution electron microscopy revealed the seamless match between core and shell, when viewed along a specific crystalline axis which allowed extremely high-quality images of core/shell particles to be produced (Figure 5.10). The core particles were found to have almost entirely (111) crystalline plane edges, suggesting an anisotropic exchange mechanism.

Other notable systems include the use of Cu_2S, which was used as both a core and a shell with CdS.[179] As the bandgaps were staggered, either combination gave a type II alignment and both structures (Cu_2S/CdS and CdS/Cu_2S) were prepared simply using existing precursor routes. By tuning precursor ratios and controlling reaction times (giving differing core sizes and shell thicknesses), the entire visible range was accessible and growth was allowed to continue until the desired emission wavelength was reached, with

Figure 5.10 High-resolution TEM images of a PbTe/CdTe core/shell QD: (a) along the (111) axis, (b) along the (110) axis. Reprinted with permission from K. Lambert, B. De Geyter, I. Moreels and Z. Hens, *Chem. Mater.*, 2009, **21**, 778. Copyright 2009 American Chemical Society.

a maximum quantum yield of 12%. To make the Cu_2S/CdS material more stable, a ZnS shell could also be deposited, which induced a significant emission blue shift, attributed to the Zn atoms diffusing into the CdS shell. A significant factor regarding these core/shell materials is that no emission was observed from just the Cu_2S cores alone, and only trap emission was observed from the uncapped CdS.

In this chapter, we have shown that core/shell systems can be used to make highly useful, resilient materials. The bandgap engineering of semiconductor nanomaterials, which was reported in earlier chapters, can be improved to a new level of complexity when the range of materials and their associated band structures is considered. Core emission can be protected, shifted and the charge carriers can be tuned to various parts of the particle by choosing the core and shell materials and their dimensions. Complicated structures can be grown that exhibit unusual and interesting optical properties, which have found application in light-emitting device, solar and bioimaging technologies.

References

1. J. V. Embden, J. Jasieniak, D. E. Gómez, P. Mulvaney and M. Giersig, *Aust. J. Chem.*, 2007, **60**, 457.
2. P. Reiss, M. Protière and L. Li, *Small*, 2009, **5**, 154.
3. F. Capasso and G. Margaritondo, *Heterojunction Band Discontinuities: Physics and Device Applications*, Elsevier, 1987.
4. W. Monch, *Semiconductor Surfaces and Interfaces*, Springer, 1995.
5. S. H. Wei and A. Zunger, *Appl. Phys. Lett.*, 1998, **72**, 2011.

6. H. Mattoussi, L. H. Radzilowski, B. O. Dabbousi, E. L. Thomas, M. G. Bawendi and M. F. Rubner, *J. Appl. Phys.,* 1998, **83**, 7965.

7. V. L. Colvin, A. P. Alivisatos and J. G. Tobin, *Phys. Rev. Lett.,* 1991, **66**, 2786.

8. H. Fu and A. Zunger, *Phys. Rev. B: Condens. Matter Mater. Phys.,* 1997, **56**, 1496.

9. S.-H. Kim, G. Markovich, S. Rezvani, S. H. Choi, K. L. Wang and J. R. Heath, *Appl. Phys. Lett.,* 1999, **74**, 317.

10. M. Green, *Small,* 2005, **1**, 684.

11. J. Li, D. Li, J. Li, J. Hu, M. Wang, J. Tang and J. Li, unpublished results.

12. H. Weller, U. Koch, M. Gutiérrez and A. Henglein, *Ber. Bunsenges. Phys. Chem.,* 1984, **88**, 649.

13. L. Spanhel, M. Haase, H. Weller and A. Henglein, *J. Am. Chem. Soc.,* 1987, **109**, 5649.

14. A. Eychmüller, A. Hässelbarth and H. Weller, *J. Lumin.,* 1992, **53**, 113.

15. A. Eychmüller, A. Mews and H. Weller, *Chem. Phys. Lett.,* 1993, **208**, 59.

16. A. Mews, A. Eychmüller, M. Giersig, D. Schooss and H. Weller, *J. Phys. Chem.,* 1994, **98**, 934.

17. D. Schooss, A. Mews, A. Eychmüller and H. Weller, *Phys. Rev. B: Condens. Matter Mater. Phys.,* 1994, **49**, 17072.

18. A. Hässelbarth, A. Eychmüller, R. Eichberger, M. Giersig, A. Mews and H. Weller, *J. Phys. Chem.,* 1993, **97**, 5333.

19. A. Mews and A. Eychmüller, *Ber. Bunsenges. Phys. Chem.,* 1998, **102**, 1343.

20. A. R. Kortan, R. Hull, R. L. Opila, M. G. Bawendi, M. L. Steigerwald, P. J. Carroll and L. E. Brus, *J. Am. Chem. Soc.,* 1990, **112**, 1327.

21. C. F. Hoener, K. A. Allan, A. J. Bard, A. Campion, M. A. Fox, T. E. Mallouk, S. E. Webber and J. M. White, *J. Phys. Chem.,* 1992, **96**, 3812.

22. J. E. Bowen Katari, V. L. Colvin and A. P. Alivisatos, *J. Phys. Chem.,* 1994, **98**, 4109.

23. M. Danek, K. F. Jensen, C. B. Murray and M. G. Bawendi, *J. Cryst. Growth,* 1994, **145**, 714.

24. M. Danek, K. F. Jensen, C. B. Murray and M. G. Bawendi, *Appl. Phys. Lett.,* 1994, **65**, 2795.

25. M. Danek, K. F. Jensen, C. B. Murray and M. G. Bawendi, *Chem. Mater.,* 1996, **8**, 173.

26. J. R. Heine, J. Rodriguez-Viejo, M. G. Bawendi and K. F. Jensen, *J. Cryst. Growth,* 1998, **195**, 564.

27. M. A. Hines and P. Guyot-Sionnest, *J. Phys. Chem.,* 1996, **100**, 468.

28. B. O. Dabbousi, J. Rodriguez-Viejo, F. V. Mikulec, J. R. Heine, H. Mattoussi, R. Ober, K. F. Jensen and M. G. Bawendi, *J. Phys. Chem. B,* 1997, **101**, 9463.

29. It is worth, at this point, commenting on the concept of a monolayer. A unit cell of a shell material is essentially two layers of semiconductor, and therefore half of the *c*-lattice parameter (for wurtzite structures) is used as the thickness of a given monolayer. Therefore a layer of both cations and anions is required to increase

the radius (not the diameter) of the particle by half a lattice parameter. Many publications also state that the amount of shell precursor is calculated once the core particles have been prepared, although few state how. Two models exist—the more common and easiest to use is the concentric shell model, first described by Dabbousi *et al.*,[28] where the volume of progressive concentric shells is estimated volumetrically, then worked backwards through the bulk densities to the number of ions required to grow the first, shell, then the second, *etc.* The unit cell model utilises bulk lattices parameters to calculate the number of atoms required for each monolayer using standard crystallography programs such as crystal maker. An excellent description of the above is given in J. L. van Embden's PhD thesis (University of Melbourne, 2008).

30. R. K. Čapek, K. Lambert, D. Dorfs, P. F. Smet, D. Poelman, A. Eychmüller and Z. Hens, *Chem. Mater.,* 2009, **21**, 1743.
31. J. Zhang, X. Zhang and J. Y. Zhang, *J. Phys. Chem. C,* 2010, **114**, 3904.
32. D. V. Talapin, A. L. Rogach, A. L. Kornowski, M. Haase and H. Weller, *Nano Lett.,* 2001, **1**, 207.
33. Q. Wang, Y. Liu, Y. Ke and H. Yan, *Angew. Chem., Int. Ed.,* 2008, **47**, 316.
34. J. Zeigler, A. Merkulov, M. Grabolle, U. Resch-Genger and T. Nann, *Langmuir,* 2007, **23**, 7751.
35. T. Mokari and U. Banin, *Chem. Mater.,* 2003, **15**, 3955.
36. L. Manna, E. C. Scher, L.-S. Li and A. P. Alivisatos, *J. Am. Chem. Soc.,* 2002, **124**, 7136.
37. J. McBride, J. Treadway, L. C. Feldman, S. J. Pennycook and S. J. Rosenthal, *Nano Lett.,* 2006, **6**, 1496.
38. H. Lee, P. H. Holloway and H. Yang, *J. Chem. Phys.,* 2006, **125**, 164711.
39. P. T. Snee, Y. Chan, D. G. Nocera and M. G. Bawendi, *Adv. Mater.,* 2005, **17**, 1131.
40. X. Peng, M. C. Schlamp, A. V. Kadanavich and A. P. Alivisatos, *J. Am. Chem. Soc.,* 1997, **119**, 7019.
41. D. V. Talapin, R. Koeppe, S. Götzinger, A. Kornowski, J. M. Lupton, A. L. Rogach, O. Benson, J. Feldmann and H. Weller, *Nano Lett.,* 2003, **3**, 1677.
42. A. Sitt, F. D. Sala, G. Menagen and U. Banin, *Nano Lett.,* 2009, **9**, 3470.
43. I. Mekis, D. V. Talapin, A. Kornowski, M. Haase and H. Weller, *J. Phys. Chem. B,* 2003, **107**, 7454.
44. M. Ristov, G. Sinadinovski, I. Grozdanov and M. Mitreski, *Thin Solid Films,* 1989, **173**, 53.
45. S. Park, B. L. Clark, D. A. Keszler, J. P. Bender, J. F. Wager, T. A. Reynolds and G. S. Herman, *Science,* 2002, **297**, 65.
46. J. J. Li, Y. A. Wang, W. Guo, J. C. Keay, T. D. Mishima, M. B. Johnson and X. Peng, *J. Am. Chem. Soc.,* 2003, **125**, 12567.
47. J. Wang, Y. Long, Y. Zhang, X. Zhong and L. Zhu, *ChemPhysChem,* 2009, **10**, 680.
48. J. van Embden, J. Jasieniak and P. Mulvaney, *J. Am. Chem. Soc.,* 2009, **131**, 14299.

49. Y. Chen, J. Vela, H. Htoon, J. L. Casson, D. J. Werder, D. A. Bussian, V. I. Klimov and J. A. Hollingsworth, *J. Am. Chem. Soc.,* 2008, **130**, 5026.

50. F. Garcia-Santamaria, Y. Chen, J. Vela, R. D. Schaller, J. A. Hollingsworth and V. I. Klimov, *Nano Lett.,* 2009, **9**, 3482.

51. D. Pan, Q. Wang, S. Jiang, X. Ji and L. An, *Adv. Mater.,* 2005, **17**, 176.

52. J. S. Steckel, J. P. Zimmer, S. Coe-Sullivan, N. E. Stott, V. Bulović and M. G. Bawendi, *Angew. Chem., Int. Ed.,* 2004, **43**, 2154.

53. S.-H. Wei, S. B. Zhang and A. Zunger, *J. Appl. Phys.,* 2000, **87**, 1304.

54. P. Reiss, J. Bleuse and A. Pron, *Nano Lett.,* 2002, **2**, 781.

55. B. Schreder, T. Schmidt, V. Ptatschek, U. Winkler, A. Materny, E. Umbach, M. Lerch, G. Müller, W. Kiefer and L. Spanhel, *J. Phys. Chem. B,* 2000, **104**, 1677.

56. S. F. Wuister, I. Swart, F. Van Driel, S. G. Hickey and C. de Mello Donegá, *Nano Lett.,* 2003, **3**, 503.

57. J. M. Tsay, M. Pflughoefft, L. A. Bentolila and S. Weiss, *J. Am. Chem. Soc.,* 2004, **126**, 1926.

58. M. Lomascolo, A. Creti, G. Leo, L. Vasanelli and L. Manna, *Appl. Phys. Lett.,* 2003, **82**, 418.

59. H.-S. Chen, B. Lo, J.-Y. Hwang, G.-Y. Chang, C.-M. Chen, S.-J. Tasi and S.-J. Jassy Wang, *J. Phys. Chem. B,* 2004, **108**, 17119.

60. A. D. Lad, P. Prem Kiran, D. More, G. Ravindra Kumar and S. Mahamuni, *Appl. Phys. Lett.,* 2008, **92**, 043126.

61. A. D. Lad, P. Prem Kiran, G. Ravindra Kumar and S. Mahamuni, *Appl. Phys. Lett.,* 2007, **90**, 133113.

62. N. Pradhan, D. Goorskey, J. Thessing and X. Peng, *J. Am. Chem. Soc.,* 2005, **127**, 17586.

63. N. Pradhan and X. Peng, *J. Am. Chem. Soc.,* 2007, **129**, 3339.

64. R. Thakar, Y. Chen and P. T. Snee, *Nano Lett.,* 2007, **7**, 3429.

65. V. Wood, J. E. Halpert, M. J. Panzer, M. G. Bawendi and V. Bulović, *Nano Lett.,* 2009, **9**, 2367.

66. H. Shen, H. Wang, X. Li, J. Z. Niu, H. Wang, X. Chen and L. S. Li, *Dalton Trans.,* 2009, 10534.

67. J. Zheng, X. Yuan, M. Ikezawa, P. Jing, X. Liu, Z. Zheng, X. Kong, J. Zhao and Y. Masumoto, *J. Phys. Chem. C,* 2009, **113**, 16969.

68. S. Wang, B. R. Jarrett, S. M. Kauzlarich and A. Y. Louie, *J. Am. Chem. Soc.,* 2007, **129**, 3848.

69. W. Jiang, A. Singhal, J. Zheng, C. Wang and W. C. W. Chan, *Chem. Mater.,* 2006, **18**, 4845.

70. T. Pons, N. Lequeux, B. Mahler, S. Sasnouski, A. Fragola and B. Dubertret, *Chem. Mater.,* 2009, **21**, 1418.

71. W. K. Bae, M. K. Nam, K. Char and S. Lee, *Chem. Mater.,* 2008, **20**, 5307.

72. X. Wang, X. Ren, K. Kahen, M. A. Hahn, M. Rajeswaran, S. Maccagnano-Zacher, J. Silcox, G. E. Cragg, A. L. Efros and T. D. Krauss, *Nature,* 2009, **459**, 686.

73. J. Wang, M. Lin, Y. Yan, Z. Wang, P. C. Ho and K. P. Loh, *J. Am. Chem. Soc.,* 2009, **131**, 11300.

74. Z.-Q. Tian, Z.-L. Zhang, P. Jiang, M.-X. Zhang, H.-Y. Xie and D.-W. Pang, *Chem. Mater.,* 2009, **21**, 3039.
75. Z.-Q. Tian, Z.-L. Zhang, J. Gao, B.-H. Huang, H.-Y. Xie, M. Xie, H. D. Abruña and D.-W. Pang, *Chem. Commun.,* 2009, 4025.
76. H. Kim, M. Achermann, L. P. Balet, J. A. Hollingsworth and V. I. Klimov, *J. Am. Chem. Soc.,* 2005, **127**, 544.
77. J. Gao, B. Zhang, Y. Gao, Y. Pan, X. Zhang and B. Xu, *J. Am. Chem. Soc.,* 2007, **129**, 11928.
78. J.-S. Lee, M. I. Bodnarchuk, E. V. Shevchenko and D. V. Talapin, *J. Am. Chem. Soc.,* 2010, **132**, 6382.
79. J. Zhang, Y. Tang, K. Lee and M. Ouyang, *Science,* 2010, **327**, 1634.
80. J. Zhang, Y. Tang, L. Weng and M. Ouyang, *Nano Lett.,* 2009, **9**, 4061.
81. Y. Jin and X. Gao, *Nat. Nanotechnol.,* 2009, **4**, 571.
82. S. Kim, Y. T. Lim, E. G. Soltesz, A. M. De Grand, J. Lee, A. Nakayama, J. A. Parker, T. Mihaljevic, R. G. Laurence, D. M. Dor, L. H. Cohn, M. G. Bawendi and J. V. Frangioni, *Nat. Biotechnol.,* 2004, **22**, 93.
83. J. Bang, B. Chon, N. Won, J. Nam, T. Joo and S. Kim, *J. Phys. Chem. C,* 2009, **113**, 6320.
84. S. Kim, B. Fisher, H.-J. Eisler and M. G. Bawendi, *J. Am. Chem. Soc.,* 2003, **125**, 11466.
85. K. Yu, B. Zaman, S. Romanova, D.-S. Wang and J. A. Ripmeester, *Small,* 2005, **1**, 332.
86. P. T. K. Chin, C. de Mello Donegá, S. S. Van Bavel, S. C. J. Meskers, N. A. J. M. Sommerdijk and R. A. J. Janssen, *J. Am. Chem. Soc.,* 2007, **129**, 14880.
87. H. Zhong and G. D. Scholes, *J. Am. Chem. Soc.,* 2009, **131**, 9170.
88. B. Blackman, D. M. Battaglia, T. D. Mishima, M. B. Johnson and X. Peng, *Chem. Mater.,* 2007, **19**, 3815.
89. H.-S. Sheu, U.-S. Jeng, W.-J. Shih, Y.-H. Lai, C.-H. Su, C.-W. Lei, M.-J. Yang, Y.-C. Chen and P.-T. Chou, *J. Phys. Chem. C,* 2008, **112**, 9617.
90. J. E. Halpert, V. J. Porter, J. P. Zimmer and M. G. Bawendi, *J. Am. Chem. Soc.,* 2006, **128**, 12590.
91. S. Kumar, M. Jones, S. S. Lo and G. D. Scholes, *Small,* 2007, **3**, 1633.
92. M. Jones, S. Kumar, S. S. Lo and G. D. Scholes, *J. Phys. Chem. C,* 2008, **112**, 5423.
93. H. McDaniel, J.-M. Zuo and M. Shim, *J. Am. Chem. Soc.,* 2010, **132**, 3286.
94. F. Shieh, A. E. Saunders and B. A. Korgel, *J. Phys. Chem. B,* 2005, **109**, 8538.
95. L. Xi, C. Boothroyd and Y. M. Lam, *Chem. Mater.,* 2009, **21**, 1465.
96. Z. A. Peng and X. Peng, *J. Am. Chem. Soc.,* 2002, **124**, 3343.
97. C.-Y. Chen, C.-T. Cheng, J.-K. Yu, S.-C. Pu, Y.-M. Cheng, P.-T. Chou, Y.-H. Chou and H.-T. Chiu, *J. Phys. Chem. B,* 2004, **108**, 10687.
98. N. N. Hewa-Kasakarage, N. P. Gurusinghe and M. Zamkov, *J. Phys. Chem. C,* 2009, **113**, 4362.
99. J.-Y. Chiang, S.-R. Wang and C.-H. Yang, *Nanotechnology,* 2007, **18**, 345602.

100. H. Seo and S.-W. Kim, *Chem. Mater.,* 2007, **19**, 2715.
101. A. M. Smith, A. M. Mohs and S. Nie, *Nat. Nanotechnol.,* 2009, **4**, 56.
102. A. Nemchinov, M. Kirsanova, N. N. Hewa-Kasakarage and M. Zamkov, *J. Phys. Chem. C,* 2008, **112**, 9301.
103. D. Dorfs, S. Salant, I. Popov and U. Banin, *Small,* 2008, **4**, 1319.
104. L. P. Balet, S. A. Ivanov, A. Piryatinski, M. Achermann and V. I. Klimov, *Nano Lett.,* 2004, **4**, 1485.
105. R. Xie, X. Zhong and T. Basché, *Adv. Mater.,* 2005, **17**, 2741.
106. R. Xie, U. Kolb and T. Basché, *Small,* 2006, **12**, 1454.
107. J. Bang, J. Park, J. H. Lee, N. Won, J. Nam, J. Lim, B. Y. Chang, H. J. Lee, B. Chon, J. Shin, J. B. Park, J. H. Choi, K. Cho, S. M. Park, T. Joo and S. Kim, *Chem. Mater.,* 2010, **22**, 233.
108. S. A. Ivanov, J. Nanda, A. Piryatinski, M. Achermann, L. P. Balet, I. V. Bezel, P. O. Anikeeva, S. Tretiak and V. I. Klimov, *J. Phys. Chem. B,* 2004, **108**, 10625.
109. X. Zhong, R. Xie, Y. Zhang, T. Basché and W. Knoll, *Chem. Mater.,* 2005, **17**, 4038.
110. During the synthesis of CdSe/ZnS particles, the formation of separate ZnS particles can often be observed as a feature in the emission spectra at *ca.* 400 nm.
111. P. Reiss, S. Carayon, J. Bleuse and A. Pron, *Synth. Met.,* 2003, **139**, 649.
112. J. W. M. Chon, P. Zijlstra, M. Gu, J. van Embden and P. Mulvaney, *Appl. Phys. Lett.,* 2004, **85**, 5514.
113. J. Lim, S. Jun, E. Jang, H. Baik, H. Kim and J. Cho, *Adv. Mater.,* 2007, **19**, 1927.
114. S. Jun and E. Jang, *Chem. Commun.,* 2005, 4616.
115. D. V. Talapin, I. Mekis, S. Götzinger, A. Kornowski, O. Benson and H. Weller, *J. Phys. Chem. B,* 2004, **108**, 18826.
116. D. V. Talapin, R. Koeppe, S. Götzinger, A. Kornowski, J. M. Lupton, A. L. Rogach, O. Benson, J. Feldmann and H. Weller, *Nano Lett.,* 2003, **3**, 1677.
117. R. Xie, U. Kolb, J. Li, T. Basché and A. Mews, *J. Am. Chem. Soc.,* 2005, **127**, 7480.
118. P. Jing, J. Zheng, M. Ikezawa, X. Liu, S. Lv, X. Kong, J. Zhao and Y. Masumoto, *J. Phys. Chem. C,* 2009, **113**, 13545.
119. C.-T. Cheng, C.-Y. Chen, C.-W. Lai, W.-H. Liu, S.-C. Pu, P.-T. Chou, Y.-H. Chou and H.-T. Chiu, *J. Mater. Chem.,* 2005, **15**, 3409.
120. B. Blackman, D. Battaglia and X. Peng, *Chem. Mater.,* 2008, **20**, 4847.
121. J. Z. Niu, H. Shen, C. Zhou, W. Xu, X. Li, H. Wang, S. Lou, Z. Du and L. S. Li, *Dalton Trans.,* 2010, **39**, 3308.
122. C.-Y. Chen, C.-T. Cheng, C.-W. Lai, Y.-H. Hu, P.-T. Chou, Y.-H. Chou and H.-T. Chiu, *Small,* 2005, **1**, 1215.
123. D. Battaglia, J. J. Li, Y. Wang and X. Peng, *Angew. Chem., Int. Ed.,* 2003, **42**, 5035.
124. J. Xu, M. Xiao, D. Battaglia and X. Peng, *Appl. Phys. Lett.,* 2005, **87**, 043107.

125. P. K. Santra, R. Viswanatha, S. M. Daniels, N. L. Pickett, J. M. Smith, P. O'Brien and D. D. Sarma, *J. Am. Chem. Soc.,* 2009, **131**, 470.
126. M. Braun, C. Burda and M. A. El-Sayed, *J. Phys. Chem. A,* 2001, **105**, 5548.
127. D. Dorfs and A. Eychmueller, *Nano Lett.,* 2001, **1**, 663.
128. D. Battaglia, B. Blackman and X. Peng, *J. Am. Chem. Soc.,* 2005, **127**, 10889.
129. E. A. Dias, A. F. Grimes, D. S. English and P. Kambhampati, *J. Phys. Chem. C,* 2008, **112**, 14229.
130. W. Zhang, G. Chen, J. Wang, B.-C. Ye and X. Zhong, *Inorg. Chem.,* 2009, **48**, 9723.
131. S. Kim, W. Shim, H. Seo, J. H. Bae, J. Sung, S. H. Choi, W. K. Moon, G. Lee, B. Lee and S.-W. Kim, *Chem. Commun.,* 2009, 1267.
132. Y.-W. Cao and U. Banin, *Angew. Chem., Int. Ed.,* 1999, **38**, 3692.
133. Y.-W. Cao and U. Banin, *J. Am. Chem. Soc.,* 2000, **122**, 9692.
134. O. Millo, D. Katz, Y. W. Cao and U. Banin, *Phys. Rev. B: Condens. Matter Mater. Phys.,* 2001, **86**, 5751.
135. O. Millo, D. Katz, Y. W. Cao and U. Banin, *Phys. Status Solidi,* 2001, **224**, 271.
136. R. Xie and X. Peng, *Angew. Chem., Int. Ed.,* 2008, **47**, 7677.
137. J. P. Zimmer, S.-W. Kim, S. Ohnishi, E. Tanaka, J. V. Frangioni and M. G. Bawendi, *J. Am. Chem. Soc.,* 2006, **128**, 2526.
138. H. S. Choi, B. I. Ipe, P. Misra, J. H. Lee, M. G. Bawendi and J. V. Frangioni, *Nano Lett.,* 2009, **9**, 2354.
139. P. M. Allen, W. Liu, V. P. Chauhan, J. Lee, A. Y. Ting, D. Fukumura, R. K. Jain and M. G. Bawendi, *J. Am. Chem. Soc.,* 2010, **132**, 470.
140. J. Zhang and D. Zhang, *Chem. Mater.,* 2010, **22**, 1579.
141. A. Aharoni, T. Mokari, I. Popov and U. Banin, *J. Am. Chem. Soc.,* 2006, **128**, 257.
142. S.-W. Kim, J. P. Zimmer, S. Ohnishi, J. B. Tracy, J. V. Frangioni and M. G. Bawendi, *J. Am. Chem. Soc.,* 2005, **127**, 10526.
143. T. Mokari, A. Aharoni, I. Popov and U. Banin, *Angew. Chem., Int. Ed.,* 2006, **45**, 8001.
144. O. I. Mićić, B. B. Smith and A. J. Nozik, *J. Phys. Chem. B,* 2000, **104**, 12149.
145. S. Haubold, M. Haase, A. Kornowski and H. Weller, *ChemPhysChem,* 2001, **5**, 331.
146. H. Borchert, S. Haubold, M. Haase and H. Weller, *Nano Lett.,* 2002, **2**, 151.
147. L. Langof, L. Fradkin, E. Ehrenfreund, E. Lifshitz, O. I. Micic and A. J. Nozik, *Chem. Phys.,* 2004, **287**, 93.
148. R. Xie, D. Battaglia and X. Peng, *J. Am. Chem. Soc.,* 2007, **129**, 15432.
149. D. J. Bharali, D. W. Lucey, H. Jayakumar, H. E. Pudavar and P. N. Prasad, *J. Am. Chem. Soc.,* 2005, **127**, 11364.
150. K.-T. Yong, H. Ding, I. Roy, W.-C. Law, E. J. Bergey, A. Maitra and P. N. Prasad, *ACS Nano,* 2009, **3**, 502.
151. A. Narayanaswamy, L. F. Feiner and P. J. van der Zaag, *J. Phys. Chem. C,* 2008, **112**, 6775.

152. A. Narayanaswamy, L. F. Feiner, A. Meijerink and P. J. van der Zaag, *ACS Nano,* 2009, **3**, 2539.

153. S. Xu, J. Ziegler and T. Nann, *J. Mater. Chem.,* 2008, **18**, 2653.

154. J. Zeigler, S. Xu, E. Kucur, F. Meister, M. Batentschuk, F. Gindele and T. Nann, *Adv. Mater.,* 2008, **20**, 4068.

155. L. Li and P. Reiss, *J. Am. Chem. Soc.,* 2008, **130**, 11588.

156. K. Huang, R. Demadrille, M. G. Silly, F. Sirotti, P. Reiss and O. Renault, *ACS Nano,* 2010, **4**, 4799.

157. U. T. D. Thuy, P. Reiss and N. Q. Liem, *Appl. Phys. Lett.,* 2010, **97**, 193104.

158. E. Ryu, S. Kim, E. Jang, S. Jun, H. Jang, B. Kim and S.-W. Kim, *Chem. Mater.,* 2009, **21**, 573.

159. L. Li, M. Protière and P. Reiss, *Chem. Mater.,* 2008, **20**, 2621.

160. A. Joshi, M. O. Manasreh, E. A. Davis and B. D. Weaver, *Appl. Phys. Lett.,* 2006, **89**, 111907.

161. M. G. Sandros, M. Behrendt, D. Maysinger and M. Tabrizian, *Adv. Funct. Mater.,* 2007, **17**, 3724.

162. M. C. Hanna, O. I. Mićić, M. J. Seong, S. P. Ahrenkiel, J. M. Nedeljković and A. J. Nozik, *Appl. Phys. Lett.,* 2004, **84**, 780.

163. K. Somaskandan, G. M. Tsoi, L. E. Wenger and S. L. Brock, *J. Mater. Chem.,* 2010, **20**, 375.

164. S.-H. Wei and A. Zunger, *Phys. Rev. B: Condens. Matter Mater. Phys.,* 1997, **55**, 13605.

165. A. C. Bartnik, F. W. Wise, A. Kigel and E. Lifshitz, *Phys. Rev. B: Condens. Matter Mater. Phys.,* 2007, **75**, 245424.

166. A. Sashchiuk, L. Langof, R. Chaim and E. Lifshitz, *J. Cryst. Growth,* 2002, **240**, 431.

167. M. Brumer, A. Kigel, L. Amirav, A. Sashchiuk, O. Solomesch, N. Tessler and E. Lifshitz, *Adv. Funct. Mater.,* 2005, **15**, 1111.

168. E. Lifshitz, M. Brumer, A. Kigel, A. Sashchiuk, M. Bashouti, M. Sirota, E. Galun, Z. Burshtein, A. Q. Le Quang, I. Ledous-Rak and J. Zyss, *J. Phys. Chem. B,* 2006, **110**, 25356.

169. J. W. Stouwdam, J. Shan, F. C. J. M. van Veggel, A. G. Pattantyus-Abraham, J. F. Young and M. Raudsepp, *J. Phys. Chem. C,* 2007, **111**, 1086.

170. D. V. Talapin, H. Yu, E. V. Shevchenko, A. Lobo and C. B. Murray, *J. Phys. Chem. C,* 2007, **111**, 14049.

171. J. Yang, L. Levina, E. H. Sargent and S. O'Kelley, *J. Mater. Chem.,* 2006, **16**, 4025.

172. T. Mokari, S. E. Habas, M. Zhang and P. Yang, *Angew. Chem., Int. Ed.,* 2008, **47**, 5605.

173. J. Zhang, K. Sun, A. Khumbhar and J. Fang, *J. Phys. Chem. C,* 2008, **112**, 5454.

174. J. M. Pietryga, D. J. Werder, D. J. Williams, J. L. Casson, R. D. Schaller, V. I. Klimov and J. A. Hollingsworth, *J. Am. Chem. Soc.,* 2008, **130**, 4879.

175. D. H. Son, S. M. Hughes, Y. Yin and A. P. Alivisatos, *Science,* 2004, **306**, 1009.

176. D. C. Lee, I. Robel, J. M. Pietryga and V. I. Klimov, *J. Am. Chem. Soc.,* 2010, **132**, 9960.

177. H. Zhao, D. Wang, T. Zheng, M. Chaker and D. Ma, *Chem. Commun.,* 2010, **46**, 5301.

178. K. Lambert, B. De Geyter, I. Moreels and Z. Hens, *Chem. Mater.,* 2009, **21**, 778.

179. X. Li, H. Shen, S. Li, J. Z. Niu, H. Wang and L. S. Li, *J. Mater. Chem.,* 2010, **20**, 923.

CHAPTER 6

Ligand Chemistry

6.1 The Functions of Ligands

The capping agent is a key parameter when synthesising nanomaterials using organometallic and inorganic precursors. Depending on your point of view, the capping agent can be thought of primarily as a stabilising agent, providing colloidal stability, hindering uncontrolled growth and agglomeration. This simplistic view can be expanded; the passivating ligands are, as described in Chapter 1, also intimately linked to the nucleation process, tuning the availability of monomers, and are important when considering the growth process. The ligands control the rate of growth, particle morphology, reaction pathways[1] and the particle size distribution.

At a deeper level, the electronic structure of the passivating ligands contribute to the overall electronic and optical profiles of the nanoparticles, blocking surface states and hence affecting emission yields and spectral position, while ligand exchange has also been shown to shift energy levels in quantum dots (QDs).[2] Surfactants may also block catalytic sites, reducing the catalytic ability of some nanomaterials,[3] and are even intimately linked to the particle's magnetic properties.[4,5] Some capping agents are chemically labile, and react with the nanoparticles producing new materials.[6]

When considering biological applications, the solubility of the particles is directly linked to the surface functionalities. Such applications must also take into account the overall particle size and the hydrodynamic diameter of the particle, which is again dictated by the surface decorations and functional groups. How surfactants interact with particles is a key concept and is surprisingly complex. For example, complicated stripe-like patterns on particle surfaces have been both theoretically predicted and allegedly observed in some nanoparticle systems[7] (although some controversy remains

RSC Nanoscience & Nanotechnology No. 33
Semiconductor Quantum Dots: Organometallic and Inorganic Synthesis
By Mark Green
© Mark Green 2014
Published by the Royal Society of Chemistry, www.rsc.org

regarding this hypothesis[8]). Here, we will discuss the key surfactants as determined by their linking functional moiety. It is worth noting that the actual backbone of the ligand offers another feature that can be manipulated, for example in the preparation of superhydrophobic materials by surface exchange with fluorinated ligands.[9] This list is not exhaustive— a wealth of possible ligands exists (including ligands not normally associated with nanoparticle passivation, *e.g.* silanes[10] and azoles,[11] that we do not cover here). The chemistry of the surface is not limited to the addition or removal of ligands; chemical treatment, such as the addition of $NaBH_4$ to a solution of QDs, has been reported to significantly improve the emission quantum yield, by removing the surface ligand and inducing the formation of a stable surface oxide layer.[12]

6.2 Phosphine Oxide-Based Ligands

6.2.1 Tri-*n*-Octylphosphine Oxide (TOPO)

In Chapter 1, the evolution of surfactants for the passivation of nanoparticles was described, through phosphate-based polymers, to monomers and eventually the phosphine oxide-based molecules such as tri-*n*-octylphosphine oxide (TOPO), which became a standard passivating ligand. The advantages of using TOPO as a surfactant, solvent and capping agent include the high boiling point, allowing reactions to proceed routinely at temperatures up to 350 °C, facilitating high-temperature annealing unavailable to aqueous-based routes, and the compatibility with organic solvents allowing a completely inert reaction environment and hence the use of air-sensitive precursors. Once coordinated to the particle, the long alkyl chains impart solubility[13] to a normally insoluble solid-state material and allow it to be manipulated like a common organic reagent (although the solvents used must possess a significantly high dielectric constant to overcome the van der Waals attraction between the colloidal particles). The ligand shell is generally robust in the case of most semiconductors, able to stand several rounds of dispersion/precipitation before losing any degree of solubility as the surfactant is gradually removed (although it is has been found that trialkylphosphine oxides do not bind as strongly to metal particles,[14] highlighting the need to choose specific surfactants for differing nanoparticles). The steric properties of the alkyl groups also further affect particle growth, controlling shape and morphology.[15,16] In semiconducting systems, the ligand also blocks the electronic surface trap sites that are normally responsible for broad emission and non-radiative charge carrier recombination, allowing clean, near band edge luminescence.[17] Surfactants also play a key role in the carrier relaxation between excited intraband states, where soft ligands are responsible for slow relaxation rates.[18]

In the case of TOPO-capped CdSe, the TOPO binds to the surface cadmium sites through the lone pairs of electrons on the phosphine oxide moiety, forming dative bonds,[19,20] sometimes referred to as an L-type ligand.[21] It is reported however, that TOPO does not remain at cadmium sites and can in fact shift to selenium sites upon illumination, forming TOPO–Se complexes that are intimately linked to the photobrightening process where the emission quantum yield of the nanoparticles temporarily increases after synthesis.[22] The interaction of surfactants and the nanoparticle surface is a key parameter, as both optical and structural properties are dominated by the interaction between the interfaces. *Ab initio* calculations have predicted that phosphine oxide species bind preferably to cadmium sites rather than selenium sites, and always through the oxygen molecule (Figure 6.1). TOPO is suggested to bind preferentially to the $(01\bar{1}0)$ and $(11\bar{2}0)$ facets on the short axis ('side') of the particle, with binding energies of 1.23 and 1.37 eV respectively. TOPO binds to the cadmium-terminated $(000\bar{1})$ face ('top') with a binding energy of 0.85 eV, but significantly less tightly to the selenium-terminated (0001) face ('bottom') as demonstrated by the smaller binding energy of 0.63 eV (assuming a relaxed cluster of wurtzite $Cd_{33}Se_{33}$ as a model). As this indicates, both these facets are the axis of growth for anisotropic particles. Phosphonic acid species, present in technical-grade TOPO and often included in reactions to induce anisotropic growth, are also found to bind preferentially through the oxygen atom onto the $(01\bar{1}0)$ and $(11\bar{2}0)$ facets, but through the hydrogen atom on the OH group to the less favoured selenium-terminated (0001) face. Phosphonic acids were also found to bind stronger than TOPO to the nanoparticle surface in all cases, except on the $(11\bar{2}0)$ facet. Later results suggested phosphine oxides had a binding energy of 3.2 eV (313.6 kJ mol^{-1}) with differences being attributed to the partial charge on the P and O atoms being overestimated.[23] As might be

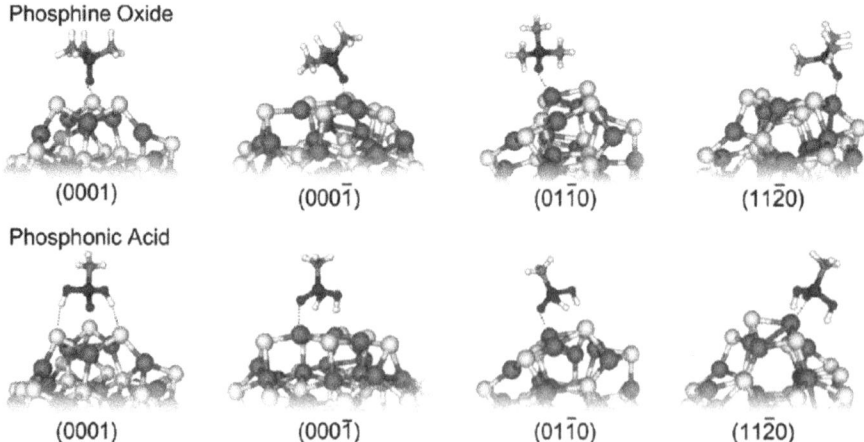

Figure 6.1 Modes of bonding for phosphine oxide and phosphonic acid species onto various crystal facets of CdSe.[25] Reprinted with permission from A. Punzder, A. J. Williams, N. Zaitseva, G. Galli, L. Manna and A. P. Alivisatos, *Nano Lett.*, 2004, **4**, 2361. Copyright 2004 American Chemical Society.

expected, the presence of phosphine-based acids in technical-grade TOPO also results in the cadmium enrichment of CdSe QDs.[24]

Calculations on the binding strength of phosphonic acid-based ligands need to take into account that the species is often deprotonated (as part of the cadmium precursor) when used in reactions. Assuming a single charge increased the binding energy substantially to 3 eV when considering the interaction of the acid with the $(000\bar{1})$ facet, as compared to 1.12 eV for the neutral complex. Despite this significant increase, the binding is comparable on all other facets. When calculations were carried out for the double negative charged acid species, it was found that the ligand bound to all surfaces tightly and even to subsurface cadmium atoms in the selenium-terminated (0001) layer.[25,26]

Phosphinic acids are usually present as impurities in reagent-grade TOPO. Whereas TOPO binds through a dative bond, phosphinic/phosphonic acids bind in an ionic manner to a surface metal site, and such ligands are some-times referred to as X-ligands.[21] A set of ligand transfer experiments by Owen *et al.* confirmed this, and raised the interesting question of whether dative bonds are enough to stabilise a particle.[21] Other impurities in TOPO have been identified and examined (the essential presence of impurities in the role of growing anisotropic particles is described in Chapter 1), notably by Wang,[27] who identified the main impurities in TOPO and their role in the reproducible growth of certain anisotropic structures. Di-*n*-octylphosphine oxide (DOPO) was found to aid in the growth of spherical QDs, di-*n*-octylphosphinic acid (DOPA) and mono-*n*-octylphosphinic acid (MOPA) were found to aid quantum rod growth, and DOPA was also found to aid wire growth. A separate study highlighted the important role played by chain length in phosphinic acid-capped QDs; shorter phosphonic acids resulted in a more branched and elongated rod structure. When mixtures were used, more branching and elongated structures were observed when a high molar fraction of the shorter ligand was employed.[28] Other studies reported that short-chain phosphonic acids stabilised CdSe particles in the zinc blende phase, whereas longer-chain phosphonic acids stabilised CdSe particles in the wurtzite phase.[29] An in-depth examination of CdSe QDs synthesised in technical-grade TOPO, purified using non-solvents, revealed that almost all the datively bound ligands were removed, leaving just *n*-octylphosphonate (OPA), an impurity in technical-grade TOPO and *P'-P'*-(di-*n*-octyl)pyrophosphonate, a self-condensation product of OPA, with small amount of stearate molecules from the cadmium precursor.[30]

6.2.2 Tri-*n*-Octylphosphine (TOP)

It is worth noting that tri-*n*-octylphosphine (TOP) is routinely used both as a surfactant and as a chalcogen delivery solvent when in the form of a trioctylphosphine selenide/sulfide/telluride (TOPSe/TOPS/TOPTe) solution during the preparation of TOPO-capped nanocrystals. The use of tri-alkylphosphine chalcogenides in materials synthesis was pioneered by

Steigerwald, who described the precursors as 'masked atoms'.[31] The general suitability of trialkylphosphine selenides as precursors was originally questioned because the P–chalcogen bond is stronger than in the analogous telluride compunds,[32] which resulted in phosphine selenides requiring further heating to yield selenium in the synthesis of, for example, NiSe.[33] Although TOPTe is also routinely used in the synthesis, of for example, CdTe, it is found to be relatively unreactive when compared to TOPSe. This made synthesis of materials such as ZnTe difficult, unless the TOPTe was reduced by superhydride giving an Te^{2-} intermediate.[34] TOP, when heated to relatively high temperatures (>300 °C) decomposes to yield phosphorus and can therefore be used as a phosphorus precursor,[35,36] as outlined in Chapter 2.

TOP is generally thought to coordinate to the surface of CdSe QDs *via* the selenium sites, supplying a more complete surface passivation.[19] Modelling has also confirmed that TOP preferentially binds to selenium-terminated sites such as the selenium-terminated $(000\bar{1})$, whereas most other ligands prefer the non-polar $(11\bar{2}0)$ surface.[37,38] The role of TOP in the electronic structure of CdSe nanocrystals is not clear; Kalyuzhny *et al.* attributed the deep trap emission observed in CdSe particles to the complex of TOP with the surface selenium sites, leading to a lower energy chemical state,[39] although Jasieniak showed that deep trap emission occurs in both TOP-passivated and TOP-free nanoparticles and that the lower energy states are surface related.[40] The presence of TOP on selenium-rich surfaces has also been shown to be the source of enhanced photoluminescence,[41] and a major contributing factor to the hexagonal crystal character of CdSe particles.[42] A related ligand, tributyl phosphate (TBP), has been shown to remove surface selenium adatoms while passivating selenium dangling bonds.[43] TOPSe is also microwave-absorbing, allowing for the synthesis of QDs under microwave irradiation.[44] TOP also plays a pivotal role in the formation mechanism of QDs, as discussed in Chapter 1.

Few other phosphines are used routinely, other than TBP as discussed in Chapter 1, which appeared to make little difference when used as a replacement for TOP,[45,46] although 1,3,5-triaza-7-phosphaadamantane (PTA) has been used as the reaction solvent for the synthesis of ruthenium and platinum nanoparticles using organometallic precursors.[47] The resulting particles were water-soluble because the PTA coordinated to the particle surface.

6.3   Amines

TOPO is not an ideal solvent, however; heating it above 300 °C is known to induce decomposition,[48] the product of which is unidentified but is known to luminesce (albeit weakly). This decomposition product emission can in some cases be mistaken for semiconductor emission, especially where the parent semiconductor is expected to exhibit a low-emission quantum yield.[49] Long-chain amines are found to be more suitable surfactants for II–VI-based semiconducting systems. Nanoparticles of CdSe prepared using long-chain primary amines are generally found to have emission quantum yields of 60%

without the need for an inorganic shell.[50,51] This has been attributed to the closer packing of the ligands on the nanoparticle surface and the etching of surface defects,[52,53] and amines have also been shown to contribute to the oxygen etching process.[54] The use of amines as capping agents on CdSe particles has also been shown to result in a surface reconstruction,[55] specifically a lattice contraction during growth, which may have contributed to the elevated emission.[56] Further evidence for a ligand-directed reconstructed surface was obtained by measuring the surface strain of CdSe capped with differing capping agents, where TOPO-capped CdSe exhibited tensile stress, whereas amine-capped CdSe exhibited compressive stress.[57]

Mulvaney examined the role of primary, secondary and tertiary amines on the emission quantum yield of CdSe particles, and noted that primary amine strongly enhanced the emission whereas secondary and tertiary amines had a negligible effect.[58] The primary amines have, theoretically, close to 100% surface coverage when assuming a footprint of 0.25 nm, as compared to a theoretical maximum of 30% with TOPO. The addition of primary amines to a solution of CdSe was also found to blue-shift the emission position, which was attributed to an electronic contribution from the amine but may also have an origin in the etching observed by Woo where the absorbance edge was also found to blue-shift and was assigned to a slight reduction in particle size.[52] Notably, the addition of amides has no effect on the emission intensity, attributed to the poor electron-donating properties. Even post-treatment of TOPO-capped CdSe particles with primary amines improved the emission quantum yield by an order of magnitude to 50%, although this was accompanied by a slight blue shift in the absorption spectra consistent with a small decrease in particle size. (This was confirmed by Foos *et al.* who also reported that secondary amines slowed the growth rate of CdSe QDs.[59]) Interestingly, other reports state that addition of amines quenches QD emission,[60,61] although work by Munro *et al.* has highlighted an interesting concentration dependence, with a bright point at a certain concentration, either side of which resulted in quenched emission, thus possibly explaining the contradictory reports.[62] It was suggested that the enhancement of the emission was not necessarily based on improved surface passivation, rather a surface reconstruction, as described earlier. Primary amines have also been shown to remove surface cadmium adatoms in CdSe QDs, and passivate cadmium dangling bonds.[43] The role of amines in the synthesis of III–V QDs is unclear. Initially, amines were thought to be activators for the reaction, notably activating the phosphorus precursor;[63,64] however, later results have questioned that particular assignment, suggesting that amines actually inhibit the reaction.[65] Yu *et al.* have suggested that the linear shape of some capping agents (amines) allows oxygen to reach the particle surface, oxidising the surface and decreasing the emission quantum yield, whereas capping agents with branched structures protect the surface in more satisfactory manner.[66] Notably, oleylamine (OAm) has also been identified as a strong reducing agent as well as a capping agent in this synthesis of Fe_3O_4, an excellent replacement for the usual diol type of synthesis.[67]

A detailed study of the dynamics of amines on CdSe nanoparticles identified several key features; the chain length of the ligand, which determined
ligand–ligand interactions on the particle surface, and notably the boiling
and melting point of the ligand. Reaction temperatures above the boiling
point of the amine resulted in a quasi-gas state in which the reaction proceeded too quickly, and too low a reaction temperature resulted in growth
that was too slow, meaning shortening of the hydrocarbon chain could result
in the reduction of the reaction temperature while still yielding high-quality
crystals.[68] Modelling of the coordination of amines on to a QD surface has
highlighted that the particle can accommodate more than one ligand per
surface cadmium ion, suggesting what is referred to as indirect adsorption,
where a second ligand would bind to an already occupied cadmium ion or
hydrogen bond with a less favourable site such as selenium.[23] This over-
loading has been suggested to be a cause of the concentration-induced
changes in optical properties.

Ab initio calculations suggest that the amine mode of bonding is through
the lone pair of electrons on the nitrogen atom, and in the case of CdSe,
the amine binds to both selenium sites and cadmium sites. Surprisingly,
amines are expected to bind to selenium sites preferentially over cadmium
sites (binding energy of 1.05 eV for the selenium-terminated (0001) face as
compared to 0.91 eV for the cadmium-terminated (000$\bar{1}$) face).[23,25] There is
an argument that the acidic protons in the NH_2 group on a primary amine
coordinate to surface metal sites.[69] The use of tertiary amines in nanoparticle synthesis appears to contradict this,[70,71] and it is possible that the
coordination of amines to nanoparticle surfaces may actually occur
through both the lone pair of electrons and protons. Fourier transform
infrared (FTIR) spectra on OAm-capped FePt particles confirm the presence
of an N–H feature and OAm appeared to bind through electron donation
from the nitrogen atom.[72] Studies using nuclear magnetic resonance (NMR)
have determined the equilibrium constant for octylamine (OA) and CdSe,
and also the kinetic constants for the desorption and adsorption
processes.[73] Interestingly, although amine-based alternatives to TOPO
are common, there are few alternative to the TOP-based system even
though 'phosphine-free' routes are becoming more popular. There are
several reasons for this. Primarily, TOP is used as a delivery system for
chalcogenide precursors, and finding a liquid alternative that dissolves
elemental chalcogens is difficult. Octadecene (ODE) is effective, although it
appears to need heating with the chalcogen, and prolonged heating is
often detrimental,[40] and ionic liquids have been used although the final
product was not as luminescent as materials prepared with TOP.[74] An
amine-related alternative, *N,N'*-dimethyl-oleyl amide, was found to dissolve
selenium easily and was used in the synthesis of CdSe[75] and ZnSe[76] in
a phosphine-free synthesis. The other notable TOP replacement was hex-
apropylphosphorus triamide used during the synthesis of CdTe particles[77]
as described in Chapter 1, although this is clearly less 'green' than
N,N'-dimethyl-oleyl amide.

Other nitrogen-based ligands have also been used as surfactants, such as pentadecanenitrile, which has been used in place of OAm in the preparation of FePt nanoparticles using standard precursors. Interestingly, the reaction provided a bimodal size distribution, and it was hypothesised that the nitrile functionality bonded to surface platinum sites.[78]

6.4 Thiols

Long-chain thiols are common surfactants and appear to be efficient capping agents for most semiconducting and metal nanoparticles. The origin of the popular use of thiols as passivating agents can be traced back to the work of Brust on the preparation of gold nanoparticles;[79,80] he chose the surfactant with reference to the well-known interaction of gold and sulfur.[81] Mićić published the first useful reports of thiol-stabilised CdTe particles[82] and described an in-depth analysis. Studies have been carried out investigating the potential use of related long-chain selenium and tellurium analogues such as dodecyl diselenide and dodecyl ditelluride as ligands for passivating gold nanoparticles (selenols and tellurols are relatively air sensitive and hence not considered good candidates for passivating ligands).[83] The use of selenium- and tellurium-based ligands is of interest because of the extended orbitals relative to the sulfur, possibly resulting in unusual electronic coupling. However, alkaneselenide-passivated gold particles were found to be slightly less stable than the almost indefinitely stable thiolate-passivated materials, and the alkanetelluride-capped particles were clearly unstable, with particles precipitating from solution in a matter of days. Medintz has highlighted that monodentate thiol-based capping agents used in biological applications of QDs coordinate through dative thiol bonds and are only stable for days. Bidentate thiols are, however, much more stable, on the order of years rather than days, and are discussed later.[84] Thiols have been found to have a binding energy of 0.36 eV (34.7 kJ mol^{-1}) whereas thiolates have a binding energy of 13.2 eV (1283 kJ mol^{-1}).[23]

Thiols are therefore generally the surfactant of choice for simple routes to inert metal nanoparticles that do not utilise organometallic precursors, although the use of thiols in organometallic-based routes is also becoming more popular. The actual modes of bonding with the gold surface in particles prepared by the relatively mild Brust route has been investigated by density functional theory and this suggests a gold–sulfur alloy is formed at the surface.[85,86] NMR investigations into thiol-capped gold nanoparticles suggest that the thiol attaches *via* the sulfur atom, and can either keep the sulfur-bound hydrogen or lose it (forming a thiolate species) depending upon reaction conditions.[87] Computational modelling of thiols on QD surfaces confirms that thiolated species bond to surface metal (*i.e.* zinc) and also to surface sulfur atoms *via* a much weaker bond.[88] Thiols are, however, extremely reactive and have been known to react with the metal particles in high-temperature routes, yielding metal sulfide particles.[6] This reactivity of the thiol has been utilised efficiently in aqueously prepared CdTe particles, where the thiol group hydrolyses, reportedly inducing a thin CdS buffer layer on the CdTe particle and effectively resulting in a wide-bandgap protective shell.[89] This is a common explanation given for the strong emission from thiol-capped CdTe, but it fails to explain why CdSe particles capped with thiols usually display low-emission quantum yields[90] despite the fact we have already described CdSe/CdS core/shell materials as highly luminescent. Clearly, the electronic structure of the thiol, the position of the energy levels and their interaction with charge carriers plays a significant role.[91] Almost all reports of thiols as capping agents utilise ligands with the thiol groups at the end of the chain, although the use of isomers where the thiol group is elsewhere in the molecule has been shown to have advantages. For example, the use of 2-mercaptopropionic acid in the synthesis of CdS QDs has been shown to increase colloidal stability and increase photoluminescence relative to similar materials formed using 3-mercaptopropionic acid as a stabiliser, due to secondary carboxylate stabilisation and the steric crowding of the methyl group.[92]

The exposure of preformed CdSe particles prepared by organometallic routes to thiols resulted in a complicated and usually detrimental interaction, notably quenching the luminescence and reducing stability in solution.[93] A notable study on CdSe QDs with octadecanethiol has demonstrated that a single thiol molecule quenched the emission by 50%.[94] A red shift in the emission peak has also been observed upon thiol addition, attributed to an electronic contribution from the adsorbed ligands.[58] By way of comparison, CdSe particles prepared by aqueous routes using thiols alone as stabilisers have extremely low quantum yields, highlighting the unsuitability of thiols as surfactants for CdSe.[95] Aldana has described in depth the instability of aqueously dispersed thiol-capped CdSe in the presence of oxygen and UV light,[96] reporting several distinct stages of photoinstability. The first phase was the photo-oxidation of the ligand, catalysed by the nanocrystals, yielding disulfides. Where the disulfides were water-soluble, the nanoparticles precipitated out of solution as they lost the capping agent. In the presence of excess free thiols in solution, the disulfides were expelled from the ligand shell by free ligands in solution and resulted in the nanoparticles remaining in solution a while longer. Where the disulfides were insoluble in water a micelle formed within which the nanoparticle oxidised, became smaller and precipitated out when the micelle became too unstable to support the particle.

The introduction of small thiol molecules to polymer-coated CdSe/ZnS core/shell particles also displayed a remarkable suppression of fluorescence intermittency (blinking).[97,98] This was attributed to the electron-rich species donating electrons to surface trapping states, reducing the traps ability to accept electrons from the nanoparticle and reducing the rate of blinking. This inspired a further study to determine the nature of QD/thiol interactions, where polymer-coated CdSe/ZnS particles were exposed to β-mercaptoethanol (BME).[99] It was observed that after exposure of QDs to low concentrations of the thiol, the emission quantum yield of the particles almost doubled.

For distinctly low concentrations, the effect was slow to manifest, but remained stable for over 12 days once observed. Large concentrations of thiol resulted in quenching of the photoluminescence in less than 1 day. The initial increase of the photoluminescence was attributed to the thiol reducing the number of electron traps, while the higher concentration resulted in the formation of hole traps. To further investigate the identification of the active species, pH dependency studies of a set concentration of BME were undertaken. At high pH, the thiol primarily dissociated into the thiolate, which was found to decrease photoluminescence quantum yields, again attributed to the presence of hole traps, whereas at low pH, exposure of the protonated species to the nanoparticle was found to increase emission quantum yield. Control experiments, exposing the nanoparticles and thiols to environments where thiolates could not form, resulted in no change in optical properties. These results suggest that thiolates and not thiols are the active species responsible for both positive and negative effects of thiol-related species. Therefore it is more accurate to say that at low concentrations, thiolates provide electrons for the trapping sites, while at higher concentrations, thiolates act as hole traps. Interestingly, the transfer of a hole to a thiolate ion results in the formation of a thiyl radical, RS, which could react with another thiyl radical to form RSSR, the disulfide observed in thiol-capped CdSe. The thiolate-trapped hole can also recombine with an electron from the nanocrystal, giving deep trap emission, or decay non-radiatively, either way reducing band edge emission. The charge transfer and blocking of surface sites was confirmed by theoretical work[88] and further experimentation with mercaptopropanoic acid[100] which concurred that thiols can be used without reducing the emission quantum yield if the thiol concentration is optimised. In related work, the quenching of the band edge emission of phosphine/amine-capped CdSe by 3-mercaptopropionic acid has also been accompanied by an enhancement of trap emission of up to 5%, but substitution of the amine with the thiol created selenium vacancies.[101] Interestingly, this phenomenon was not observed in QDs capped solely with phosphine/phosphine oxide.

Aldana has also shown that thiolate-capped nanoparticles of CdE (E = S, Se, Te) undergo ligand dissociation from the particle surface when exposed to relatively low pH systems, precipitating (undamaged) particles. The particles were redispersed in solutions with a significantly higher pH, displaying a distinct hysteresis curve, attributed to the differing processes governing precipitation and redispersion, namely ligand protonation and deprotonation.[102] This process was found to be size (and hence bandgap) dependent, showing the Gibbs reaction energy ($\Delta_r G_o$) for the formation of surface-ligand bonds for materials of differing diameters. This size/binding energy relationship has, however, been suggested to be coincidental, and the changing binding energies attributed to the differing crystal facets on the nanoparticle surface.[103] In related work, a thiolated spin trap has been coordinated to TOPO-capped CdSe QDs, and the nature of disulfide bond cleavage investigated.[104]

The role of thiols in other systems, such as the III–V materials is much more beneficial. As-synthesised InP QDs capped with TOPO have a low quantum yield (<1% at room temperature), which can be greatly improved by treatment with octanethiol, increasing to *ca.* 30%.[105]

6.5   Carboxylic Acids

The use of carboxylic acids as capping agents and surfactants in colloids predates the organometallic route by over 20 years; oleic acid was used as a capping agent for Fe_3O_4 as far back as 1971.[106] Oleic acid, $CH_3(CH_2)_7CH=CH(CH_2)_7COOH$, is the standard carboxylic acid to be used as a surfactant, the double bond and associated 'kink' in the alkyl chain is found to be an essential feature for imparting colloidal stability;[107] the closely related stearic acid does not stabilise magnetic colloids, but is often used in the capping of QDs. Long-chain fatty acids have also been described as inhibitors in nanoparticles reactions, slowing down the synthesis.[108,109] What is notable about the use of carboxylic acids is the fact that they are Lewis acids, unlike the majority of other capping agents which are generally Lewis bases. One might arguably state that oleic acid was first used as a stabiliser in an organometallic route in the preparation of an iron colloid, where a solution of $Fe(CO)_5$ was sonicated with oleic acid, giving particles of iron *ca.* 8 nm in diameter with a narrow size distribution (notably narrower than when the polymer polyvinylpyrolidone (PVP) was used instead).[110] Oleic acid was then used in the seminal paper describing the preparation of cobalt nanoparticles, where it was found to be an extremely tightly binding ligand, and required the use of a weaker binding co-ligand (an alkylphosphine) to control the particle size.[111] The use of oleic acid alone inhibited particle growth. X-ray photoelectron spectroscopy (XPS) and Fourier transform infrared (FTIR) spectroscopy studies concluded that oleic acid chemisorbed to cobalt nanoparticles in a symmetrical bidentate manner onto a single cobalt atom, forming Co–O bonds.[112] In a similar study on oleic acid-capped FePt particles, two bonding modes were observed; monodentate (RC(=O)–O–M) and bidentate (RCO_2–M).[72] A change in the ligand stereochemistry, from the oleyl (*cis*) to the elaidyl (*trans*) isomer, was also observed, attributed to the high-temperature synthesis, and is important when self-assembly of the nano-particles is considered. Interestingly, oleic acid appears to be able, in some cases, to coordinate to a nanoparticle surface through the double bond, making the nanoparticle soluble in polar rather than non-polar solvents, and allowing phase-transfer reaction without changing the surfactant.[113] This feature, having an interchangeable binding point, exposing either the alkyl group or the alkyl group and carboxylic acid group, makes oleic acid extremely unusual among standard capping agents. The formation of bila-yers of oleic acid on a particle surface has been reported, where the ligand has been used to cap the particles, and use as a phase-transfer reagent has shown that the ligand is extremely versatile and can be used effectively with magnetic and luminescent materials, and is, to date almost unique in its use in both polar and non-polar solvents.[114] Since oleic acid in particular has been extensively used in magnetic nanoparticle synthesis, and materials such as oleic acid-capped FePt need to be heated to 560 °C to be converted to the face-centred cubic phase for useful applications, the thermal behaviour of these ligands at such temperatures has been explored, identifying that the excess secondary surface ligands desorb from the particle surface at *ca.* 200 °C, and dehydrogenates at 400 °C yielding a graphitic layer with oleic acid alkyl fragments underneath.[115]

A particularly useful application of carboxylic acids is in the phosphine-free synthesis of QDs, as mentioned in Chapter 1. In a typical example,[40] the synthesis of CdSe was achieved by dissolving the cadmium precursor (CdO) in a solution of oleic acid and ODE, followed by the injection of ODE/Se. It was found that TOP was not essential for the production of high-quality CdSe, although the use of ODE/Se instead of TOPSe resulted in poorer nucleation, producing approximately half as many particles. It does, however, allow the study of the nature of the selenium precursor, which is thought to convert from the polymeric elemental form to various allotropes over 2 hours of heating in ODE, unlike TOPSe, in which the selenium is thought to exist as the elemental monomer. Excessive heating of ODE/Se (>4 hours) results in a polymeric species unsuitable for QD synthesis (the inclusion of a phosphonic acid in the synthesis allowed the formation of the wurtzite-structured crystalline form rather than the zinc blende-structured, oleic acid-capped CdSe). Little difference was found in the optical quality between the QDs prepared using TOP and those prepared using just oleic acid. It is also worth noting than trap emission was observed in both synthetic pathways, irrespective of whether TOP was used. Other methods of producing phosphine-free QDs includes the use of olive oil as a solvent and oleic acid as a capping agent, as outlined in Chapter 1, giving high-quality QDs of CdSe. Olive oil is known to contain a high proportion of triglyceride esters of oleic acid, so might be thought of as a trimer of oleic acid. The kinetics of QD growth using carboxylic acids have also been reported.[116,117] Related to carboxylic acids is the use of high boiling point esters and ketones, such as hexadecyl hexadecanoate and benzophenone, as replacements for TOPO.[118] Notably, the initial nuclei growth and particle growth rate at high temperatures were hindered in hexadecyl hexadecanoate, allowing potentially smaller nanoparticles with a narrower size distribution.

6.6 Surfactant Exchange

The key aspect of most surfactants is the dynamic character associated with ligands bound to a nanoparticle surface. Most surfactants are interchangeable, importantly allowing the exchange of pendant functionalities or the switch from organic to aqueous phases, or *vice versa*. Most surfactant exchange reactions are driven by mass action where the exposure of the QDs to a large excess of the new surfactant drives the exchange. A study of some typical phase-transfer protocols to prepare QDs for biological applications and exposure reaction has been reported by Smith *et al.*, who highlighted the problems associated with such systems.[119]

With the development of the original synthetic method came the first ligand-exchange reactions. Murray elegantly described the use of pyridine as a labile intermediate when replacing TOPO with other Lewis base species such as TBP and tributyl phosphine oxide (TBPO), pyrazine and tricetyl-phosphate.[120] The use of pyrazine, a molecule with two ring nitrogens, induced bridging of the nanoparticles, resulted in flocculation. Heating pyridine-capped particles under vacuum reportedly removed all capping agents, yielding particles with no passivation at all. Pyridine was also used as an intermediate for the deposition of an inorganic shell. In early work describing the synthesis of CdSe/CdS core/shell particles, the core particles had the TOPO ligands replaced with pyridine prior to the epitaxial growth of the CdS shell,[121] although this procedure appears to be extraneous and the deposition of the shell can be achieved with the original ligands. This, however, highlights that the coordination of ligands to a nanoparticle surface is clearly a dynamic process, in which ligands absorb and desorb, allowing further interactions such as further shell growth. Later work demonstrated that up to 80% of surface ligands could be removed from oleic acid-capped CdSe by repeated exposure to pyridine, although solar cells made from the resulting QDs decreased in efficiency due to the resulting increased number of trap states and aggregation.[122] Murray also alluded to the use of Lewis acid species, such as tributylborane and trioctylaluminium, as capping agents, although further details were not reported.[120] A more in-depth report investigating surface exchange reactions with 4-picoline, 4-(trifluromethyl)thiophenol and tris(2-ethylhexyl)phosphate revealed that despite rigorous processing, 10–15% of the original capping agent remained on the particle surface after surfactant exchange.[123] To be an ideal phase-transfer agent, the resulting transferred materials should exhibit colloidal stability in the chosen solvent (water for biological applications, over a wide range of pH and over an extended time period), maintain its inherent optical or magnetic properties, maintain its original size, provide opportunities for further functionalisation, and in the case of biological applications, exhibit non-specific binding and avoid unwanted side reactions.

Not all exchange reactions are mass action driven; surface-ligand substitution can be achieved by reaction-based exchange.[21] For example, addition of bis(trimethylsilyl)selenide/sulfide to alkylphosphonate-passivated CdSe/ZnSe resulted in the elimination of *O,O*-bis(trimethylsilyl)octadecylphosphonic acid and the precipitation of the nanoparticles as the new surface cap failed to stabilise them. Adding related reagents such as *S*-trimethylsilyl-2,5,8,11-tetraoxatridecane-13-thiol resulted in the cap exchange where the phosphonate was replaced with the thiolate, giving QDs passivated with a long polar chain making them soluble in polar solvents, although the emission was quenched. Similarly, addition of chlorotrimethylsilane and tridecyltrimethylammonium chloride resulted in chlorine-terminated QDs, although the emission was quenched only slightly. The use of silylated chalcogenide compounds as reactive ligand-exchange reagents has been expanded, resulting in the capping of QDs and metal oxides with more usual ligands such as long-chain mercaptocarboxylic acids.[124]

Here we cover some exchange reactions based on the linking molecules, the majority of which are explored with specific applications in mind.

6.6.1 Thiol-Based Surfactant Exchange

Surfactant exchange has been used to produce designer structures; the simple labile capping agents such as TOPO can be removed and replaced with ligands that provide functionalities that may undergo further reactions. In one of the earliest reports of this type, TOPO-capped CdSe particles were subject to surfactant exchange for a thiol-based ligand, *N*-methyl-4-sulfanylbenzamide (Figure 6.2). The thiol-capped particles were then dissolved in methanol, and cross-linked with bis(acyl hydrazide). After removal of large aggregates and size-selective precipitation, dimers, trimers and multimers of nanoparticles were clearly observed by electron microscopy. Further purification resulted in the removal of the larger linked particles, leaving the dimeric structures.[125] Similar work has been carried out on CdTe particles.[126]

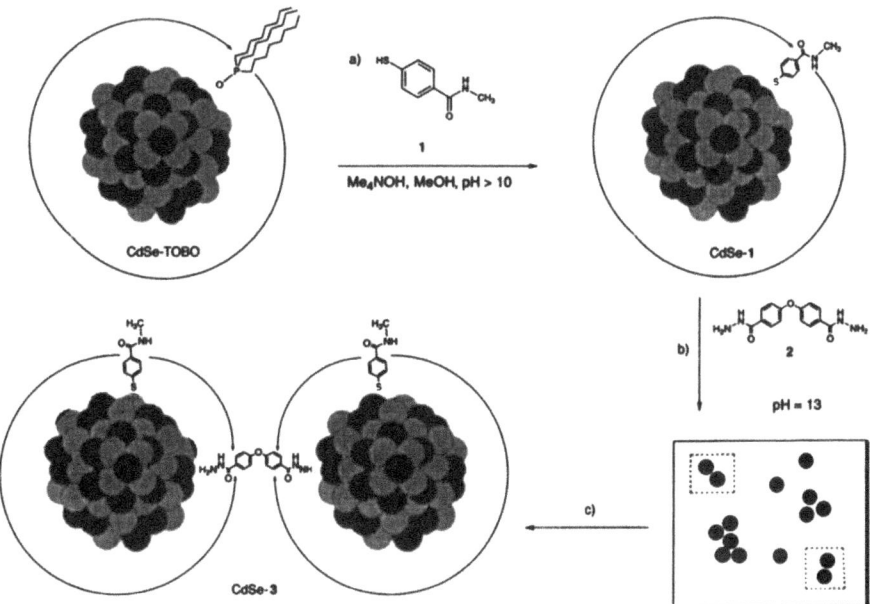

Figure 6.2 Diagram of the preparation of homodimers of CdSe particles.[125] Reproduced with permission from X. Peng, T. E. Wilson, A. P. Alivisatos and P. G. Schultz, Synthesis and Isolatin of a Homodimer of Cadmium Selenide Nanocrystals, *Angew. Chem., Int. Ed. Engl.*, 1997, **36**, 145. Copyright Wiley-VCH Verlag GmbH & Co. KGaA.

The foremost benefit of surfactant exchange is the ability to prepare particles that are suitable for differing environments and applications. The obvious example is the use of semiconductor nanoparticles in biological labelling, where particles are required to be water-soluble. Nie described the simple reaction of TOPO-capped CdSe/ZnS nanoparticles and mercaptoacetic acid in a chloroform solution. The thiol group coordinated to the nanoparticle surface and the deprotonated carboxylic acid group imparted water solubility. The resulting mercaptoacetic acid-capped particles were then isolated by extraction into water.[127] This simple reaction was one of the seminal phase-transfer methods for preparing QDs for use in biology. The particles exhibited significantly reduced emission, and further investigations into the size and charge of thiols have shown that smaller thiols increase emission quenching.[128] The availability of the pendant carboxylic acid functional group provided a simple anchor point for further functionalities to be attached to the luminescent particle, a common starting point for the synthesis of more complex biological labels.[129] It is worth noting that the coupling of a carboxylic acid group with an amine containing biological entity is a simple way of functionalising QDs with biological vectors, although the common use of 1-ethyl-3-(3-dimethylaminopropyl)carbodiimide (EDC) as coupling reagent is sometimes problematic, occasionally resulting in the irreversible precipitation of the particles due to the interaction of the dimethylamine functionality with the nanoparticles. The use of polyethylene glycol (PEG) carbodiimide coupling agents is an elegant way of circumventing this problem, as the particles are stable even at high loading of the reagent, and can even be used as a method of adding PEG to a nanoparticle surface.[130] PEG has become arguably one of the most useful molecules available to nanoparticle chemists interested in biological applications, as materials functionalised with PEG are found to be non-antigenic, non-immunogenic and protein resistant;[131] and, importantly, nanoparticles capped with PEG show limited non-specific binding[132] (a major problem with early labelling studies), prolonged circulation times in biological systems and reduced toxicity.[133]

In related work mercaptoundecanoic acid-capped particles have had the terminal carboxylic acid groups cross-linked with lysine, giving a stable hydrophilic shell composed of carboxylic acid and amine functional groups available for further conjugation.[134] In this way, the thiolated molecule can be thought of as a bridge, linking the particle to further entities such as proteins[135] or antibodies,[136] which are usually capable of further interaction with the biological material of interest. It is worth noting that pendant carboxylic acid groups are not exclusively used; thiols with protonated amine groups have also been used to transfer TOPO-capped CdTe to water.[137]

An important part of bioconjugation is rendering the particle hydrophilic while maintaining the smallest possible hydrodynamic diameter; Liu *et al.* have prepared CdSe/ZnCdS QDs which were made water-soluble by phase-transfer using cysteine, a thiol-based amino acid zwitterion. Interestingly, the resulting material possessed a quantum yield of 40%, due to the effective passivation of the alloy ZnCdS shell (materials with a ZnS shell had quantum yields of only 13%). The cysteine-capped core/shell QDs had an inorganic core of 3.6 nm as determined by transmission electron microscopy (TEM), and a hydrodynamic diameter of only 5. 9 nm, notably small for a QD biolabel.[138] The preparation of such cysteine-capped QDs has provided particles small enough to be cleared through the renal system, due to the small size (ideally <5.5 nm) and zwitterionic charge which prevented serum protein adsorption.[139] Interestingly, the cysteine-capped particles required the addition of dithiothreitol (DTT) to prevent the dimerisation of the capping agent; DTT has itself been used as a capping agent[140] and its role as an additive in this case was not discussed. Cysteine-capped QDs have also been further conjugated with dye molecules and small-molecule targeting ligands for tumour imaging while maintaining a small enough hydrodynamic diameter to be cleared. This required the coordination of only 5–10 targeting molecules per QD.[141]

Alivisatos also reported the use of thiols in the preparations of water-soluble QDs for labelling applications. In this work, (3-mercaptopropyl)trimethoxysilane was coordinated to the particle surface as an anchor point for the growth of a silica shell.[142] The silica shell was added in a long procedure based on Mulvaney's seminal work on silica shell deposition on various nanoparticles, notably gold.[143] In this case, the thiol surfactant can be considered as a precursor for the final shell of SiO_2, which could be further decorated with various functional groups.[144,145] The emission of silica-passivated QDs was partially quenched, although quantum yields of up to 18% were reported. The use of silica-capped CdSe/ZnS particle was one of the first nanomaterials used for biolabelling applications although this has generally been surpassed by simpler, more reproducible methods. Related routes to siloxane-capped QDs have, however, been developed for use in electroluminescent devices.[146]

Thiolated PEG[147] has become a notable ligand used in nanoparticle biolabelling, as attachment of the long-chain PEG molecule results in excellent water solubility. Functionalised PEG groups can also be used as precursors for further structures: a notable study includes the use of nitriloacetic acid-based ligands which incorporated a nickel complex. The nickel complex was found to bind to histidine, and used to visualise 5HT2C serotonin receptors. In this case, the particles were initially capped with PEG with terminal amine groups, which were then cross-linked to the nitriloacetic acid ligand.[148]

In most cases, the monodentate thiol ligands impart only a temporal stability to the nanoparticle dispersion, in some cases lasting only a matter of hours or days.[72,119] Dithiol and multidentate-based ligands extend the stability in water to, in some cases, years.[149] Notably, one of the earliest examples using dithiols was the phase transfer of CdSe/ZnS QDs using dithiothreitol (DTT), which was followed by linking oligonucleotides to the ligands which were then used in fluorescent *in situ* hybridization (FISH) experiments.[140] The DTT-capped QDs were reportedly stable for 3–4 weeks, substantially longer than monodentate thiol-capped QDs. This simple surfactant exchange has been improved by the use of dihydrolipoic acid (DHLA), a bidentate thiol with a carboxylic acid group that is more stable than the monodentate ligand. Particles of CdSe/ZnS capped with the bidentate ligand were found to retain quantum yields of up to 20% in water.[150-152] The one drawback is the agglomeration of DHLA-capped nanoparticles in acidic media (due to the carboxylic acid group requiring deprotonation by a base to impart solubility in water). This has been overcome by the use of engineered ligands, based on a DHLA anchoring group linked to a hydrophilic PEG unit (see Figure 6.3A),[153,154] which can also be followed by a terminal functional group for further bioconjugation.[149,155,156] QDs passivated with such ligands have been successfully used in cellular imaging applications and displayed stability over a range of pH values and ionic concentrations. As expected with thiol ligand substitution reactions, a drop in quantum yield from *ca.* 70% to *ca.* 30% was routinely observed. The colloidal stability of QDs could be impressively enhanced even further, over a wider range of pHs (although quenching was observed after just a few hours at pH values <2) for several months by the use of bis(DHLA)PEGylated ligands which provided four thiol anchoring points per PEG chain for coordination to the particle surface.[157] QDs capped with the bis(DHLA) ligands could be conjugated to peptides and used in cell imaging studies.

Earlier studies which showed the use of zwitterions (such as cysteine) as effective capping agents yielding particles with small hydrodynamic diameters[138] inspired the use of DHLA-based sulfobetaine zwitterions as

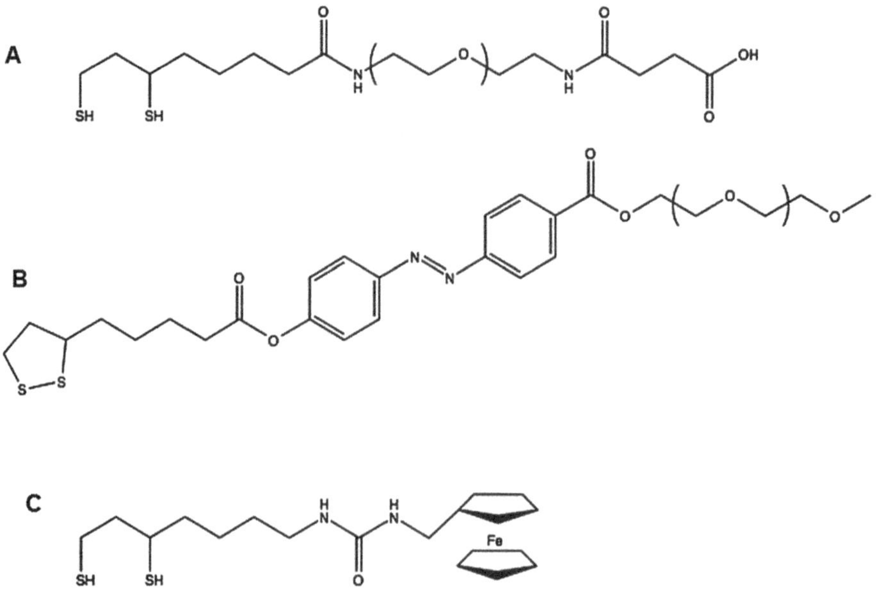

Figure 6.3 A range of dithiol-based ligands: for biological applications (A), as a photoreactive switch (B) and as a ferrocene-based photoswitch (C).

phase-transfer agents, and the resulting phase-transferred QDs were stable for months (unlike cysteine-capped materials) but increased the hydrodynamic diameter to a larger degree than cysteine.[158] Quantum yields of up to 70% using multishell QDs were obtained, and significantly, the resulting materials were insensitive to solution pH and the materials could be used routinely in cell imaging applications with non-specific adhesion, a problem normally addressed by the use of PEG-based ligands.

The use of dithiol-based ligands is not restricted to biological applications. The addition of an azobenzene chromophore to the PEG group (Figure 6.3B) on a dithiol-based ligand, which switched conformation from *trans* to *cis* upon activation with UV light and could be reversed by thermal reisomerisation has led to a simple photoswitch. The polymeric analogue of the ligand (Figure 6.4) was prepared, and when attached to the QDs, significantly reduced the photoluminescence quantum yield when transformed to the *cis* isomer, which was again reversed upon restoration of the configuration.[159] Similar ligands, which shift absorption characteristics at differing pH values, have been coordinated to QD surfaces by dithiol groups. The emission from the QDs can be either reversibly increased by electron transfer or quenched across a pH range of 3–11.[160,161] The concept of a photoswitch using a QD as the signal has been extended further, using dithiol ligands containing a ferrocene unit (Figure 6.3C). Addition of the ligand to CdSe/ZnS QDs quenched emission, which could be restored by addition of fluorine ions that facilitated photoinduced electron transfer.[162] Dithiol-based linking molecules connected to a silica substrate through a terminal siloxy group have also been used to tether QDs to an optical fibre.[163]

Other molecular species with two sulfur groups have also been explored as phase-transfer agents; dithiocarbamates, prepared *in situ* by addition of the required amine and carbon disulfide to a QD solution, have been shown to be effective surfactant exchange reagents.[164] QDs with multiple shells, such as CdSe/CdS/CdZnS/ZnS which had the initial hydrophobic shell exchanged for a glycine-based dithiocarbamate, exhibited a bathochromic shift in emission and absorption spectra, along with a substantial drop in emission quantum yield of *ca.* 40%. A very similar reaction was reported using simple TOPO-capped CdSe QDs, where the initial quantum yield of *ca.* 3% was, unusually, found to increase to 15% upon phase transfer using a glycine-based dithiocarbamate; this was attributed to the improved surface passivation.[165] The addition of phenyldithiocarbamate to a pre-existing solution of TOPO-capped CdSe QDs has been found to shift the optical bandgap by up to 220 meV, which was attributed to relaxation of the exciton hole into the ligand shell.[166] Addition of *N*-2,4,6-trimethylphenyl-*N*-methyldithiocarbamate at room temperature to a solution of oleic acid-capped PbS QDs resulted in ligand exchange, blue-shifting and broadening the absorption profile due to surface etching. The resulting particles were found to be more resistant to oxidation, allowing the fabrication of efficient solar cells using a simple solution process.[167] The phase transfer of Fe_3O_4 nanoparticles to water (for MRI applications) using dimercaptosuccinic acid, a ligand with two thiol

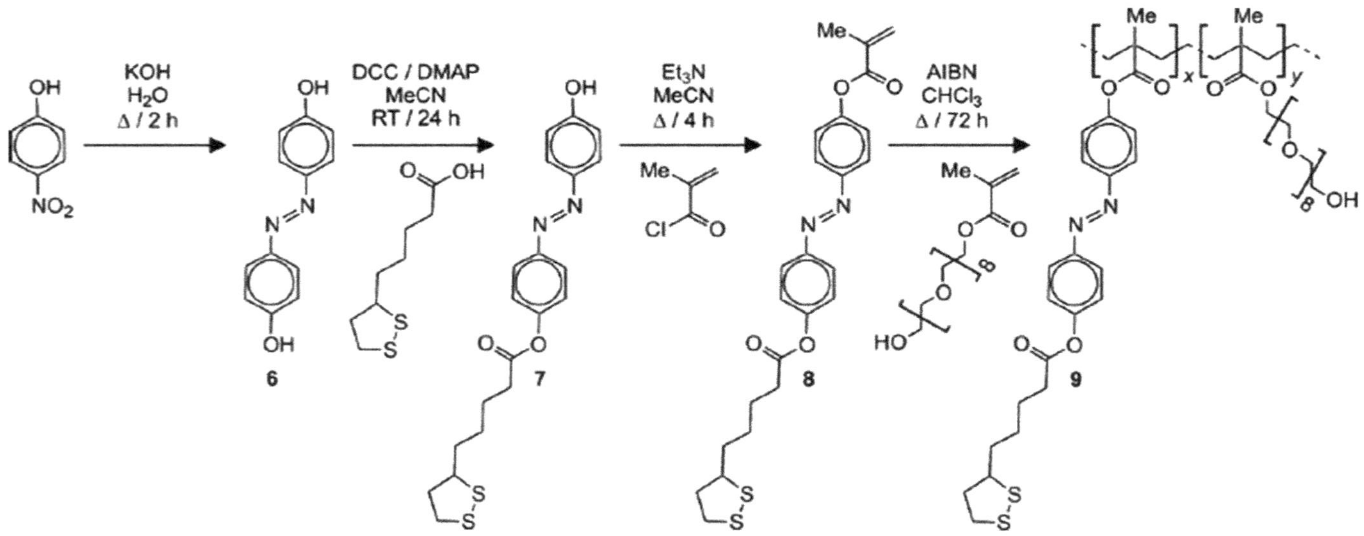

Figure 6.4 Preparation of a polymeric dithio-ligand. Reprinted with permission from I. Yildiz, S. Ray, T. Benelli and F. M. Raymo, *J. Mater. Chem.*, 2008, **18**, 3940. Copyright 2008 The Royal Society of Chemistry.

groups and two carboxylic acid groups, has been reported. In this case, the ligand bound through the carboxylic acid groups leaving the thiol groups pendant in solution and available for bioconjugation.[168]

Despite being initially designed as ligands for bioimaging, the inherent stability and ease of manipulation of thiol- and dithiol-based ligands has resulted in their application in other areas. The treatment of QD surfaces with thiols usually (as mentioned above) leads to at least some degree of emission quenching, which is generally considered detrimental. The use of most surfactants during synthesis enhances emission as a result of effective steric passivation, and also because most surfactants have a wide bandgap and are electronically insulating. Thiols, however, have band structures that overlap with QD bandgaps, and, although this is usually problematic in cellular labelling, in applications such as photovoltaics the transfer of charge by such means is essential. Therefore, QDs passivated by thiols, which allow hole transfer from excited semiconductor QDs to ligands, may be of immense benefit.[169] For example, CdSe nanorods linked to an oligothiophene *via* a pendant thiol group on the polymer quenched the polymer emission; this was attributed to an energy transfer mechanism.[170] Similarly, TOPO-capped CdSe rods have been passivated with *tert*-butyl *N*-(2-mercaptoethyl)carbamate, which, in the presence of a photoacid generator and under UV irradiation, was modified to give shorter ligands[171] which were used in the manufacture of polymer photovoltaic composites.[172] Nanoparticulate CdSe has also been linked to thymine, a DNA base, *via* an alkyl thiol linkage, and this has been coordinated to a polythiophene *via* an aminopyridine side group,[173] an interesting mix of molecular recognition based on DNA base interactions and potential photovoltaic applications. Similarly, a bifunctional carbodithioic acid molecule has been used to replace TOPO on CdSe nanoparticles, linking to the particle surface by the dithiol units leaving a pendant carboxylic acid group which was then further reacted with an aniline tetramer, giving an overall structure with possible applications in photovoltaics (Figure 6.5).[174] Attachment of the carbodithioic molecule resulted in enhanced photostability for the nanoparticle, although all emission was quenched, and the attached ligand retained its electrochemical activity.[175]

Other applications have also been explored since attaching reactive groups *via* a thiol became routine: for example, linking a photoactive spiropyran molecule to CdSe/ZnS nanoparticles has been achieved, where the QD emission can be switched off by UV photoinduced ring-opening of the functional group and reversibly 'switched back on' by heat or irradiation by visible light.[176]

The phase transfer of QDs and nanoparticles is not always organic → aqueous. In applications where materials require processing in a water-free environment, an aqueous → organic phase transfer is required, and this can be achieved by using a ligand such as 1-dodecanethiol,[177] where the long simple alkyl chain made nanoparticles passivated with such a ligand hydrophobic. In this case, acetone was found to be an important constituent to reduce the interface boundaries and induce efficient interactions.

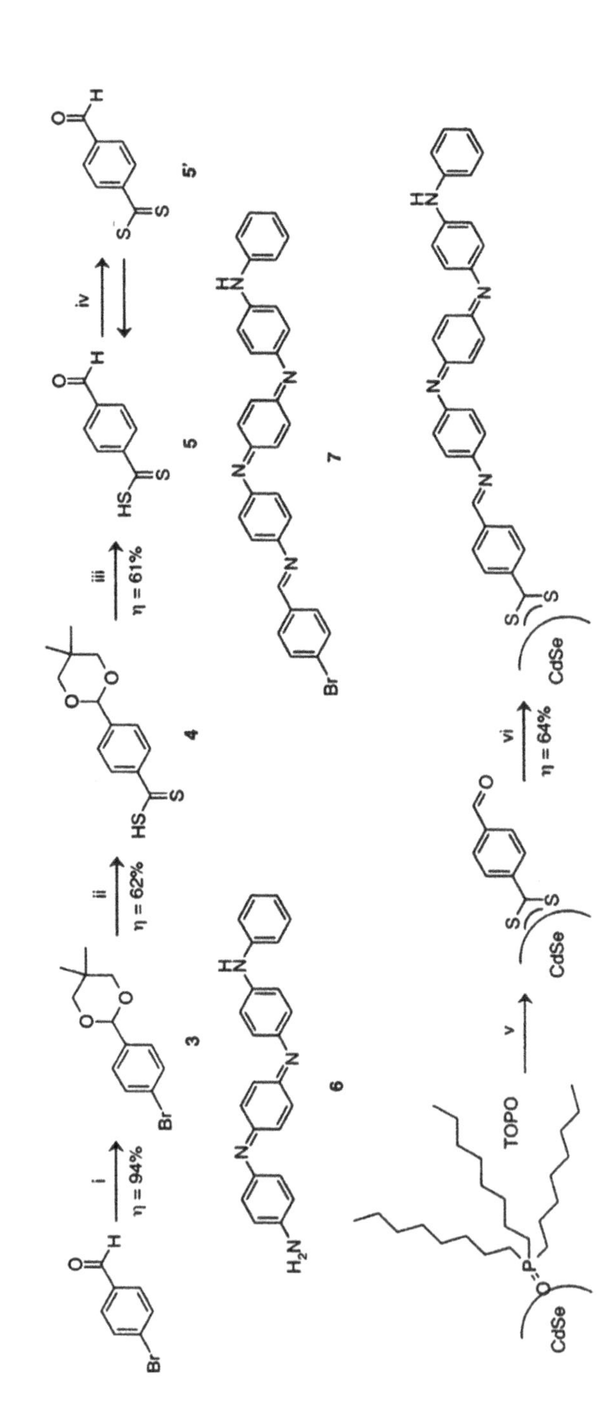

Figure 6.5 The preparation of a carbodithioic capping agent, the passivation of a CdSe particle and the further reaction of the ligand giving an aniline related structure. Reprinted with permission from C. Querner, P. Reiss, J. Bleuse and A. Pron, *J. Am. Chem. Soc.*, 2004, **126**, 11574. Copyright 2004 American Chemical Society.

The phase-transferred materials exhibited a reduced emission quantum yield, and some materials (HgTe) displayed a shift in emission profile.

Another benefit of using thiol compounds is in the monitoring of shell growth on emitting cores. As thiols almost immediately quenched the emission of a 'naked' (unshelled) CdSe QD and only reduced the emission on shelled particles, addition of a thiolated compound such as thiophenol is a simple and effective test for the efficiency of a shelling method (although it is dependent on the solution used).[178] Phenothiazine, another such compound used for optimising the shell thickness of a core/shell system, also worked by a similar method, although this ligand was suggested to coordinate through an amine group.[179]

6.6.2 Phosphine-Based Surfactant Exchange

Although thiolated ligands are extremely versatile for facilitating surfactant exchange, their usually detrimental effects on particle luminescence have driven investigations towards less disruptive capping agents. Naturally, as the majority of synthetic routes use surfactants related to phosphine and phosphine oxide, more complicated, engineered phosphine-based ligands have been developed that can be used in exchange reactions giving the desired functionality while retaining the efficient linkage.

As mentioned previously, a common problem in the incorporation of TOPO-capped particles into electronic devices is the monolayer of surfactant on the surface, which significantly reduces electron transfer.[180] To circumvent this problem, Milliron *et al.* have designed an electroactive oligothiophene ligand with a phosphinic acid functionality to allow strong coordination to a nanoparticle surface,[181] a similar concept to the thiolated aniline capping agent mentioned above. Simple coordination to the CdSe particle surface, in this case, was not the only factor; correct energy level alignment between the particle and the organic group and solubility in a solvent convenient for further processing are all requirements for ligands designed for such applications. The energy level alignment was addressed by the modular synthesis of the ligand that enabled specifically engineered conjugation lengths, and the usually poor solubility of the oligothiophene was rectified by the inclusion of alkyl chains on the thiophene units inducing the required solubility. The free oligothiophene ligands fluoresced with quantum yields in the region of 15%, but quenched when linked to CdSe nanoparticles. The emission of the nanoparticle/ligand complex varied with the length of the oligo backbone, as the electronic structure was staggered and hence the nature of the charge/energy transfer varied with electronic alignment, either slightly enhancing or reducing emission. Likewise, branched (dendron) oligothiophene/phosphinic acid ligands have been used to cap CdSe nanoparticles, using the branched structure to fill the ligand sphere around the particle. When using branched oligothiophene, the QD emission was completely quenched; this was attributed to the successful electron transfer between the ligand and nanoparticle.[182]

The incorporation of nanoparticles into differing matrices without the presence of extraneous capping agents is a major driving force in the search for designer capping agents. Modified tetramethylpiperidinyloxy/phosphinic acid-based capping agents have been used to incorporate particles into polystyrene, reportedly growing the polymer from the nanoparticle surface.[183] In similar work, PVP ligands were linked to CdSe nanoparticles *via* phosphonate groups. Initially, TOPO-capped CdSe QDs were surface-exchanged with a related phosphine oxide type ligand with one pendant bromobenzyl group instead of a long alkyl chain. This anchored group was then reacted further using a palladium-catalysed Heck reaction to give poly(*p*-phenylenevinylene) (PPV)-functionalised CdSe particles, with a quantum yield of 65%.[184] Capping with PPV was also carried out with CdSe/ZnS.[185] Interesting optical properties were observed in such structures, such as 1D emission and absorption spectra attributed to photoinduced charge transfer from the ligand to the nanoparticles.[186] Blinking (fluorescence intermittency) was also found to be suppressed in such structures, again attributed to charge transfer.[187]

Again, the suitability of phosphine oxide system as a capping agent has been exploited and combined with the use of a PEG group to impart water solubility and compatibility with a cellular environment to Fe_3O_4 nanoparticles. In a simple reaction, phosphoryl trichloride was reacted with three equivalents of poly(ethylene glycol)methyl ether to give a range of substituted PEGylated phosphine oxides. The dissolution of oleic acid-capped Fe_3O_4 and the PEGylated phosphine oxide in tetrahydrofuran ensured the ligand substitution reaction, and allowed the resulting nanoparticles to be dispersed in water. Ligands could also be prepared that possessed an amine group such as 1,2-ethylenediamine, using a larger excess of the amine and only two equivalents of the poly(ethylene glycol)methyl ether during the ligand synthesis. This functional pendant amine group could then be reacted further; for example, coupling the particle to a dye molecule. The resulting nanoparticles were easily transfected into cells and were found to be non-toxic.[188]

To overcome the instability of monodentate ligands, oligophosphines have been developed that can be tailored to suit specific environments by altering the functionality on a pendant side arm,[189] as shown in Figure 6.6. The oligophosphines were described as having three distinct units; a phosphine layer that coordinated to the particle surface and a second linking layer that connected to the final layer, which consisted of the surface functionality. Example of functionalities as shown in Figure 6.6 include octyl groups for allowing particle dispersion in non-polar hydrocarbons, carboxylic acids, which when deprotonated allowed dispersion in polar solution and further functionalisation (such as cross-linking, enhancing particle stability) for bioapplication, and methacrylate for processing into polymeric matrices. Comparison of the oligophosphine-capped CdSe/ZnS particles and the monodentate analogues showed increased stability and reduced emission quenching for the oligomeric species. In similar work, a phosphine oxide-based polymer has been developed that facilitated the effective phase

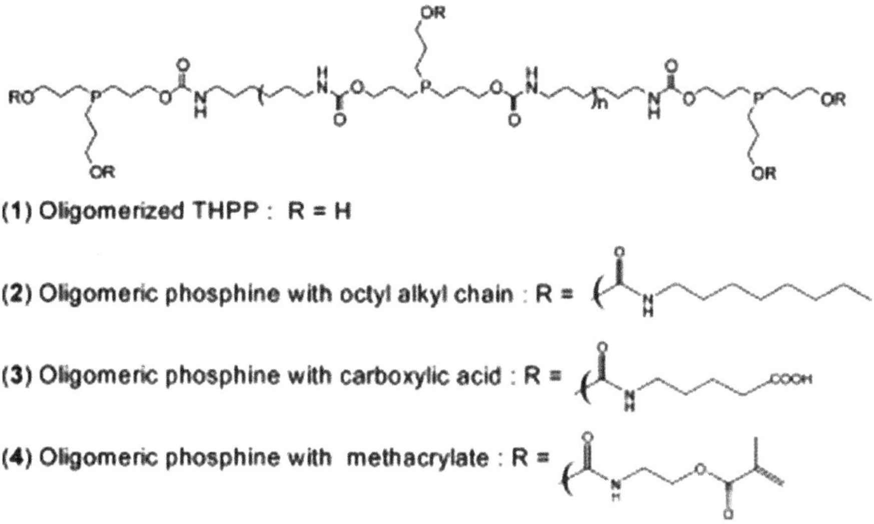

(1) Oligomerized THPP : R = H

(2) Oligomeric phosphine with octyl alkyl chain : R =

(3) Oligomeric phosphine with carboxylic acid : R =

(4) Oligomeric phosphine with methacrylate : R =

Figure 6.6   The structure of oligomeric phosphine capping agents with different functionalities. Reprinted with permission from S. Kim and M. G. Bawendi, *J. Am. Chem. Soc.*, 2003, **125**, 14652. Copyright 2003 American Chemical Society.

transfer of a variety of materials, from QDs, to metals, and metal oxides.[190] The resulting products retained the majority of their original optical properties and were proven in magnetic resonance imaging (MRI), labelling and catalytic applications.

6.6.3   Amine-Based Surfactant Exchange

As amines make excellent capping agents for QDs, it makes sense that they are used as versatile phase-transfer reagents too. Indeed, pyridine was considered one of the first phase-transfer reagents, which initially was thought to be essential for the deposition of a shell layer, as mentioned earlier.[121] The use of dodecylamine (DDA) as a phase-transfer reagent for metal ions and its subsequent use in metal and semiconductor nanoparticle synthesis highlights the versatility of amines.[191] Following the above reports, 4-polyethyleneglycol pyridine has been used to transfer TOPO-capped CdSe QDs to water, either by first removing the initial ligand with pyridine, or by direct addition of the aqueous solution of the substituted pyridine ligand. The resulting particles were soluble in water for months, unlike monothiol analogues, which were found to be unstable due to disulfide formation. The optical spectra showed that the position of the features was maintained after phase transfer, although the quantum yield was not reported.[192] Pyridine-type ligands have also been prepared that, when attached to the surface of CdSe QDs, can be used as an iodide sensor by detecting the linear luminescence quenching with a detection limit of 1.5×10^{-9} M.[193] The overall system is very sensitive and specific but is not water-soluble.

The exchange of surface ligands has also been used in attempts to make linked nanocomposites. Ligands such as 2,2'-bipyrimidine and related molecules were used to link QDs of CdSe or CdS with PbS with the resulting cross-linked materials examined by electron microscope, infrared, absorption and emission spectroscopies.[194,195] Similar bipyridine-type ligands have also been used to develop a new type of photochromic system, by linking a bipyridinium dication to a nanoparticle surface (although this time through a thiol group), which can transfer electrons from the particle to the accepter ligand by irradiation with visible light.[196] The large extinction coefficient of the nanoparticles allowed efficient harvesting of photons and hence the injection into the bipyridinium ligand resulted in a colour change. The large absorption cross-section of a QD has also been exploited with an amine-based dye, where a single dye molecule attached to a CdSe QD surface reportedly quenched the excitonic emission of the entire nanoparticle, transferring the energy to the dye molecule, enhancing the emission characteristics of the dye.[197]

6.6.4   Carboxylic Acid-Based Surfactant Exchange

Although not as common as thiol, amine or phosphine-based transfer ligands, the stability of carboxylic acid groups bound to a nanoparticle surface make then ideal phase-transfer reagents and this appears to be a popular method of modifying nanoparticulate metal and metal oxide surfaces. In this case, as mentioned above, it seems that oleic acid is almost unique in that, depending upon reaction conditions, it can be used to render particles soluble in either polar or non-polar solvents,[113] arguably making surfactant exchange unnecessary in some cases where the solubility can be altered chemically rather than by actual ligand replacement. The exchange of surface-bound oleic acid on Fe_3O_4 for other carboxylic acids with reactive hydroxyl groups has been reported by Lattuada and Hatton, and highlights the simplistic nature of phase transfer using carboxylate species.[198] Phase transfer, however, is not simply a case of changing solvents—the phase-transfer ligand often requires a pendant functional group, especially if the ligand is designed for a biological application. One might therefore design a ligand with a carboxylic acid group to coordinate to a nanoparticle surface, leaving a further group exposed. Kasture *et al.* have prepared glycolipid-terminated cobalt nanoparticles, using a capping agent obtained by a biosynthetic reaction between oleic acid and glucose. It was reported that the ligand coordinated to the nanoparticle surface through both the double bond and the carboxylate group.[199]

The strength of carboxylic acid binding has also been exploited to manufacture electronically coupled assemblies. Carboxy-type species have been used to phase-transfer Fe_3O_4 into water leaving the particles either positively or negatively charged. In this case, either (3-carboxypropyl)trimethylammonium chloride or 2-carboxyethyl phosphonate was simply attached to the nanoparticle surface, giving either a negative or positively charged surface species respectively, with the cationic species used to label stem cells.[200] CdSe QDs have been linked to carboxylated polypyridine ruthenium complexes, which were oxidised upon transfer of a photoexcited hole.[201]

In related work, phenyl groups were grafted to CdSe QDs *via* a family of carboxylic acids with differing chain lengths. It was found that ligands with shorter chain lengths quenched emission more than the longer-chain acids, which was attributed to charge transfer to the phenyl ring.[202] CdSe QDs have also been functionalised with electroactive carboxy-terminated thiophene-based molecules, such as 10-((3-methyl-3,4-dihydro-2*H*-thieno[3,4-*b*][1,4] dioxepin-3-yl)methoxy)-10-oxodecanoic acid, for inclusion in solar energy conversion applications.[203] Notably, the particles required pretreatment with pyridine before ligand substitution.

Related to carboxylate-induced phase transfer is the transfer using hydroxyl groups. Dopamine-based molecules, with a triazine central unit and a terminal PEG group, have been linked to Fe_3O_4 particles *via* the diol functionality, making the initially oleic acid-capped particles water-soluble.[204,205] Similarly, hyaluronic acid/dopamine conjugates have been attached to the surface of Fe_3O_4 nanoparticles. In this case, the nanoparticles were transferred to water prior to the attachment of the conjugate, and the ligand attached afterwards, again *via* the diol functionality. The particles were then used in labelling experiments with the hyaluronan receptor CD44, a receptor whose overexpression is linked with various tumours.[206] Dopamine has also been used to link PEG to oleic acid-capped Fe_3O_4 particles, giving water-soluble materials with a terminal carboxylic acid group.[207] The resulting particles were stable in water and were found not to be taken up by macrophage cells in labelling experiments. Other molecules with ketone groups have been used to exchange ligands on Fe_3O_4 nanoparticles, such as deprotonated quinone-based compounds, which have been used to form organometallic coordination polymers of nanoparticles.[208]

Polymer-based carboxylic acids such as poly(acrylic acid), PAA, have also been found to be effective phase-transfer reagents, and can be used to exchange with oleic acid on Fe_2O_3 at high temperatures.[209] The benefits of a polymeric capping agent are numerous, and are discussed in the next section.

6.6.5 Polymers and Dendrimer-Based Surfactant Exchange

As stated, monodentate ligands are relatively unstable, and if the bond to the particle surface breaks, the functionality is lost. A way round this is to attach the functional groups to a polymer. Polymers are fast becoming the favoured way of protecting QDs in biological applications,[210] despite the fact they increase the hydrodynamic diameter. Numerous papers explore the interaction of QDs and polymers (normally conjugated) for applications such as solar energy conversion or electronic device manufacture; one of the earliest reports of linking QDs to a polymer with a pendant linking functionality, [(norbornenylmethoxy-carbonyl) biphenyl-yl-*tert*-butylphenyl oxadiazole]$_{150}$ [norbornene-2-yl-CH$_2$O(CH$_2$)$_5$P(oct)$_2$]$_{10}$, was carried out to allow fabrication of a light-emitting device.[211] We will also concentrate on polymers that have been added after nanoparticle synthesis, to replace the ligands described above, although we appreciate that some polymers are included during synthesis and are an inherent part of the resulting materials, such as the addition of block copolymers during the synthesis of Co particles[212] or conjugated polymers such as poly(3-hexylthiophene) which have been used instead of TOPO in the synthesis of CdSe QDs, potentially allowing charge transfer between the polymer and the nanoparticle without hindrance from a native capping agent.[213] Closely related to this is the inclusion of [(4-bromophenyl)methyl]dioctylphosphine oxide as a capping agent during the synthesis of CdSe, which then facilitated the grafting of a conjugated polymer onto the capping agent by a palladium-catalysed Heck coupling.[214] The resulting composites have since found use in photovoltaic devices.[215] Other reports that are worthy of note include the use of poly(*N*-isopropyl-*co-t*-butylacrylamide) as an *in situ* reaction capping agent for cobalt and Fe$_2$O$_3$.[216] The polymer has a sharp coil-to-globule phase transition at a low critical solution temperature below which the polymer is water-soluble but above which it is organically soluble. Nanoparticles can therefore be prepared by the thermolysis of metal carbonyl complexes at relatively high temperatures in organic solvents, yet yields particles that are water-soluble on cooling and hence require no further phase-transfer procedures. Related to this is the use of (poly-4-vinylpyridine-*co*-acrylamidoethylamine-*co*-polyethylene-glycol-*co*-polyethyeleneglycolic acid) which has three pendant groups—amine, PEGy-lated hydroxy and a PEGylated carboxylic acid—as an *in situ* stabiliser for the synthesis of amphiphilic Fe$_3$O$_4$ nanoparticles.[217]

The chemistry does not necessarily need to be complicated; simple polymers, such as PVP and poly(ethylenimine) (PEI) have been used to replace oleic acid and OAm on magnetic nanoparticles such as FePt at room temperature changing the solubility in differing solvents which allowed controllable self-assembly on a substrate.[218] This use of simple polymers such as PEI has been extended to induce phase transfer in organically soluble QDs, using the pendant amine groups both as attachment points to the particles, and potentially as linking points to biological entities. The phase transfer appeared to induce photo-oxidation, but the particles in water appeared stable if stored in the dark.[219] PEI has also been used as a transfer agent when linked to PEG, which provided a branched capping agent with the amine linking functionalities supplied by PEI, and the benefits of the pendant long-chain PEG. This work was also extended to link PEG to diethylenetriamine (DETA) giving a linear capping agent again with amine binding units and a pendant PEG group.[220] Interestingly, when DETA–PEG was used to phase-transfer QDs, chain-like aggregates were observed leading to an increased hydrodynamic radius. A triblock polymer based on PEI, polycaprolactone and terminal PEG has been found to be an extremely effective phase-transfer agent for QDs, giving very stable solutions due to the intermediate hydrophobic polycaprolactone moiety shielding the binding group from solution.[221]

Other simple polymers used include poly(dimethylaminoethyl methacrylate) (PDMAEMA), which was used to replace TOPO on CdSe/ZnS QDs, making them soluble in toluene and methanol.[222] Notably, the radius of the particle/polymer conjugate almost doubled in diameter from 3 to 5.9 nm, and the emission dropped 30% in intensity. Pyrene-functionalised PDMAEMA has also been used to determine the number of polymer strands attached to a QD using size exclusion chromatography, by monitoring the amount of free polymer removed from solution upon addition of TOPO-capped QDs.[223] QDs of CdSe which emitted at 550 nm were found to have on average 12.5 ± 1.9 polymer molecules attached, whereas the smaller particles of CdSe which emitted at 520 nm had on average 4.6 ± 0.6 polymer molecules attached.

Poly(allylamine), poly(sodium styrene sulfonate) and PAA have all been used to transfer hydrophobic nanoparticles to water using a high-temperature solution method.[224] A diethylene glycol solution of the polymer was heated under an inert atmosphere, followed by injection of a toluene solution of the nanoparticles, forming a cloudy solution. This was followed by heating at *ca.* 240 °C until the solution cleared, indicating a successful phase transfer. The particles were isolated by addition of a dilute solution of acid causing the nanoparticles to precipitate, although they could be redispersed in water by deprotonating the carboxylic acid groups. Another simple polymer, poly(*N*-isopropylacrylamide) (PNiPAm), a thermoresponsive material, has also been used to phase-transfer luminescent QDs, by simply adding a chloroform solution and shaking, followed by addition of the resulting solid to water and storage in a refrigerator.[225] The quantum yields dropped only slightly although the overall size did increase. The resulting material could also be cross-linked. Not all phase-transfer protocols with polymers result in simple ligand exchange; some protocols result in thick polymer shells or polymer particles doped with smaller nanoparticles.[226–232]

There are, however, only a few polymers that are commercially available and present the desired range of linking functionalities, solubility and potential for immediate use. Most polymers have to be prepared specifically for use with nanoparticles. Again, the chemistry does not have to be complicated; the simple grafting of mercaptoethylamine onto a PAA backbone yielded a polymer with pendant thiol groups available to coordinate to a QD surface, with the remaining carboxylic acid groups available for further conjugation.[233] When used to phase-transfer QDs with multiple shells, the resulting water-soluble material exhibited emission quantum yields that were actually higher than the initial hydrophobic sample, although the hydrodynamic diameter increased notably, suggesting the polymer had a thickness of about 3 nm. In another example,[234] as shown in Figure 6.7, a polymeric ligand of *ca.* 290 repeat units (esters) was functionalised with

Nanocrystal modification

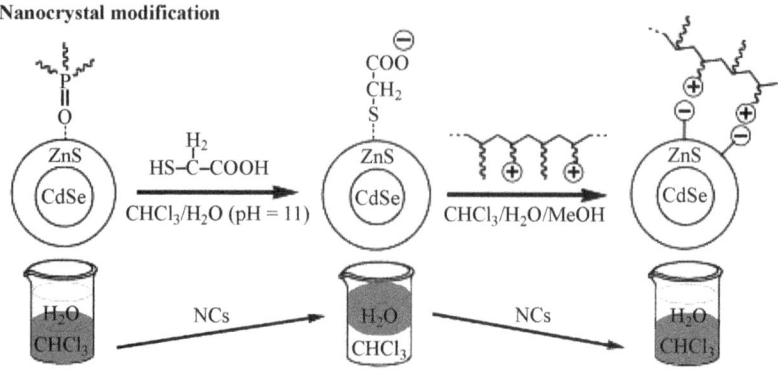

Figure 6.7 Surface modification of nanoparticles and the resulting phase transfer. Reprinted with permission from I. Potapova, R. Mruk, S. Prehl, R. Zentel, T. Basché and A. Mews, *J. Am. Chem. Soc.*, 2003, **125**, 320. Copyright 2003 American Chemical Society.

a mixture of octadecylamine (ODA) and aliphatic phosphonium salts, giving a positive charge to link to QDs with a negatively charged functional group. Mixing an aqueous solution of the nanoparticles with a chloroform solution of polymer resulted in the coordination of the positively charged groups, leaving the long alkyl chain to induce solubility in non-polar solvents, hence inducing another phase transfer back to an organic solution. It was noted that the phase transfer would only work with 30–50% of charged side chains; fewer charged side chains resulted in insufficient linking with the dots whereas too many resulted in a polymer that was insoluble in chloroform.

Another polymeric capping agent designed specifically to address solubility problems with thiolated monomeric species has been described by Liu *et al.*[235] The polymer was engineered and prepared by reversible addition–fragmentation chain-transfer mediated polymerisation to yield a material with three distinct subunits—an imidazole group for linking to QDs, a PEG group to induce water solubility, and a primary amine/biotin group for biofunctionalisation. The exchange of hydrophobic ligands on CdSe/ZnCdS with the multifunctional polymer was easily achieved by simple dissolution and precipitation, followed by dialysis. Phase exchange of CdSe/CdS particles (prepared by the successive ion layer adsorption and reaction (SILAR) route described in Chapter 5) was achieved and was notable for the material's high quantum yield in water (65%). The hydrodynamic diameter was equivalent to DHLA–PEG functionalised QDs (*ca.* 11 nm), but exhibited superior colloidal stability of at least 2 months. The multifunctional polymer could then be further functionalised with dyes such as 5-carboxy-X-rhodamine, and Förster resonance energy transfer (FRET) experiments were undertaken. Further experiments with QDs capped with the multifunctional polymer showed almost no non-specific binding in labelling experiments, and conjugation to streptavidin for targeted imaging was simply achieved by coupling chemistry. It is worth noting at this point that aminopolymers with dyes attached have previously been reported and FRET experiments explored.[236] Multifunctional designer polymers have also been used with magnetic iron oxide nano-particles for biological applications.[237] In this example, a polymer with various subunits has been utilised, with a dopamine group for attachment to the iron oxide particle, a dye molecule to allow cellular imaging and a free amine group for further conjugation to other species including double strands of RNA/polymer conjugates, which allowed the particles to mark the cells wall of a Caki-1 cell line. The preparation of such polymeric species allowed nanoparticles to retain their inherent optical or magnetic properties while providing additional groups that induced water solubility and allowed conjugation to differing molecules of interest to biologists, while inhibiting non-specific binding and maintaining a small hydrodynamic diameter, all key factors in biological labelling.

One notable feature of the general use of polymers is the resulting increase in hydrodynamic diameter (in some cases up to 40 nm [119])—an undesirable side effect of phase transfer, as the renal clearance limit of nanoparticles is suggested to be 5.5 nm.[139] An elegant way round this has been described by Smith and Nie, who replaced *ca.* 35% of the carboxylic acid groups on PAA with amine and thiol functionalities.[238] QDs of CdTe were then initially phase-transferred to water using thioglycerol, which was then replaced with the polymer. This combination of thiol and amine reportedly resulted in the polymer wrapping round the QD, giving a shell thickness of 1.5–2 nm, significantly smaller than if the parent PAA polymer alone was used. In this case, the photoluminescent stability of the CdTe was maintained (probably due to the use of thiol groups) with quantum yields of up to 50% if the optimum capping ratio was used.

A related family of structures, dendrons, have also been used as capping agents for semiconductor QDs.[239] Hydroxy-terminated dendrons (genera-tions 1–3) with a thiol focal point were prepared from the parent disulfide dendrimer and were linked to TOPO-capped CdSe particles by refluxing an alkaline solution of methanol, nanoparticles and dendron in the dark.[96] A two-phase procedure was also found to be effective, where the hydrophobic nanoparticles were dissolved in ethyl ether, the dendrons dissolved in water at pH 10.3 and both solvents mixed, effecting a phase transfer *via* surfactant exchange. The resulting structures, where the TOPO is replaced by the den-drons, were water-soluble and extremely stable to chemical processing and resistant to photo-oxidation. The use of higher-generation dendrons resulted in increased protection, attributable to the steric effects afforded by the increasingly branched structures. The dendron terminal groups could then be functionalised further by an amide coupling reaction, resulting in struc-tures that were soluble in either water or organics, depending on the func-tionality. This is significant, as the dendrons present a wide range of coupling chemistries comparable to organic reagents used in biochemistry.

Another type of third-generation dendron investigated was notable for the eight alkene terminal functionalities (Figure 6.8), which were cross-linked by ring closing metathesis (RCM) reactions which resulted in a dendron

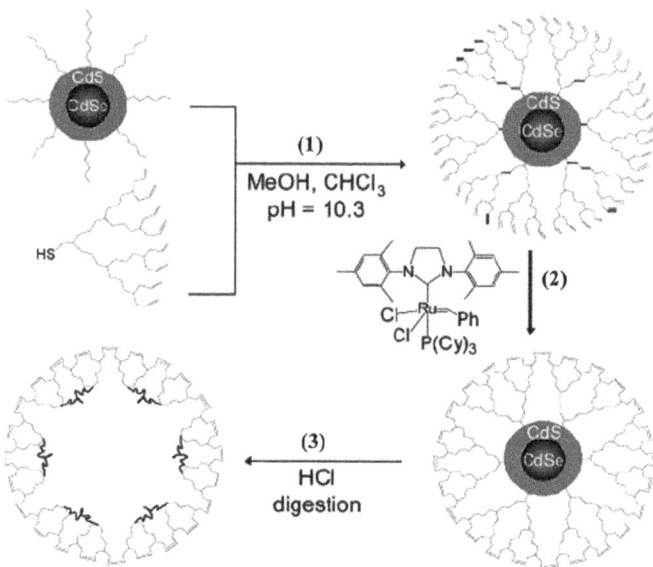

Figure 6.8   Passivation of CdSe/CdS particles with alkene-terminated dendrons, box formation and etching reactions. Reprinted with permission from W. Guo, J. J. Li, Y. A. Wang and X. Peng, *J. Am. Chem. Soc.*, 2003, **125**, 3901. Copyright 2003 American Chemical Society.

box.[240,241] The dendron boxes were found to be soluble only in aromatic or polar organic solvents, whereas the parent dendrons were soluble in a number of non-polar organic solvents.

The dendron boxes exhibited increased stability against HCl etching, oxidation with H_2O_2, photo-oxidation and heating when compared to the simple dendron-stabilised particles, although the core could be removed by digestion with concentrated HCl, leaving 'empty boxes'. Despite the increased stability, replacement of TOPO for the thiolated dendron almost completely quenched the emission from CdSe particles. This was overcome using CdSe/CdS core/shell particles, whose emission remained although quenched to *ca.* 20% of the original value. Photobrightening, linked to oxidation, was also observed and could, in some cases return the emission back to its initial level.

The excellent stability afforded by the RCM reaction limited the applications as the resulting material was only soluble in organic solvents. To avoid this, a dendron box structure was developed based on the water-soluble third-generation hydroxyl-terminated dendron, passivated with another second-generation dendron resulting in an amine-terminated dendron box, the stability of which was comparable to the box prepared by the RCM route. The amine functional group allowed numerous coupling reactions to be carried out (Figure 6.9), and investigations into the material's biocompatibility were also undertaken.[242] Similarly, carboxylate-terminated dendrons

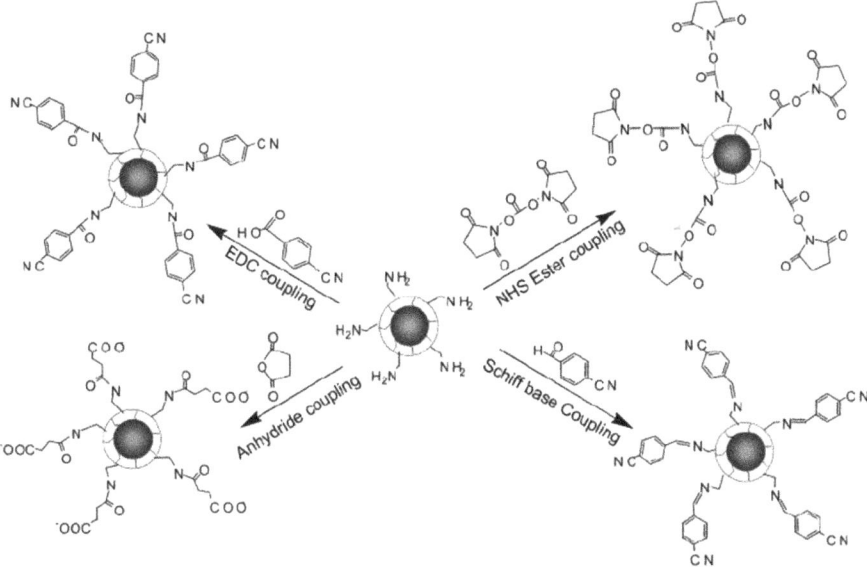

Figure 6.9 Coupling reactions of amine box dendron-passivated particles. Reprinted with permission from W. Guo, J. J. Li, A. Wang and X. Peng, *Chem. Mater.*, 2003, **15**, 3125. Copyright 2003 American Chemical Society.

have also been used on CdSe/CdS particles.[243] In a related study, second-generation polyamidoamine (PAMAM) dendrimers were used as capping agents, although the terminal amine groups caused irreversible precipitation by cross-linking on the CdSe surface. Capping the dendrimer's terminal groups with 1,2-epoxyethane resulted in a long-chain alkyl that did not interact with the nanoparticle surface. The dendrimers coordinated to the particle surface through the internal amines in the branched structure, with dendrimers consisting of secondary and tertiary amines providing the better surface protection.[244] Similarly, PAMAM dendrimers of various generations were used to phase-transfer QDs of CdSe/ZnS using a simple solution method, where the ester terminated dendrimers were dissolved in water with a surfactant (poloxamer 188) and the chloroform solution of the QDs was added and stirred until the chloroform evaporated. In this case, the resulting material was not dendrimer-capped nanoparticles, but in fact a composite of QDs, dendrimers and surfactant, with a final particle size much larger than that of the parent QDs. The use of amine-terminated dendrimers resulted in a precipitate that could not be dissolved in most solvents, and the generation of dendrimers used also dictated emissive properties, with the use of a 4.5G dendrimer resulting in a material with a quantum yield of *ca.* 40%, which is notably high for QD materials in water. Lower-generation dendrimers had notably lower quantum yields, which limited their use in labelling.[245]

Similarly, a second-generation polybenzyl ether dendron with a phospho-nate group has been used to cap CdSe nanoparticles. In this case, the dendron was added during the synthesis rather than used afterwards as a surface exchange reagent. The resulting material exhibited enhanced emission. Electropolymerization of the pendant carbazole groups was readily achieved, but red-shifted the QD absorption and quenched the emission, attributed to the polymerisation step, lifting the ground state of the carbazole groups above the ground state of the CdSe.[246]

6.7   Inorganic Ligands

The majority of exchange reactions utilise organic-based ligands as a vast number of compounds are available, providing coordinating groups and pendant functionalities. A drawback with the use of organic ligands is the insulating properties this induces in the final nanomaterials, which is problematic if the final application requires the unhindered passage of a charge carrier. A simple, yet elegant way to circumvent this problem is to use inorganic complexes. Kovalenko *et al.* have reported the use of metal chalcogenide complexes (MCCs) in effective surfactant exchange reactions, in which compounds such as $(N_2H_5)_4Sn_2S_6$ completely replaced the organic surface cap on a variety of nanoparticles with ions such as $Sn_2S_6^{4-}$.[247,248] The exchange reaction proceeded smoothly in solution, leaving the particles essentially physically and optically unchanged, and soluble in a wide range of solvents yielding all-inorganic colloids. When deposited as films, QDs and gold nanoparticles both showed impressive conductivity values, consistent with the inorganic ligands facilitating efficient charge transfer, which led to the fabrication of QD based field-effect transistors. Although MCCs are themselves precursors for chalcogenide materials, their primary use in these cases was as molecular ligands, although gentle heating resulted in the preparation of the relevant semiconducting phase, producing composite materials. A wide range of QDs and metals of various types and structures were reported with numerous MMCs, giving a potentially enormous of variation of conducting nanoparticle solids. A range of nanomaterials (PbS, PbTe, Bi_2S_3) have been functionalised with the Zintl ion Sb_2Te_3 MCCs, and used to prepare thin films of nanostructured thermoelectrics with relatively large values of the figure of merit, *ZT*.[249]

6.8   Encapsulation Driven by Hydrophobic Interactions

Although surfactant exchange is a versatile proven chemistry, this process is not always ideal. As the capping ligands are an inherent part of a nano-particle's structure, changing the surface capping chemistry alters the electronic properties, and applications such as solar energy conversion rely on the intimate electronic link between solid-state material and organic ligand.

In biological applications, this is usually viewed as detrimental as surfactant exchange is known to reduce emission quantum yield, for example, and the nanoparticle's magnetic or luminescent attributes should ideally be maintained.

One method of phase-transferring particles to water is to encapsulate the entire particle, including the original ligand, in a micelle structure.[250] This has the disadvantage of significantly increasing the hydrodynamic diameter,[251] but the increased protection provided by the extra layer of surfactant and the preservation of the interface integrity by avoiding damage to the surface by surfactant exchange results in the stabilisation of the nanoparticles' properties, notably the optical characteristics. Micelle formation is also relatively inexpensive, quick and simple, unlike many other methods of preparing water-soluble luminescent materials.

In the seminal work in QD encapsulation reported by Dubertret *et al.* CdSe/ZnS QDs capped with TOPO were overcoated with phospholipids, specifically *n*-poly(ethylene glycol) phosphatidylethanolamine (PEG-PE) and phosphatidylcholine (PC), an extremely simple and effective method.[252,253] The PEG-PE was found to interdigitate with ligands on the nanoparticle surface, leaving the PEG groups pendant in solution, and PC was found to effectively control the effective spacing of the PEG-PE molecules. The resulting particles were stable in aqueous solution for months, with each micelle usually containing just one nanoparticle with emission quantum yields of 24%. The initial nanoparticles were 3–4 nm in diameter, ideally suited for encapsulation within PEG-PE, but were found to be 10–15 nm in diameter as determined by microscopy after encapsulation. Replacement of PEG-PE with an amino-functionalised PEG-PE allowed further linking to DNA, opening the potential for targeted delivery. Investigations into labelling *Xenopus* embryos were undertaken, with essentially no toxicity observed for what was considered normal injection concentrations (2×10^9 QDs per cell). At concentrations in excess of 5×10^9 QDs per cell, abnormalities were observed. Other biological molecules, such as bovine serum albumin (BSA) have been attached to the lipid, demonstrating the versatility of the capping layer.[254]

This technique has been extended to the use of 1,2-dioctanoyl-*sn-glycero*-3-phosphocholine to phase-transfer CdSe/CdS QDs, which were used to label rat hippocampal neurons.[255] A nice example of the use of QDs encapsulated in phospholipids is in their use as a model for the uptake of vasoactive intestinal peptide (VIP)-grafted phospholipids as drug delivery platforms for breast cancer cells.[256]

Although the use of phospholipids appears to be an excellent way of stabilising particles, the structure was only maintained by hydrophobic interactions. Micelles with a more stable shell were achieved by using specially designed phospholipids that included a norbornene group, which was then oligomerised to form a phospholipid hexamer.[257] The hexamers were then used in the same way as simple monomeric phospholipids to stabilise hydrophobic nanoparticles, although encapsulation required significantly more hexameric material to ensure phase transfer. Stability tests showed that

QDs transferred using the hexamer were stable in buffer solution at pH 6 and 9, whereas the same materials transferred using the monomer displayed a reduction in emission. QDs transferred by both materials showed a significant reduction of emission in 5 M NaCl solution, although the species transferred by oligomer showed less of a drop-off. Similar work has been reported by Kim *et al.*[258] where a PEG-10,12-pentacosadiynoic acid conjugate was used to phase transfer QDs in a reverse micelle. Exposure to UV light resulting in cross-linking across numerous pentacosadiynoic acid groups, resulting in stabilised micelles, *ca.* 150 nm in diameter.

The use of phospholipids was not limited to semiconductor QDs. Magnetic nanoparticles such as $MnFe_2O_4$ were encapsulated in either 1,2-distearoyl-*sn-glycero*-3-phosphoethanolamine-*N*-[biotinyl(poly(ethylene glycol))2000], 1,2-distearoyl-*sn-glycero*-3-phosphoethanolamine-*N*-[methoxy(poly(ethylene glycol))2000] or 1,2-distearoyl-*sn-glycero*-3-phosphoethanolamine-*N*-[maleimide(poly(ethylene glycol))2000] by simple dissolution in chloroform along with the nanoparticles.[259] Once the chloroform had evaporated, the solid was redispersed in phosphate buffer solution and heated to affect a clear stable solution. Strands of DNA then could be attached to the maleimide-functionalised phospholipid. The particles, *ca.* 12 nm in diameter, were substantially larger than the QDs described above and might initially be considered too large for PEG-PE type micelles, although particles up to 15 nm in diameter were found to be effectively passivated. To demonstrate that the DNA-linked and biotinylated particles were still biologically active, they were attached to substrates through the functional groups, while the nanoparticles retained their magnetic properties. Fe_3O_4 particles, capped with oleic acid have also been phase-transferred using PEG-PE-based phospholipids, specifically 1,2-distearoyl-*sn-glycero*-3-phosphoethanolamine-*N*-[methoxy(poly(ethylene glycol))2000], and used in MRI experiments.[260] In this case, the non-toxic particles were 8–10 nm in diameter without the PEG-PE capping, and determined to be *ca.* 22 nm by light scattering after phase transfer. Upon incubation in 1 M NaCl solution, the PEG-PE capped particles showed no increase in size, whereas iron oxide particles capped with other polymeric ligands almost doubled in diameter, indicating the stability of particles capped with the PEG-PE system.

One of the advantages of the ligand system is that numerous phospholipids and mixtures thereof can be easily utilised, allowing the preparation of multifunctional materials. The phospholipid system has also been extended to incorporate a gadolinium-containing lipid, Gd-DTPA-bis(stearylamide).[261] Inclusion of this ligand with 1,2-distearoyl-*sn-glycero*-3-phosphoethanolamine-*N*-[methoxy-(poly(ethylene glycol))] (PEG-DSPE) in the phase transfer of CdSe/ZnS QDs gave a luminescent water-soluble material with paramagnetic character and also having the benefits associated with the encapsulation of the particles in PEG-terminated micelles. The resulting particles had a high relaxivity of almost 2000 mM^{-1} s^{-1}. The ligand could be further conjugated to Arg-Gly-Asp (RGD) peptides and used in imaging specific endothelial cells using both MRI and optical techniques. Another method of preparing multifunctional particles using phospholipids is based on the ability to encapsulate more than one particle at a time, which is usually considered an unwanted occurrence. Park *et al.* have incorporated near-IR-emitting QDs and magnetic iron oxide nanoparticles in a phospholipid micelle.[262] It is noteworthy that the QDs and iron oxide were encapsulated together, and the ratio of QDs to metal oxide particles could be altered controllably. An anticancer drug, doxorubicin, was also included in the heterostructure, and a peptide known to target nucleolin in endothelial cells was conjugated to the particle surface. The targeted structures were then used to image mice *in vivo* with MDA-MB-435 tumours. After harvesting the tumours, examinations by both MRI and optical microscopy confirmed the presence of the nanostructures, although significant levels of particles were also found in the liver. An interesting related method of preparing multifunctional materials is the inclusion of iron oxide and QDs in the lipid cores of lipoprotein micelles,[263] giving what was termed 'nanosomes' with hydrodynamic diameters of 250 nm. These materials were functionalised with apolipoprotein and lipoprotein lipase and used in examining liver cells *in vivo* using both MRI and optical imaging. An similar interesting example is the incorporation of CdSe/ZnS QDs into glyconanospheres with a diameter of *ca.* 190 nm.[264] The QDs were initially phase-transferred using mercaptosuccinic acid, and linked electrostatically with polylysine inside the sphere. The spheres were inherently unstable and dissociated after *ca.* 10 hours, although this was addressed by coupling the particles to avidin. The nanospheres used dextran molecules on the surface to examine interactions with carbohydrate-binding proteins.

A similar method of encapsulation has been widely adopted, using an engineered amphiphilic polymer with pendant side chains that resulted in interdigitation with the nanoparticle capping group. The usual polymer chosen was PAA modified with 40% OA.[265] Phase transfer was achieved in the same manner as phospholipid phase transfer; CdSe/ZnS QDs were dissolved in chloroform along with the polymer, and left to evaporate, and the resulting solid was redispersed in water. The polymer could be further cross-linked by EDC-mediated cross-coupling to lysine, and then further conjugated to proteins or antibodies as needed. In a nice example, the negatively charged polymer has also been used to electrostatically coordinate to a cationic dye, which then formed J-aggregates.[266] In most applications, however, cross-linking by lysine was not necessary as most applications only needed phase transfer. A detailed study by Anderson[267] highlighted that 40% modification appeared to be the optimum degree of modification, although ODA appeared to be a better side chain for efficient phase transfer. In a specific example, particle diameter measured by dynamic light scattering was found to increase from 9.7 to 17.2 nm upon addition of the polymer. Interestingly, the size of the final particle/polymer conjugate appeared to be approximately the same regardless of the size of the initial particle. The emission of encapsulated QDs appeared to drop initially after transfer, although the luminescence was found to recover to some degree on standing for several days.

PAA could also be amended to include a 'chemical handle'—a single functional group on the polymer chain such as a thiol that could be used to attach a further functionality, which therefore did not require the usual EDC coupling chemistry.[268] This handle could be used to attach dye molecules and other biological moieties, notably any material that contains a maleimide group as the double bond is known to form a stable C–S bond.

When one considers the structure of an amphiphilic polymer, essentially carboxylic acid groups along a polymer backbone with hydrophobic tails, comparisons to the long-chain carboxylic acids are obvious—it could be suggested that the polymers are simply a polymeric version of fatty acid capping agents. Smith and Nie have developed a new synthetic pathway based on previous work done with amphiphilic polymers as the *in situ* capping agent rather than a reagent added after nanoparticle synthesis.[269] In the case of QD synthesis, metal salts were mixed with the polymer (termed 'amphipols', a term given to amphiphilic polymers used to stabilise integral membrane proteins that predates the use to phase-transfer QDs[270]), a non-coordinating solvent (low molecular weight PEG) and a chalcogen source such as TOPTe/PEG, and injected into the reaction solution at elevated temperatures. The amphipol released the metal ions during the synthesis yielding the nanoparticle, to which the amphipol bound through the carboxylic acid groups. The particles could then be directly added to non-polar solvents as the pendant alkyl chains allowed dissolution. The particles could also be directly added to polar solvents, as the amphipols released from the metal ions during the reaction formed a bilayer structure, exposing the polar carboxylic acid groups to solution and again allowing dissolution. The particles were, however, only soluble in one medium at a time and were not truly amphiphilic—once isolated from a polar solvent, they could not be redispersed in a non-polar solvent, and *vice versa* (without further processing). The quality of the nanoparticles prepared was comparable to the standard organometallic route and could be extended to metals and metal oxides. An interesting observation from what was termed the 'amphibious bath reaction' was the precipitation of particle from the reaction mixture when methoxy-terminated PEG was used and the reaction mixture cooled to below 50 °C. This potentially allowed easy isolation of the particles. Similar reactions with poly(maleic anhydride) have been explored where cysteamine groups have been grafted onto the polymer backbone, yielding a polymer with a pendant thiol group available for coordination to the particle.[271] Interestingly, chemical modification of 40% of the maleic anhydride groups was found to be optimum (similar to the modified PAA described above), giving *ca.* 20 thiol groups per polymer chain. QDs with standard ligands (such as TOPO and long-chain amines) could have them exchanged for the thiolated polymer with mild heat treatment, and further treatment of the polymer/QD with a PEG ligand containing amine and hydroxyl groups resulted in efficient phase-transfer to water, forming a tightly bound polymer layer. The resulting particles were notably smaller than those encapsulated in a similar poly(maleic anhydride) micelle using long alkyl chains to inter-digitate between the existing capping ligands.

This use of interdigitating polymers was extended to the application of poly(maleic acid anhydride-*alt*-1-tetradecene), where the long alkyl chain again intercalated with the surface ligands of the nanoparticle of various materials, from semiconductors, to metals, alloys and metal oxides (Figure 6.10).[272] Once the polymer had intercalated with the surface capping agents, the anhydride groups required hydrolysation to induce water solubility. The polymer chains could again be cross-linked with bis(6-aminohexyl) amine, and hydrodynamic diameters increased from 10 nm or less to more than 20 nm upon phase transfer. Similarly, poly(maleic anhydride-*alt*-1-ODE), a (then) commercially available polymer, was used to phase-transfer iron

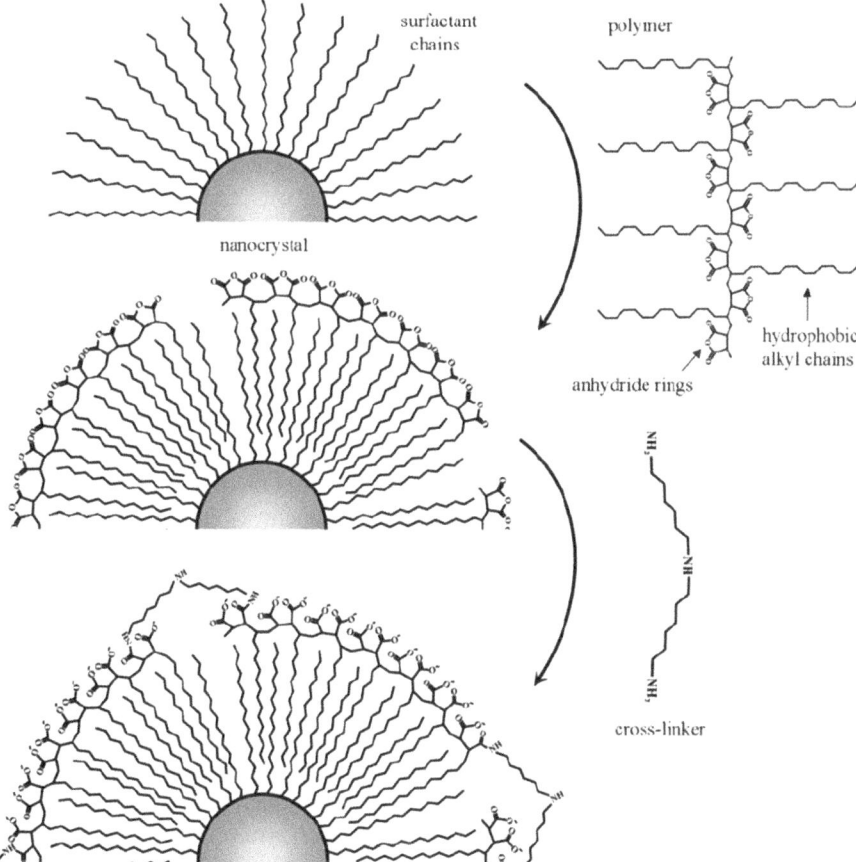

Figure 6.10 Scheme for polymer coating of nanocrystals. Reprinted with permission from T. Pellegrino, L. Manna, S. Kudera, T. Liedl, D. Koktysh, A. L. Rogach, S. Keller, J. Rädler, G. Natile and W. J. Parak, *Nano Lett.*, 2004, **4**, 703. Copyright 2004 American Chemical Society.

oxide nanoparticles to water.[273] In this case, a monomer of the maleic anhydride group was functionalised with PEG by reacting the polymer with amine-terminated PEG methyl ether, leaving a pendant PEG group and a single carboxylic acid group for further functionalisation to biological molecules of interest. The resulting phase-transferred nanoparticles were stable in water over a wide range of pH, and found to be up to 60 nm in size (the parent nanoparticles were up to 40 nm in diameter). Further studies on this system included the use of poly(maleic anhydride-*alt*-1-ODE) to encapsulate a range of QDs,[274] and it was found that the hydrodynamic diameter could be tuned by altering the molecular weight of the PEG side chain, with sizes ranging from 24 to 46 nm. Impressively, the quantum yields of the encapsulated QDs were found to be 34.5–54.7% when in water. The particles were also extremely stable to further harsh reaction conditions, such as sonication at 70 °C. The particles were successfully used in labelling breast cancer cells, with no non-specific binding observed. The toxicity of the particles was explored in human breast cancer cells, and a relationship between cell death and particle concentration was found, although the presence of the PEG group was found to significantly reduce toxicity. Other groups also explored the use of poly(maleic anhydride-*alt*-1-ODE) to phase-transfer iron oxide nano-particles.[275] In this case, the particles were slightly larger than the QDs described above (*ca.* 16–21 nm) therefore a larger molecular weight polymer was used (30–50 kDa). Again, individual polymer chains were found to interact only with a single nanoparticle.

Other functional groups have been added to poly(maleic anhydride-*alt*-1-ODE) when used with iron oxide, such as the thermoresponsive polymer poly(*N*-isopropylacrylamide).[276] In this case, the water-soluble particles of iron oxide capped with the polymer grafted onto the maleic anhydride backbone showed a large difference in size at different temperatures, consistent with the collapse of the polymer chains at lower temperatures. The particles could easily be isolated with a bar magnet at higher temperatures, whereas the smaller particle/polymer composites stayed in solution at lower temperatures.

Work on poly(maleic anhydride-*alt*-1-ODE)[277] was also carried out because of the commercial discontinuation of the tetradecene-based polymer, but highlighted a new purification step based on ultracentrifugation as a key part of the phase transfer, as the previous use of size exclusion chromatography did not work to the same degree. The need for a more dilute solution of nanoparticles and an increased amount of polymer to ensure encapsulation of a single nanoparticle was also discussed and other factors such as the need for prolonged sonication were highlighted. The hydrodynamic diameter of several materials, before and after encapsulation, was investigated and as might be expected, the use of the ODE-based polymer almost doubled the diameter of the particle, although the use of the tetradecene-based polymer increased the diameter even more.

The use of a maleic anhydride group to functionalise a polymer backbone was extended by Lees *et al.*, producing water-soluble QDs with a final hydrodynamic diameter similar to that of a surface-exchanged species.[278] In this study, poly(styrene-*co*-maleic anhydride) (PSMA) was adding to a chloroform solution of QDs, and then the polymer converted to an amphiphilic species *in situ* by addition of either ethanolamine or a poly-ether amine (essentially an amine-terminated PEG) as ring-opening agents, leaving pendant carboxylic acid groups and either an alcohol-terminated short chain or, in the case of the polyether amine, a PEG group. Following this, the particles could simply be transferred to water. Small QDs phase-transferred using ethanolamine as a ring-opening agent showed a reduction in emission quantum yields, whereas larger particles actually displayed a significant increase in emission. The QDs passivated with the polyether amine/PSMA displayed exceptional stability across a wide range of pHs, whereas the ethanolamine/PSMA particles appeared stable primarily at pH 5–7. The ethanolamine-functionalised particles were, however, extremely small; the hydrodynamic diameter of CdSe/ZnS particles (displaying an emission peak at 551 nm) exhibited a diameter of 14.8 nm when phase-transferred using a PEGylated DHLA as described earlier. Phase transfer of the same material with PSMA/ethanolamine resulted in a diameter of 13.4 nm, which is unusually small. However, phase transfer using the more stable polyether amine resulted in a diameter of 17.8 nm, and ethanolamine-functionalised poly(maleic anhydride-*alt*-1-ODE) resulted in notably larger particles of 24.5 nm. This highlights that PSMA functionalised with the shorter chain can in fact rival phase-transferred material produced by surfactant exchange. The exchanged material does, however, maintain the PEG pendant group, so one must choose between size and surface functionality. Particles with the polyether amine group were also found to exhibit reduced non-specific binding in cell labelling, again attributable to the PEG group.

The use of a poly(maleic anhydride) backbone was extended by slightly varying the structure of the materials by reacting a fraction of the anhydride rings (75%) on poly(isobutylene-*alt*-maleic acid) with DDA to form the hydrophobic intercalating arms of the polymer, rather than relying on the polymer already having a hydrophobic pendant side chain attached to the backbone.[279] This left 25% of the maleic acid groups available for further conjugation, which was carried out to PEG, proteins, biotin, dyes and sugars. Importantly, the hydrophobic side chains and other groups could be added to the polymer backbone before the polymer was wrapped around a nanoparticle.

An interesting and unusual use of QDs within a micelle is the preparation of a QD–inorganic chelate composite with extreme sensitivity to nitric oxide, and important biological free radicals. This composite is unusual as it does not directly involve a surface modification as TOPO-capped QDs are phase-transferred into a cetyltrimethylammonium cationic micelle, which was electrostatically linked with tris(*N*-(dithiocarboxy)sarcosine)iron(III). After synthesis of the composite, QD fluorescence was found to have quenched due to overlapping electronic spectra of the ligand with the QD emission. Addition of a NO-donating compound resulted in restoration of the QD emission, which was stable for an hour, after which time it dropped to 75% of the initial value. The limit of detection was found to be *ca.* 9.0 μM.[280]

6.9 Calixarenes and Related Macrocyclic Compounds

Amphiphilic polymers and phospholipids are not the only species to induce phase transfer by interdigitation; unfunctionalised cyclodextrins also induced water solubility in oleic acid-capped nanoparticles of metals and metal oxides by forming inclusion complexes between the hydrophobic surface species and the hydrophobic cavity of the molecule.[281] Calix[4]arene carboxylic acids have also been used to phase-transfer QDs into water extremely effectively,[282] while retaining the high quantum yields of *ca.* 30%. In this case, the carboxylic acids were deprotonated to induce water solubility, so the mode of bonding was not clear, although hydrophobic interactions were assigned as the coordination mechanism. The fact that the carboxylic acids required deprotonation suggested intercalation of the molecule interior with the surface ligand, although addition of large concentrations of Na^+ or K^+ quenched the emission by 10%, suggesting sensing behaviour. Interestingly, the emission of the QDs could be significantly shifted towards the red end of the visible spectrum simply by using a larger calixarene,[283] and the exchange of surface sulfur atoms for hydroxyl groups was suggested as a possible explanation based on similar work with cyclodextrins.[284] This does seem to point to the direct coordination of the carboxylic acid group with the QD surface, so clearly the mode of bonding remains unresolved.

More typical for calixarenes and related macrocyclic compounds is the use of functional groups, as demonstrated by the use of monothiolated β-cyclodextrin on CdSe/CdS particles to induce water solubility through ligand exchange,[285] which has the advantage of freeing the interior of the bowl-shaped molecule for potential sensing applications. Crown compounds (15-crown-5) have been attached to CdSe/ZnS particles by dithiol linkers, and detected potassium ions at levels as low as 10^{-6} M.[286] Similarly, thioethoxy-functionalised calix[4]arene has been attached to the surface of CdSe/ZnS QDs, and been used to sense mercury ions *via* fluorescence quenching with a limit of detection of 15 nM.[287] More complicated structures have also been reported, such as tris(phenylureido)calix[6]arene with three pendant long-chain thiols to facilitate coordination to QDs. This structure was found to decrease the emission quantum yield by 30% when attached to CdSe/ZnS QDs, and induced a bathochromic shift of 5 nm in the emission.[288] The paper also reported the quenching effect of bipyridinium cations on the QDs emission. Related to this is the use of pyridyl-substituted free-base porphyrins which have been coordinated to CdSe/ZnS QDs through amine groups, quenching the emission by a FRET mechanism.[289]

In an example of a more complicated structure based on calixarenes,[290] CdSe/ZnS QDs were phase-transferred with glutathione, and linked with (3-aminophenyl)boronic acid and then β-cyclodextrin. Rhodamine B was then inserted into the cyclodextrin cavity. The dye absorbed at the emission wavelength of the QD, so excitation of the QD resulted in emission of the dye *via* a FRET process. In the presence of a specific analyte, the dye was displaced and emission from the QD dominated upon excitation. The QD/ cyclodextrin without the dye could also be used as a sensor for various organic groups, as long as the analyte acted as an electron transfer quencher.

Related to this is the reversible phase transfer of QDs based on the formation of complexes between ferrocene and β-cyclodextrin.[291] In this reaction, TOPO-capped CdSe/ZnS QDs were dissolved in an organic solution containing either 6-ferrocenyl-1-hexanethiol or 11-ferrocenyl-1-undecanethiol and gently heated under an inert atmosphere. Reaction of a chloroform solution of the ferrocene-modified QDs with a phosphate buffer solution of cyclodextrin resulted in phase transfer as the ferrocene unit threaded into the hydrophilic cyclodextrin, and QD emission was quenched by the cyclodextrin. This was reversed by adding a solution of sodium-2-naphthalene sulfonate and adamantine carboxylate, known to displace ferrocene from the hydrophobic interior of the cyclodextrin. Similarly, adamantaneacetic acid-capped nanoparticles of the up-converting phosphor $NaYF_4$:Yb,Er were phase-transferred from chloroform into water using simple β-cyclodextrin, exhibiting no notable loss of emission and with the final water-soluble material showing colloidal stability over the pH range 3–10.[292] Interestingly, the phase transfer was not successful when α-cyclodextrin was used.

6.10 Biological Molecules as Capping Agents

In some cases, when a new ligand is added to a nanoparticle surface, it is with the aim of attaching a further biological molecule. However, the use of a capping agent as a bridge to a biological molecule[135] is not always necessary. The successful use of thiols has resulted in direct ligand exchange for thiolated biological reagents such as peptides,[293] lipids,[294] glycoconjugates,[295] glycopeptide antibiotics,[296,297] neurotransmitters,[298–301] chemotherapeutic drugs,[302] hormones[299] and DNA,[303] and has now become a common and simple method of bioactivating semiconductor QDs, based on previous studies of gold nanoparticle/DNA interactions.[304–306] How the biological molecule of interest orientates itself once linked to a particle is obviously a key parameter, and reports exist on how for, example, DNA bases interact with a nanoparticle surface.[307] The coordination of a single strand of DNA onto a particle surface opens obvious possibilities—attachment of, for example, a dye molecule to a complementary strand of DNA allowed investigations in FRET.[308,309]

Other biological molecules are notable for their ability to act as a linking molecule in the same manner as simple mercaptoacids in early studies.[127] Glutathione, a tripeptide, has also been successfully used as a thiol-based surfactant which has a further two carboxylic acid groups and an amine group available for cross-linking once on the nanoparticle surface.[310] Glutathione has been used to phase-transfer CdSeTe/CdS QDs into water. In this case, a macrocycle, 1,4,7,10-tetraazacyclododecane-1,4,7,10-tetra-acetic acid (DOTA) was linked to one of the available amine groups which then trapped a gadolinium ion while maintaining the QD luminescence. This was then demonstrated to be MRI active, and hence the particles could potentially be used as a bimodal imaging agent.[311] A related ligand, glutathione tetra-methyl-ammonium salt, has also been used to phase-transfer metal nano-particles and QDs into water.[312] In this case, a methanolic solution of the glutathione was necessary because of the lack of mutual solubility. Addition of the ligand initially caused the particles to precipitate, although they dissolved back into solution as the polarity of the solvent increased, and finally, rapid phase transfer was affected by addition of water. The phase transfer was reversible upon further addition of hexadecyltrimethylammonium bromide, which coordinated to the positive charge of the glutathione ligand, resulting in long alkyl chains exposed to solution. The hexadecyl-trimethylammonium bromide could then be removed by addition of tetra-methylammonium decanoate, allowing the particles to return to the aqueous phase.

Ai *et al.* have examined the role of a combination of amino acids in capping CdSe/ZnS QDs, and found that surface exchange of TOPO for a combination of histidine and *N*-acetyl-cysteine (NAC) at pH 8 red-shifted the absorption spectra and increased the photoluminescence quantum yield.[313] Interestingly, the addition of the amino acids on their own was found to be ineffective. The use of histidine takes advantage of the well-known interactions between the imidazole group and zinc (the increase in photoluminescence from amine species has already been discussed earlier), while the use of the thiol-containing NAC introduces thiolates that have been known to increase the emission intensity at low concentration (again, discussed earlier). The use of histidine-rich peptides as capping agents for QDs has also been explored by Sapsford *et al.* who described the binding affinity of such materials to QD surfaces in assays, with peptides with more than four monomer units binding better, notably HIS_6.[314] (Similarly, histidine-rich genetically modified proteins have been used to assemble QDs.[315]) The presence of the native ligands on the QD surface also affected the ability of the peptide to attach to the surface; peptides with a small molecular mass could easily attach to DHLA and DHLA–PEG passivated particles, but larger proteins rich in histidine were not able to penetrate the ligand shell. These peptides could also be engineered to provide spacing units and pendant functional groups such as a 4-formylbenzoyl, which were then available for further reactions such as aniline-catalysed hydrazone ligation.[316]

These reactions allowed simple conjugation of, for example, further luminescent peptides onto the QD for FRET studies and enzymatic assays. Polyhistidine-based peptides have also been used to attach dopamine to QDs, forming QD–dopamine–peptide bioconjugates that acted as pH sensors.[317]

Related polypeptides, polymers of amino acids, which can form micelle-like structures that encapsulate a number of nanoparticles to form a large cluster of nanoparticles have also been used in phase-transfer procedures. A block copolypeptide such as poly(EG_2-lys)$_{100}$-b-poly(asp)$_{30}$ linked to organically soluble Fe_2O_3 particles *via* the aspartic acid residue resulted in a cluster of 20 particles, exhibiting enhanced stability in physiological conditions.[318] In related work, a peptide/polymer hybrid material consisting of a peptide backbone with PEG side groups and thiol/amine groups for coordination, produced to a very specific accuracy with regard to chain length and functional groups, has been used to phase-transfer CdSe/ZnCdS core/shell QDs, with a drop in quantum yield from 48% to 16%.[319] Cysteine-functionalised polyaspartic acid has also been used to cap QDs,[320] although the hydrophobic particles required phase transfer into a micelle prior to polymer capping. The polyaspartic acid-capped QDs could them be functionalised with antibodies, and used in protein detection studies and cell labelling. Upon phase transfer, no emission quenching was observed and the materials appeared stable for months.

Cysteine-rich α-peptides (phytochelatins) have also been used in the phase transfer of QDs and labelling studies. The peptides were engineered to consist of 20 amino acids, 10 of which formed a cysteine-rich hydrophobic domain that coordinated to the QDs, while the other 10 formed a polar, negatively charged hydrophilic linking domain that could be engineered to include functional groups, such as biotin or binding sequences.[321] Interestingly, the adhesive cysteine-rich domain required an alanine group between every two cysteine groups, as a cysteine-only chain would not stabilise the particle (although it was hypothesised that the alanine group, in the form of bulky 3-cyclohexylalanine, could also have been involved in the bonding process). Here, the cysteine groups were converted to cysteine thiolates that coordinated to the zinc sulfide shell by addition of a base. The resulting materials showed excellent stability in buffer solution, with a minimal size increase consistent with the attachment of a single layer of peptide, while displaying no obvious change in the position of absorption or emission, although a reduction in quantum yield was observed (the emission quenching in buffer solutions could reportedly be addressed by UV irradiation[322]). Incubation of biotin-functionalised particles with HeLa cells that expressed GPI-anchored avidin-CD14 chimeric proteins resulted in targeted cell labelling with little non-specific labelling observed.

In this chapter, we have shown how the synthesis of the inorganic core is only part of the inherent chemistry required for the preparation of useful QDs. Functionality can be induced by using differing surface ligands and a range of applications can be realised by the exploitation of engineered surfactants.

References

1. R. M. de Silva, V. Palshin, K. M. Nalin de Silva, L. L. Henry and C. S. S. R. Kumar., *J. Mater. Chem.,* 2008, **18**, 738.
2. M. Soreni-Harari, N. Yaacobi-Gross, D. Steiner, A. Aharoni, U. Banin, O. Millo and N. Tessler, *Nano Lett.,* 2008, **8**, 678.
3. R. M. Rioux, H. Song, S. Habas, K. Niesz, J. D. Hoefelmeyer, P. Yang and G. A. Somorjai, *Top. Catal.,* 2006, **39**, 167.
4. R. W. Meulenberg, J. R. I. Lee, S. K. McCall, K. M. Hanif, D. Haskel, J. C. Lang, L. J. Terminello and T. van Buuren, *J. Am. Chem. Soc.,* 2009, **131**, 6888.
5. M. S. Seehra, P. Dutta, S. Neeleshwar, Y.-Y. Chen, C. L. Chen, S. W. Chou, C. C. Chen, C.-L. Dong and C.-L. Chang, *Adv. Mater.,* 2008, **20**, 1656.
6. T. Teranishi, Y. Inoue, M. Nakaya, Y. Oumi and T. Sano, *J. Am. Chem. Soc.,* 2004, **126**, 9914.
7. C. Singh, P. K. Ghorai, M. A. Horsch, A. M. Jackson, R. G. Larson, F. Stellacci and S. C. Glotzer, *Phys. Rev. Lett.,* 2007, **99**, 226106.
8. Y. Cesbron, C. P. Shaw, J. P. Birchall, P. Free and R. Lévy, *Small,* 2012, **8**, 3714.
9. Y. Ofir, B. Samanta, P. Arumugan and V. M. Rotello, *Adv. Mater.,* 2007, **19**, 4075.
10. R. De Palma, S. Peeters, M. J. Van Bael, H. Van den Rul, K. Bonroy, W. Laureyn, J. Mullens, G. Borghs and G. Maes, *Chem. Mater.,* 2007, **19**, 1821.
11. S. V. Voitekhovich, D. V. Talapin, C. Klinke, A. Kornowski and H. Weller, *Chem. Mater.,* 2008, **20**, 4545.
12. E. Jang, S. Jun, Y. Chung and L. Pu, *J. Phys. Chem. B,* 2004, **108**, 4597.
13. The term 'solubility' is used to describe the dispersion of nanoparticles in a compatible solvent. Whether the particles actually dissolve in a classic sense is debatable, since they form a colloid. However, most TOPO-capped particles are easily dispersible in solvents such as hexane and toluene, giving optical clear 'solutions' that are almost indefinitely stable in a similar manner to organic compounds. Addition of a more polar solvent such as methanol induces precipitation. This gives the particles a definite set of solvents and non-solvents and allows the materials to be manipulated like a common organic reagent. We therefore refer to the process as 'dissolving' or 'dispersing' and the terms are often used interchangeably.
14. M. Green, N. Allsop, G. Wakefield, P. J. Dobson and J. L. Hutchison, *J. Mater. Chem.,* 2002, **12**, 2671.
15. C. B. Murray, D. J. Norris and M. G. Bawendi, *J. Am. Chem. Soc.,* 1993, **115**, 8706.
16. S. Sun and C. B. Murray, *J. Appl. Phys.,* 1999, **85**, 4325.
17. P. Guyot-Sionnest, M. Shim, C. Matranga and M. Hines, *Phys. Rev. B: Condens. Matter Mater. Phys.,* 1999, **60**, R2181.

18. P. Guyot-Sionnest, B. Wehrenberg and D. Yu, *J. Chem. Phys.,* 2005, **123**, 074709.
19. L. R. Becerra, C. B. Murray, R. G. Griffin and M. G. Bawendi, *J. Chem. Phys.,* 1994, **100**, 3297.
20. J. K. Lorenz and A. B. Ellis, *J. Am. Chem. Soc.,* 1998, **120**, 10970.
21. J. S. Owen, J. Park, P.-E. Trudeau and A. P. Alivistatos, *J. Am. Chem. Soc.,* 2008, **130**, 12279.
22. H. Asami, Y. Abe, T. Ohtsu, I. Kamiya and M. Hara, *J. Phys. Chem. B,* 2003, **107**, 12566.
23. P. Schapotschnikow, B. Hommersom and T. J. H. Vlugt, *J. Phys. Chem. C,* 2009, **113**, 12690.
24. A. J. Morris-Cohen, M. T. Frederick, G. D. Lilly, E. A. McArthur and E. A. Weiss, *J. Phys. Chem. Lett.,* 2010, **1**, 1078.
25. A. Punzder, A. J. Williams, N. Zaitseva, G. Galli, L. Manna and A. P. Alivisatos, *Nano Lett.,* 2004, **4**, 2361.
26. L. Manna, L. W. Wang, R. Cingolani and A. P. Alivisatos, *J. Phys. Chem. B,* 2005, **109**, 6183.
27. F. Wang, R. Tang and W. E. Buhro, *Nano Lett.,* 2008, **8**, 3521.
28. W. Wang, S. Banerjee, S. Jia, M. L. Steigerwald and I. P. Herman, *Chem. Mater.,* 2007, **19**, 2573.
29. J. Huang, M. V. Kovalenko and D. V. Talapin, *J. Am. Chem. Soc.,* 2010, **132**, 15866.
30. A. J. Morris-Chen, M. D. Donakowski, K. E. Knowles and E. A. Weiss, *J. Phys. Chem. C,* 2010, **114**, 897.
31. S. M. Stczynski, Y.-U. Kwon and M. L. Steigerwald, *J. Organomet. Chem.,* 1993, **449**, 167.
32. M. Steigerwald, *Polyhedron,* 1994, **13**, 1245.
33. J. G. Brennan, T. Siegrist, Y. U. Kwon, S. M. Stuczynski and M. L. Steigerwald, *J. Am. Chem. Soc.,* 1992, **114**, 10334.
34. J. Zhang, K. Sun, A. Khumbar and J. Fang, *J. Phys. Chem. C,* 2008, **112**, 5454.
35. A. E. Henkes, Y. Vasquez and R. E. Schaak, *J. Am. Chem. Soc.,* 2007, **129**, 1896.
36. A. E. Henkes and R. E. Schaak, *Chem. Mater.,* 2007, **19**, 4234.
37. J. Y. Rempel, B. L. Trout, M. G. Bawendi and K. F. Jensen, *J. Phys. Chem. B,* 2006, **110**, 18007.
38. J. Y. Rempel, B. L. Trout, M. G. Bawendi and K. F. Jensen, *J. Phys. Chem. B,* 2005, **109**, 19320.
39. G. Kalyuzhny and R. W. Murray, *J. Phys. Chem. B,* 2005, **109**, 7012.
40. J. Jasieniak, C. Bullen, J. van Embden and P. Mulvaney, *J. Phys. Chem. B,* 2005, **109**, 20665.
41. J. Jasieniak and P. Mulvaney, *J. Am. Chem. Soc.,* 2007, **129**, 2841.
42. S. Sapra, A. L. Rogach and J. Feldmann, *J. Mater. Chem.,* 2006, **16**, 3391.
43. W. Kim, S. J. Lim, S. Jung and S. K. Shin, *J. Phys. Chem. C,* 2010, **114**, 1539.
44. A. L. Washington and G. F. Strouse, *Chem. Mater.,* 2009, **21**, 2770.

45. J. E. Bowen Katari, V. L. Colvin and A. P. Alivisatos, *J. Phys. Chem.*, 1994, **98**, 4109.
46. S. J. Rosenthal, J. McBride, S. J. Pennycook and L. C. Feldman, *Surf. Sci. Rep.*, 2007, **62**, 111.
47. P.-J. Deboutiére, V. Martinez, K. Philippot and B. Chaudret, *Dalton Trans.*, 2009, 10172.
48. O. I. Mićić, J. R. Sprague, C. J. Curtis, K. M. Jones, J. L. Machol, A. J. Nozik, H. Giessen, B. Fluegel, G. Mohs and N. Peyghambarian, *J. Phys. Chem.*, 1995, **99**, 7754.
49. M. Furis, Y. Sahoo, D. J. MacRae, F. S. Manciu, A. N. Cartwright and P. N. Prasad, *J. Phys. Chem. B*, 2003, **107**, 11622.
50. D. V. Talapin, A. L. Rogach, I. Mekis, S. Haubold, A. Kornowski, M. Haase and H. Weller, *Colloids Surf., A*, 2002, **202**, 145.
51. D. V. Talapin, A. L. Rogach, A. Kornowski, M. Haase and H. Weller, *Nano Lett.*, 2001, **1**, 207.
52. W. K. Woo, PhD thesis, MIT, 2002.
53. C. Landes and M. A. El-Sayed, *J. Phys. Chem. A*, 2002, **106**, 7621.
54. R. Li, J. Lee, B. Yang, D. N. Horspool, M. Aindow and F. Papadimitrakopoulos, *J. Am. Chem. Soc.*, 2005, **127**, 2524.
55. J.-Y. Zhang, X.-Y. Wang, M. Xao, L. Qu and X. Peng, *Appl. Phys. Lett.*, 2002, **81**, 2076.
56. L. Qu and X. Peng, *J. Am. Chem. Soc.*, 2002, **124**, 2049.
57. R. W. Meulenberg, T. Jennings and G. F. Strouse, *Phys. Rev. B: Condens. Matter Mater. Phys.*, 2004, **70**, 235311.
58. C. Bullen and P. Mulvaney, *Langmuir*, 2006, **22**, 3007.
59. E. E. Foos, J. Wilkinson, A. J. Mäkinen, N. J. Watkins, Z. H. Kafafi and J. P. Long, *Chem. Mater.*, 2006, **18**, 2886.
60. C. Landes, C. Burda, M. Braun and M. A. El-Sayed, *J. Phys. Chem. B*, 2001, **105**, 2981.
61. C. F. Landes, M. Braun and M. A. El-Sayed, *J. Phys. Chem. B*, 2001, **105**, 10554.
62. A. M. Munro, I. Jen-La Plante, M. S. Ng and D. S. Ginger, *J. Phys. Chem. C*, 2007, **111**, 6220.
63. O. I. Mićić, S. P. Ahrenkiel and A. J. Nozik, *Appl. Phys. Lett.*, 2001, **78**, 4022.
64. J. M. Nedeljković, O. I. Mićić, S. P. Ahrenkiel, A. Miedaner and A. J. Nozik, *J. Am. Chem. Soc.*, 2004, **126**, 2632.
65. P. M. Allen, B. J. Walker and M. G. Bawendi, *Angew. Chem., Int. Ed.*, 2010, **49**, 760.
66. W. W. Yu, Y. A. Wang and X. Peng, *Chem. Mater.*, 2003, **15**, 4300.
67. Z. Xu, C. Shen, Y. Hou, H. Gao and S. Sun, *Chem. Mater.*, 2009, **21**, 1778.
68. N. Pradhan, D. Reifsnyder, R. Xie, J. Aldana and X. Peng, *J. Am. Chem. Soc.*, 2007, **129**, 9500.
69. Y. Bao, M. Beerman, A. B. Pakhomov and K. M. Krishnan, *J. Phys. Chem. B*, 2005, **109**, 7220 (supplementary information).
70. M. Yamamoto and M. Nakmoto, *J. Mater. Chem.*, 2003, **13**, 2064.

71. M. Aslam, L. Fu, M. Su, K. Vijayamohanan and V. P. Dravid, *J. Mater. Chem.*, 2004, **14**, 1795.
72. N. Shukla, C. Liu, P. M. Jones and D. Weller, *J. Magn. Magn. Mater.*, 2003, **266**, 178.
73. X. Ji, D. Copenhaver, C. Sichmeller and X. Peng, *J. Am. Chem. Soc.*, 2008, **130**, 5726.
74. M. Green, P. Rahman and D. Smyth-Boyle, *Chem. Commun.*, 2007, 574.
75. C. Wang, Y. Jiang, Z. Zhang, G. Li, L. Chen and J. Jie, *J. Nanosci. Nanotechnol.*, 2009, **9**, 4735.
76. L. Chen, Y. Jiang, C. Wang, X. Liu, Y. Chen and J. Jie, *J. Exp. Nanosci.*, 2010, **5**, 106.
77. F. V. Mikulec, PhD thesis, MIT, 1999.
78. V. Monnier, M. Delalande, P. Bayle-Guillemaud, Y. Samson and P. Reiss, *Small*, 2008, **4**, 1139.
79. M. Brust, M. Walker, D. Bethell, D. J. Schiffrin and R. J. Whyman, *Chem. Commun.*, 1994, 801.
80. M. Brust, J. Fink, D. Bethell, D. J. Schiffrin and C. J. Kiely, *Chem. Commun.*, 1995, 1655.
81. M. D. Porter, T. B. Bright, D. L. Allara and C. E. D. Chidsey, *J. Am. Chem. Soc.*, 1987, **109**, 3559.
82. T. Rajh, O. I. Mićić and A. J. Nozik, *J. Phys. Chem.*, 1993, **97**, 11999.
83. M. Brust, N. Stuhr-Hansen, K. Nørgaard, J. B. Christensen, L. K. Nielsen and T. Bjørnholm, *Nano Lett.*, 2001, **1**, 189.
84. I. L. Medintz, H. Tetsuo Uyeda, E. R. Goldman and H. Mattoussi, *Nat. Mater.*, 2005, **4**, 435.
85. H. Häkkinen, R. N. Barnett and U. Landman, *Phys. Rev. Lett.*, 1999, **82**, 3264.
86. I. L. Garzón, C. Rovira, K. Michaelian, M. R. Beltrán, P. Ordejón, J. Junquera, D. Sánchez-Portal, E. Artacho and J. M. Soler, *Phys. Rev. Lett.*, 2000, **85**, 5250.
87. M. Hasan, D. Bethell and M. Brust, *J. Am. Chem. Soc.*, 2002, **124**, 1132.
88. B.-K. Pong, B. L. Trout and J.-Y. Lee, *Langmuir*, 2008, **24**, 5270.
89. A. L. Rogach, *Mater. Sci. Eng., B*, 2000, **69–70**, 435.
90. A. L. Rogach, A. Kornowski, M. Gao, A. Eychmüller and H. Weller, *J. Phys. Chem. B*, 1999, **103**, 3065.
91. S. F. Wuister, C. de Mello Donegá and A. Meijerink, *J. Phys. Chem. B*, 2004, **108**, 17393.
92. H. Y. Acar, R. Kas, E. Yurtsever, C. Ozen and I. Lieberwirth, *J. Phys. Chem. C*, 2009, **113**, 10005.
93. X. Gao, W. C. W. Chan and S. Nie, *J. Biomed. Opt.*, 2002, 7, 532.
94. A. M. Munro and D. S. Ginger, *Nano Lett.*, 2008, **8**, 2585.
95. A. L. Rogach, A. Kornowski, M. Gao, A. Eychmüller and H. Weller, *J. Phys. Chem. B*, 1999, **103**, 3065.
96. J. Aldana, Y. A. Wang and X. Peng, *J. Am. Chem. Soc.*, 2001, **123**, 8844.
97. S. Hohng and T. Ha, *J. Am. Chem. Soc.*, 2004, **126**, 1324.
98. H. Chen, H. Gai and E. S. Yeung, *Chem. Commun.*, 2009, 1676.

99. S. Jeong, M. Achermann, J. Nanda, S. Ivanov, V. I. Klimov and J. A. Hollingsworth, *J. Am. Chem. Soc.,* 2005, **127**, 10126.
100. Q. B. Wang, Y. Xu, X. H. Zhao, Y. Chang, Y. Liu, L. J. Jiamg, J. Sharma, D. K. Seo and H. Yan, *J. Am. Chem. Soc.,* 2007, **129**, 6380.
101. D. R. Baker and P. V. Kamat, *Langmuir,* 2010, **26**, 11276.
102. J. Aldana, N. Lavelle, Y. Wang and X. Peng, *J. Am. Chem. Soc.,* 2005, **127**, 2496.
103. J. Schrier and L.-W. Wang, *J. Phys. Chem. B,* 2006, **110**, 11982.
104. P. S. Billone, L. Maretti, V. Maurel and J. C. Scaiano, *J. Am. Chem. Soc.,* 2007, **129**, 14150.
105. O. I. Mićić, J. Sprague, Z. Lu and A. J. Nozik, *Appl. Phys. Lett.,* 1996, **68**, 3150.
106. S. E. Khalafalla and G. W. Reimers, *US Pat.*, no. 3764540, 1973.
107. R. Tadmor, R. E. Rosensweig, J. Frey and J. Klein, *Langmuir,* 2000, **16**, 9117.
108. X. Peng and J. Thessing, *Struct. Bonding,* 2005, **118**, 79.
109. R. Xie and X. Peng, *Angew. Chem., Int. Ed.,* 2008, **47**, 7677.
110. K. S. Suslick, M. Fang and T. Hyeon, *J. Am. Chem. Soc.,* 1996, **118**, 11960.
111. S. Sun and C. B. Murray, *J. Appl. Phys.,* 1999, **85**, 4325.
112. N. Wu, L. Fu, M. Aslam, K. C. Wong and V. P. Dravid, *Nano Lett.,* 2004, **4**, 383.
113. T. Bala, A. Swami, B. L. V. Prasad and M. Sastry, *J. Colloid Interface Sci.,* 2005, **283**, 422.
114. A. Prakash, H. Zhu, C. J. Jones, D. N. Benoit, A. Z. Ellsworth, E. L. Bryant and V. L. Colvin, *ACS Nano,* 2009, **3**, 2139.
115. V. Pérez-Dieste, O. M. Castellini, J. N. Crain, M. A. Eriksson, A. Kirakosian, J.-L. Lin, C. T. Black and C. B. Murray, *Appl. Phys. Lett.,* 2003, **83**, 5053.
116. G. G. Yordanov, G. D. Gicheva, B. H. Bochev, C. D. Dushkin and E. Adachi, *Colloids Surf., A,* 2006, **273**, 10.
117. B. D. Dickerson, D. M. Irving, E. Herz, R. O. Claus, W. B. Spillman and K. E. Meissner, *Appl. Phys. Lett.,* 2005, **86**, 171915.
118. D. Wu, M. E. Kordesch and P. G. Van Patten, *Chem. Mater.,* 2005, **17**, 6436.
119. A. M. Smith, H. Duan, M. N. Rhyner, G. Ruan and S. Nie, *Phys. Chem. Chem. Phys.,* 2006, **8**, 3895.
120. C. B. Murray, PhD thesis, MIT, 1995.
121. X. Peng, M. C. Schlamp, A. V. Kadanavich and A. P. Alivisatos, *J. Am. Chem. Soc.,* 1997, **119**, 7019.
122. I. Lokteva, N. Radchev, F. Witt, H. Borchert, J. Parisi and J. Kolny-Olesiak, *J. Phys. Chem. C,* 2010, **114**, 12784.
123. M. Kuno, J. K. Lee, B. O. Dabbousi, F. V. Mikulec and M. G. Bawendi, *J. Chem. Phys.,* 1997, **106**, 9869.
124. M. A. Caldwell, A. E. Albers, S. C. Levy, T. E. Pick, B. E. Cohen, B. A. Helms and D. J. Milliron, *Chem. Commun.,* 2011, **47**, 556.

125. X. Peng, T. E. Wilson, A. P. Alivisatos and P. G. Schultz, *Angew. Chem., Int. Ed. Engl.,* 1997, **36**, 145.
126. K. Hoppe, E. Geidel, H. Weller and A. Eychmüller, *Phys. Chem. Chem. Phys.,* 2002, **4**, 1704.
127. W. C. W. Chan and S. Nie, *Science,* 1998, **281**, 2016.
128. V. V. Breus, C. D. Heyes and G. U. Nienhaus, *J. Phys. Chem. C,* 2007, **111**, 18589.
129. J. K. Jaiswal, E. R. Goldman, H. Mattoussi and S. M. Simon, *Nat. Methods,* 2004, **1**, 73.
130. H. Shen, A. M. Jawaid and P. T. Snee, *ACS Nano,* 2009, **3**, 915.
131. C. G. Golander, J. N. Herron, K. Lim, P. Claesson and J. D. Andrade, *Polyetherglycol Chemistry, Biotechnical and Biomedical Applications*, Plenum, New York, 1992.
132. E. L. Bentyen, I. D. Tomlinson, J. Mason, P. Gresch, M. R. Warnement, D. Wright, E. Sanders-Bush, R. Blakely and S. J. Rosenthal, *Bioconjugate Chem.,* 2005, **16**, 1488.
133. C. Kirchner, T. Liedl, S. Kudera, T. Pellegrino, A. Muñoz, H. E. Gaub, S. Stölzle, N. Fertig and W. J. Parak, *Nano Lett.,* 2005, **5**, 331.
134. W. Jiang, S. Mardyani, H. Fischer and W. C. W. Chan, *Chem. Mater.,* 2006, **18**, 872.
135. E. R. Goldman, E. D. Balighian, H. Mattoussi, M. K. Kuno, J. M. Mauro, P. T. Tran and G. P. Anderson, *J. Am. Chem. Soc.,* 2002, **124**, 6378.
136. J. K. Jaiswal, H. Mattoussi, J. M. Mauro and S. M. Simon, *Nat. Biotechnol.,* 2003, **21**, 47.
137. S. F. Wuister, I. Swart, F. van Driel, S. G. Hickey and C. de Mello Donegá, *Nano Lett.,* 2003, **3**, 503.
138. W. Liu, H. S. Choi, J. P. Zimmer, E. Tanaka, J. V. Frangioni and M. Bawendi, *J. Am. Chem. Soc.,* 2007, **129**, 14530.
139. H. S. Choi, W. Liu, P. Misra, E. Tanaka, J. P. Zimmer, B. I. Ipe, M. G. Bawendi and J. V. Frangioni, *Nat. Biotechnol.,* 2007, **25**, 1165.
140. S. Pathak, S.-K. Choi, N. Arnheim and M. E. Thompson, *J. Am. Chem. Soc.,* 2001, **123**, 4103.
141. H. S. Choi, W. Kiu, F. Liu, K. Nasr, P. Misra, M. G. Bawendi and J. V. Frangioni, *Nat. Nanotechnol.,* 2010, **5**, 42.
142. M. Bruchez, M. Moronne, P. Gin, S. Weiss and A. P. Alivisatos, *Science,* 1998, **281**, 2013.
143. P. Mulvaney, L. M. Liz-Marzán, M. Giersig and T. Ung, *J. Mater. Chem.,* 2000, **10**, 1259.
144. D. Gerion, F. Pinaud, S. C. Williams, W. J. Parak, D. Zanchet, S. Weiss and A. P. Alivisatos, *J. Phys. Chem. B,* 2001, **105**, 8861.
145. C. Earhart, N. R. Jana, N. Erathodiyil and J. Ying, *Langmuir,* 2008, **24**, 6215.
146. J.-J. Park, P. Prabhakaran, K. K. Jang, Y. G. Lee, J. Lee, K. H. Lee, J. Hur, J.-M. Kim, N. Cho, Y. Son, D.-Y. Yang and K.-S. Lee, *Nano Lett.,* 2010, **10**, 2310.

147. R. Hong, N. O. Fischer, T. Emrick and V. M. Rotello, *Chem. Mater.,* 2005, **17**, 4617.
148. J. Kim, H.-Y. Park, J. Kim, J. Ryu, D. Y. Kwon, R. Grailhe and R. Song, *Chem. Commun.,* 2008, 1910.
149. K. Susumu, H. T. Uyeda, I. L. Medintz, T. Pons, J. B. Delehanty and H. Mattoussi, *J. Am. Chem. Soc.,* 2007, **129**, 13987.
150. H. Mattoussi, J. M. Mauro, E. R. Goodman, G. P. Anderson, V. C. Sundar, F. V. Mikulec and M. G. Bawendi, *J. Am. Chem. Soc.,* 2000, **122**, 12142.
151. H. Mattoussi, J. M. Mauro, E. R. Goodman, T. M. Green, G. P. Anderson, V. C. Sundar and M. G. Bawendi, *Phys. Status Solidi B,* 2001, **224**, 277.
152. A. R. Clapp, E. R. Goldman and H. Mattoussi, *Nat. Protoc.,* 2006, **1**, 1258.
153. J. P. Zimmer, S.-W. Kim, S. Ohnishi, E. Tanaka, J. V. Frangioni and M. G. Bawendi, *J. Am. Chem. Soc.,* 2006, **128**, 2526.
154. H. T. Uyeda, I. L. Medintz, J. K. Jaiswal, S. M. Simon and H. Mattoussi, *J. Am. Chem. Soc.,* 2005, **127**, 3870.
155. W. Liu, M. Howarth, A. B. Greytak, Y. Zheng, D. G. Nocera, A. Y. Ting and M. G. Bawendi, *J. Am. Chem. Soc.,* 2008, **130**, 1274.
156. B. C. Mei, K. Susumu, I. L. Medintz, J. B. Delanty, T. J. Mountziaris and H. Mattoussi, *J. Mater. Chem.,* 2008, **18**, 4949.
157. M. H. Stewart, K. Susumu, B. C. Mei, I. L. Medintz, J. B. Delehanty, J. B. Blanco-Canosa, P. E. Dawson and H. Mattoussi, *J. Am. Chem. Soc.,* 2010, **132**, 9804.
158. E. Muro, T. Pons, N. Lequeux, A. Fragola, N. Sanson, Z. Lenkei and B. Dubertret, *J. Am. Chem. Soc.,* 2010, **132**, 4556.
159. I. Yidiz, S. Ray, T. Benelli and F. M. Raymo, *J. Mater. Chem.,* 2008, **18**, 3940.
160. M. Tomasulo, I. Yildiz, S. L. Kaanumalle and F. M. Rayno, *Langmuir,* 2006, **22**, 10284.
161. M. Tomasulo, I. Yildiz and F. M. Rayno, *J. Phys. Chem. B,* 2006, **110**, 3853.
162. R. C. Mulrooney, N. Singh, N. Kaur and J. F. Callan, *Chem. Commun.,* 2009, 686.
163. W. R. Algar and U. J. Krull, *Langmuir,* 2008, **24**, 5514.
164. F. Dubois, B. Mahler, B. Dubertret, E. Doris and C. Mioskowski, *J. Am. Chem. Soc.,* 2007, **129**, 482.
165. J. Wang, J. Xu, M. D. Goodman, Y. Chen, M. Cai, J. Shinar and Z. Lin, *J. Mater. Chem.,* 2008, **18**, 3270.
166. M. T. Frederick and E. A. Weiss, *ACS Nano,* 2010, **4**, 3195.
167. R. Debnath, J. Tang, D. A. Barkhouse, X. Wang, A. G. Pattantyus-Abraham, L. Brzozowski, L. Levina and E. H. Sargent, *J. Am. Chem. Soc.,* 2010, **132**, 5952.
168. Y.-W. Jun, Y.-M. Huh, J.-S. Choi, J.-H. Lee, H.-T. Song, S. Kim, S. Yoon, K.-S. Kim, J.-S. Shin, J.-S. Suh and J. Cheon, *J. Am. Chem. Soc.,* 2005, **127**, 5732.
169. I.-S. Liu, H.-H. Lo, C.-T. Chien, Y.-Y. Lin, C.-W. Chen, Y.-F. Chen, W.-F. Su and S. C. Liou, *J. Mater. Chem.,* 2008, **18**, 675.
170. B. C. Sih and M. O. Wolf, *J. Phys. Chem. C,* 2007, **111**, 17184.

171. W. J. Kim, S. J. Kim, K.-S. Lee, M. Samoc, A. N. Cartwright and P. N. Prasad, *Nano Lett.,* 2008, **8**, 3262.
172. J. Seo, W. J. Kim, S. J. Kim, K.-S. Lee, A. N. Cartwright and P. N. Prasad, *Appl. Phys. Lett.,* 2009, **94**, 133302.
173. J. De Girolamo, P. Reiss and A. Pron, *J. Phys. Chem. C,* 2007, **111**, 14681.
174. C. Querner, P. Reiss, J. Bleuse and A. Pron, *J. Am. Chem. Soc.,* 2004, **126**, 11574.
175. C. Querner, P. Reiss, M. Zagorska, O. Renault, R. Payerne, F. Genoud, P. Rannou and A. Pron, *J. Mater. Chem.,* 2005, **15**, 554.
176. L. Zhu, M.-Q. Qiang Zhu, J. K. Hurst and A. D. Q. Li, *J. Am. Chem. Soc.,* 2005, **127**, 8968.
177. N. Gaponik, D. V. Talapin, A. L. Rogach, A. Eychmüller and H. Weller, *Nano Lett.,* 2002, **2**, 803.
178. J. Ziegler, A. Merkulov, M. Grabole, U. Resch-Genger and T. Nann, *Langmuir,* 2007, **23**, 7751.
179. R. Vinayakan, T. Shanmugapriya, P. V. Nair, P. Ramamurthy and K. G. Thomas, *J. Phys. Chem. C,* 2007, **111**, 10146.
180. N. C. Greenham, X. Peng and A. P. Alivisatos, *Phys. Rev. B: Condens. Matter Mater. Phys.,* 1996, **54**, 17628.
181. D. J. Milliron, A. P. Alivisatos, C. Pitois, C. Edder and J. M. J. Frechet, *Adv. Mater.,* 2003, **15**, 58.
182. J. Locklin, D. Patton, S. Deng, A. Baba, M. Millan and R. C. Advincula, *Chem. Mater.,* 2004, **16**, 5187.
183. K. Sill and T. Emrick, *Chem. Mater.,* 2004, **16**, 1240.
184. H. Skaff, K. Sill and T. Emrick, *J. Am. Chem. Soc.,* 2004, **126**, 11322.
185. M. Y. Odoi, N. I. Hammer, K. Sill, T. Emrick and M. D. Barnes, *J. Am. Chem. Soc.,* 2006, **128**, 3506.
186. P. K. Sudeep, K. T. Early, K. D. McCarthy, M. Y. Odoi, M. D. Barnes and T. Emrick, *J. Am. Chem. Soc.,* 2008, **130**, 2384.
187. N. I. Hammer, K. T. Early, K. Sill, M. Y. Odoi, T. Emrick and M. D. Barnes, *J. Phys. Chem. B,* 2006, **110**, 14167.
188. H. B. Na, I. S. Lee, H. Seo, Y. Il Park, J. H. Lee, S.-W. Kim and T. Hyeon, *Chem. Commun.,* 2007, 5167.
189. S. Kim and M. G. Bawendi, *J. Am. Chem. Soc.,* 2003, **125**, 14652.
190. S.-W. Kim, S. Kim, J. B. Tracy, A. Jasanoff and M. G. Bawendi, *J. Am. Chem. Soc.,* 2005, **127**, 4556.
191. J. Yang and J. Y. Ying, *Nat. Mater.,* 2009, **8**, 683.
192. H. Skaff and T. Emrick, *Chem. Commun.,* 2003, 52.
193. H. Li, C. Han and L. Zhang, *J. Mater. Chem.,* 2008, **18**, 4543.
194. T. Trindade, P. O'Brien and X.-M. Zhang, *Chem. Mater.,* 1997, **9**, 523.
195. T. Trindade, P. O'Brien and X.-M. Zhang, *J. Mater. Res.,* 1999, **14**, 4140.
196. I. Yildiz and F. M. Raymo, *J. Mater. Chem.,* 2006, **16**, 1118.
197. A. M. Funston, J. J. Jasieniak and P. Mulvaney, *Adv. Mater.,* 2008, **20**, 4274.
198. M. Lattuada and T. A. Hatton, *Langmuir,* 2007, **23**, 2158.

199. M. Kasture, S. Singh, P. Patel, P. A. Joy, A. A. Prabhune, C. V. Ramana and B. L. V. Prasad, *Langmuir,* 2007, **23**, 11409.
200. H.-T. Song, J.-S. Choi, Y.-M. Huh, S. Kim, Y.-W. Jun, J.-S. Suh and J. Cheon, *J. Am. Chem. Soc.,* 2005, **127**, 9992.
201. M. Sykora, M. A. Petruska, J. Alstrum-Acevedo, I. Bezel, T. J. Mayer and V. I. Klimov, *J. Am. Chem. Soc.,* 2006, **128**, 9984.
202. H. Amiri, Z. Zhao, T. M. Dansereau, M. A. Petrukhina and M. A. Carpenter, *J. Phys. Chem. C,* 2010, **114**, 4272.
203. R. C. Shallcross, G. D. D'Ambruoso, J. Pyun and N. R. Armstrong, *J. Am. Chem. Soc.,* 2010, **132**, 2622.
204. J. Xie, C. Xu, Z. Xu, Y. Hou, K. L. Young, S. X. Wang, N. Pourmond and S. Sun, *Chem. Mater.,* 2006, **18**, 5401.
205. B. Wang, J. Hai, Z. Liu, Q. Wang, Z. Yang and S. Sun, *Angew. Chem., Int. Ed.,* 2010, **49**, 4576.
206. Y. Lee, H. Lee, Y. B. Kim, J. Kim, T. Hyeon, H. W. Park, P. B. Messersmith and T. G. Park, *Adv. Mater.,* 2008, **20**, 4154.
207. J. Xie, C. Xu, N. Kohler, Y. Hou and S. Sun, *Adv. Mater.,* 2007, **19**, 3163.
208. S. B. Kim, C. Cai, S. Sun and D. A. Sweigart, *Angew. Chem., Int. Ed.,* 2009, **48**, 2907.
209. T. Zhang, J. Ge, Y. Hu and Y. Yin, *Nano Lett.,* 2007, **7**, 3203.
210. A. F. E. Hezinger, J. Tessmar and A. Göpferich, *Eur. J. Pharm. Biopharm.,* 2008, **68**, 138.
211. H. Mattoussi, L. H. Radzilowski, B. O. Dabbousi, D. E. Fogg, R. R. Schrock, E. L. Thomas, M. F. Rubner and M. G. Bawendi, *J. Appl. Phys.,* 1999, **86**, 4390.
212. G. Liu, X. Yan, Z. Lu, S. A. Curda and J. Lal, *Chem. Mater.,* 2005, **17**, 4985.
213. S. Dayal, N. Kopidakis, D. C. Olson, D. S. Ginley and G. Rumbles, *J. Am. Chem. Soc.,* 2009, **131**, 17726.
214. J. Xu, J. Wang, M. Mitchell, P. Mukherjee, M. Jeffries-El, J. W. Petrich and Z. Lin, *J. Am. Chem. Soc.,* 2007, **129**, 12828.
215. M. D. Goodman, J. Xu, J. Wang and Z. Lin, *Chem. Mater.,* 2009, **21**, 934.
216. I. Robinson, C. Alexander, L. T. Lu, L. D. Tung, D. G. Fernig and N. T. K. Thanh, *Chem. Commun.,* 2007, 4602.
217. W. Thiessen, A. Dubavik, V. Lesnyak, N. Gaponik, A. Eychmüller and T. Wolff, *Phys. Chem. Chem. Phys.,* 2010, **12**, 2063.
218. S. Sun, S. Anders, H. F. Hamann, J.-U. Thiele, J. E. E. Baglin, T. Thomson, E. E. Fullerton, C. B. Murray and B. D. Terris, *J. Am. Chem. Soc.,* 2002, **124**, 2884.
219. T. Nann, *Chem. Commun.,* 2005, 1735.
220. M. S. Nikolic, M. Krack, V. Aleksandrovic, A. Kornowski, S. Förster and H. Weller, *Angew. Chem., Int. Ed.,* 2006, **45**, 6577.
221. E. Pöselt, S. Fischer, S. Foerster and H. Weller, *Langmuir,* 2009, **25**, 13906.
222. X.-S. Wang, T. E. Dykstra, M. R. Salvador, I. Manners, G. D. Scholes and M. A. Winnik, *J. Am. Chem. Soc.,* 2004, **126**, 7784.

223. M. Wang, T. E. Dykstra, X. Lou, M. R. Salvador, G. D. Scholes and M. A. Winnik, *Angew. Chem., Int. Ed.,* 2006, **45**, 2221.
224. T. Zhang, J. Ge, Y. Hu and Y. Yin, *Nano Lett.,* 2007, **7**, 3203.
225. Z. Ceng, S. Liu, P. W. Beines, N. Ding, P. Jakubowicz and W. Knoll, *Chem. Mater.,* 2008, **20**, 7215.
226. Y. Shibasaki, B.-S. Kim, A. J. Young, A. L. McLoon, S. C. Ekker and T. A. Taton, *J. Mater. Chem.,* 2009, **19**, 6324.
227. A. Salcher, M. S. Nikolic, S. Casado, M. Vélez, H. Weller and B. H. Juárez, *J. Mater. Chem.,* 2010, **20**, 1367.
228. M. Bradley, N. Bruno and B. Vincent, *Langmuir,* 2005, **21**, 2750.
229. B. H. Lee, K. W. Kwon and M. Shim, *J. Mater. Chem.,* 2007, **17**, 1284.
230. X. Yang and Y. Zhang, *Langmuir,* 2004, **20**, 6071.
231. J.-Y. Chang, C.-H. Yang and K.-S. Huang, *Nanotechnology,* 2007, **18**, 305305.
232. W. Yin, H. Liu, M. Z. Yates, H. Du, F. Jiang, L. Guo and T. D. Krauss, *Chem. Mater.,* 2007, **19**, 2930.
233. L. Liu, X. Guo, Y. Li and X. Zhong, *Inorg. Chem.,* 2010, **49**, 3768.
234. I. Potapova, R. Mruk, S. Prehl, R. Zentel, T. Basché and A. Mews, *J. Am. Chem. Soc.,* 2003, **125**, 320.
235. W. Liu, A. B. Greytak, J. Lee, C. R. Wong, J. Park, L. F. Marshall, W. Jiang, P. N. Curtin, A. Y. Ting, D. G. Nocera, D. Fukumura, R. K. Jain and M. G. Bawendi, *J. Am. Chem. Soc.,* 2010, **132**, 472.
236. I. Potapova, R. Mruk, C. Hubner, R. Zentel, T. Basché and A. Mews, *Angew. Chem., Int. Ed.,* 2005, **44**, 2437.
237. M. I. Shukoor, F. Natalio, N. Metz, N. Glube, M. N. Tahir, H. A. Therese, V. Ksenofontov, P. Theato, P. Langguth, J.-P. Boissel, H. C. Schröder, W. E. G. Müller and W. Tremel, *Angew. Chem., Int. Ed.,* 2008, **47**, 4748.
238. A. M. Smith and S. Nie, *J. Am. Chem. Soc.,* 2008, **130**, 11278.
239. Y. A. Wang, J. J. Li, H. Chen and X. Peng, *J. Am. Chem. Soc.,* 2002, **124**, 2293.
240. W. Guo, J. J. Li, Y. A. Wang and X. Peng, *J. Am. Chem. Soc.,* 2003, **125**, 3901.
241. W. Guo and X. Peng, *C. R. Chim.,* 2003, **6**, 989.
242. W. Guo, J. J. Li, A. Wang and X. Peng, *Chem. Mater.,* 2003, **15**, 3125.
243. Y. Liu, M. Kim, Y. Wang, Y. A. Wang and X. Peng, *Langmuir,* 2006, **22**, 6341.
244. C. Zhang, S. O'Brien and L. Balogh, *J. Phys. Chem. B,* 2002, **106**, 10316.
245. J. Liu, H. Li, W. Wang, H. Xu, X. Yang, J. Liang and Z. He, *Small,* 2006, **2**, 999.
246. Y. Park, P. Taranekar, J. Y. Park, A. Baba, T. Fulghum, R. Ponnapati and R. C. Advincula, *Adv. Funct. Mater.,* 2008, **18**, 2071.
247. M. V. Kovalenko, M. Scheele and D. V. Talapin, *Science,* 2009, **324**, 1417.
248. M. V. Kovalenko, M. I. Bodnarchuk, J. Zaumseil, J.-S. Lee and D. V. Talapin, *J. Am. Chem. Soc.,* 2010, **132**, 10085.
249. M. V. Kovalenko, B. Spokoyny, J.-S. Lee, M. Scheele, A. Weber, S. Perera, D. Landry and D. V. Talapin, *J. Am. Chem. Soc.,* 2010, **132**, 6686.

250. H. Fan, *Chem. Commun.,* 2008, 1383.
251. T. Pons, H. T. Uyeda, I. L. Medintz and H. Mattoussi, *J. Phys. Chem. B,* 2006, **110**, 20308.
252. B. Dubertret, P. Skourides, D. J. Norris, V. Noireaux, A. H. Brivaniou and A. Libchaber, *Science,* 2002, **298**, 1759.
253. O. Carion, B. Mahler, T. Pons and B. Dubertret, *Nat. Protoc.,* 2007, **2**, 2383.
254. N. Depalo, A. Mallardi, R. Comparelli, M. Striccoli, A. Agostiano and M. L. Curri, *J. Colloid Interface Sci.,* 2008, **325**, 558.
255. H. Fan, E. W. Leve, C. Scullin, J. Gabaldon, D. Tallant, S. Bunge, T. Boyle, M. C. Wilson and C. J. Brinker, *Nano Lett.,* 2005, **5**, 645.
256. I. Rubinstein, I. Soos and H. Onyuksel, *Chem.-Biol. Interact.,* 2008, **171**, 190.
257. N. Travert-Branger, F. Dubois, O. Carion, G. Carrot, B. Mahler, B. Dubertret, E. Doris and C. Mioskowski, *Langmuir,* 2008, **24**, 3016.
258. Y.-H. Kim, E. Subramanyam, J. H. Im, K. M. Huh, H. Choi, J. S. Choi, Y.-K. Lee and S.-W. Park, *J. Nanosci. Nanotechnol.,* 2010, **10**, 3275.
259. S. G. Grancharov, H. Zeng, S. Sun, S. X. Wang, S. O'Brien, C. B. Murray, J. R. Kirtley and G. A. Held, *J. Phys. Chem. B,* 2005, **109**, 13030.
260. J. Park, M. Yu, Y. Jeong, J. Kim, K. Lee, V. Phan and S. Jon, *J. Mater. Chem.,* 2009, **19**, 6412.
261. W. J. M. Mulder, R. Koole, R. J. Brandwijk, G. Storm, P. T. K. Chin, G. J. Strijkers, C. de Mello Donegá, K. Nicolay and A. W. Griffioen, *Nano Lett.,* 2006, **6**, 1.
262. J.-H. Park, G. von Maltzahn, E. Ruoslahti, S. N. Bhatia and M. J. Sailor, *Angew. Chem., Int. Ed.,* 2008, **47**, 7284.
263. O. T. Burns, H. Ittrich, K. Peldschus, M. G. Kaul, U. I. Tromsdorf, J. Lauterwasser, M. S. Nikolic, B. Mollwitz, M. Merkel, N. C. Bigall, S. Sapra, R. Reimer, H. Hohenberg, H. Weller, A. Eychmüller, G. Adam, U. Beisiegel and J. Heeren, *Nat. Nanotechnol.,* 2009, **4**, 193.
264. Y. Chen, T. Ji and Z. Rosenzweig, *Nano Lett.,* 2003, **3**, 581.
265. X. Wu, H. Liu, J. Liu, K. N. Haley, J. A. Treadway, J. P. Larson, N. Ge, F. Peale and M. P. Bruchez, *Nat. Biotechnol.,* 2003, **21**, 41.
266. J. E. Halpert, J. R. Tischler, G. Nair, B. J. Walker, W. Liu, V. Bulović and M. G. Bawendi, *J. Phys. Chem. C,* 2009, **113**, 9986.
267. R. E. Anderson and W. C. W. Chan, *ACS Nano,* 2008, **2**, 1341.
268. Y. Chen, R. Thakar and P. T. Snee, *J. Am. Chem. Soc.,* 2008, **130**, 3744.
269. A. M. Smith and S. Nie, *Angew. Chem., Int. Ed.,* 2008, **47**, 9916.
270. C. Tribet, R. Auderbert and J.-L. Popot, *Proc. Natl. Acad. Sci. U. S. A.,* 1996, **93**, 15047.
271. H. Duan, M. Kuang and Y. A. Wang, *Chem. Mater.,* 2010, **22**, 4372.
272. T. Pellegrino, L. Manna, S. Kudera, T. Liedl, D. Koktysh, A. L. Rogach, S. Keller, J. Rädler, G. Natile and W. J. Parak, *Nano Lett.,* 2004, **4**, 703.
273. W. W. Yu, E. Chang, C. M. Sayes, R. Drezek and V. L. Colvin, *Nanotechnology,* 2006, **17**, 4483.

274. W. W. Yu, E. Chang, J. C. Falkner, J. Zhang, A. M. Al-Somali, C. M. Sayes, J. Johns, R. Drezek and V. L. Colvin, *J. Am. Chem. Soc.*, 2007, **129**, 2871.

275. E. V. Shytkova, X. Huang, X. Gao, J. C. Dyke, A. L. Schmucker, B. Dragnea, N. Remmes, D. V. Baxter, B. Stein, P. V. Konarev, D. I. Svergun and L. M. Bronstein, *J. Phys. Chem. C*, 2008, **112**, 16809.

276. J. Qin, Y. S. Jo and M. Muhammed, *Angew. Chem., Int. Ed.*, 2009, **48**, 7845.

277. R. Di Corato, A. Quarta, P. Piacenza, A. Ragusa, A. Figuerola, R. Buonsanti, R. Cingolani, L. Manna and T. Pellegrino, *J. Mater. Chem.*, 2008, **18**, 1991.

278. E. E. Lees, T.-L. Nguyen, A. H. A. Clayton and P. Mulvaney, *ACS Nano*, 2009, **3**, 1121.

279. C.-A. J. Lin, R. A. Sperling, J. K. Li, T.-Y. Yang, P.-Y. Li, M. Zanella, W. H. Chang and W. J. Parak, *Small*, 2008, **4**, 334.

280. S. Wang, M.-Y. Han and D. Huang, *J. Am. Chem. Soc.*, 2009, **131**, 11692.

281. Y. Wang, J. F. Wong, X. Teng, X. Z. Lin and H. Yang, *Nano Lett.*, 2003, **3**, 1555.

282. T. Jin, F. Fujii, H. Sakata, M. Tamura and M. Kinjo, *Chem. Commun.*, 2005, 2829.

283. T. Jin, F. Fujii, Y. Nodasaka and M. Kinjo, *J. Am. Chem. Soc.*, 2006, **128**, 9288.

284. J. Feng, S.-K. Ding, M. P. Tucker, M. E. Himmel, Y.-H. Kim, S. B. Zhang, B. M. Keyes and G. Rumbles, *Appl. Phys. Lett.*, 2005, **86**, 033108.

285. K. Palaniappan, C. Xue, G. Arumugam, S. A. Hackney and J. Liu, *Chem. Mater.*, 2006, **18**, 1275.

286. C.-Y. Chen, C.-T. Cheng, C.-W. Lai, P.-W. Wu, K.-C. Wu, P.-T. Chou, Y.-H. Chou and H. T. Chiu, *Chem. Commun.*, 2006, 263.

287. H. Li, Y. Zhang, X. Wang, D. Xiong and Y. Bai, *Mater. Lett.*, 2007, **61**, 1474.

288. B. Gadenne, I. Yildiz, M. Amelia, F. Ciesa, A. Secchi, A. Arduini, A. Credi and F. M. Raymo, *J. Mater. Chem.*, 2008, **18**, 2022.

289. E. Zenkevich, F. Cichos, A. Shulga, E. P. Petrov, T. Blaudeck and C. von Borczyskowski, *J. Phys. Chem. B*, 2005, **109**, 8679.

290. R. Freeman, T. Finder and I. Willner, *Nano Lett.*, 2009, **9**, 2073.

291. D. Dorokhin, N. Tomczak, M. Han, D. N. Reinhoudt, A. H. Velders and G. J. Vancso, *ACS Nano*, 2009, **3**, 661.

292. Q. Liu, C. Li, T. Yang, T. Yi and F. Li, *Chem. Commun.*, 2010, **46**, 5551.

293. M. E. Åkerman, W. C. W. Chan, P. Laakkonen, S. N. Bhatia and E. Ruoslahti, *Proc. Natl. Acad. Sci. U. S. A.*, 2002, **99**, 12617.

294. M. J. Murcia, D. E. Minner, G.-M. Mustata, K. Ritchie and C. A. Naumann, *J. Am. Chem. Soc.*, 2008, **130**, 15054.

295. C.-T. Chen, Y. S. Munot, S. B. Salunke, Y.-C. Wang, R.-K. Lin, C.-C. Lin, C.-C. Chen and Y.-H. Liu, *Adv. Funct. Mater.*, 2008, **18**, 527.

296. H. Gu, P.-L. Ho, K. W. T. Tsang, L. Wang and B. Xu, *J. Am. Chem. Soc.*, 2003, **125**, 15702.

297. H. Gu, P.-L. Ho, K. W. T. Tsang, C. W. Yu and B. Xu, *Chem. Commun.,* 2003, 1966.

298. I. D. Tomlinson, J. Mason, J. N. Burton, R. Blakely and S. J. Rosenthal, *Tetrahedron,* 2003, **59**, 8035.

299. I. D. Tomlinson, T. Kippeny, L. Swafford, N. H. Siddiqui and S. J. Rosenthal, *J. Chem. Res., Synop.,* 2002, 204.

300. S. J. Rosenthal, I. Tomlinson, E. M. Adkins, S. Schroeter, S. Adams, L. Swafford, J. McBride, Y. Wang, L. J. DeFelice and R. D. Blakely, *J. Am. Chem. Soc.,* 2002, **124**, 4586.

301. S. J. Clarke, C. A. Hollman, F. A. Aldaye and J. L. Nadeau, *Bioconjugate Chem.,* 2008, **19**, 562.

302. S. S. Narayanan, S. S. Sinha and S. K. Pal, *J. Phys. Chem. C,* 2008, **112**, 12716.

303. G. P. Mitchell, C. A. Mirkin and R. L. Letsinger, *J. Am. Chem. Soc.,* 1999, **121**, 98122.

304. A. P. Alivisatos, K. P. Johnsson, X. Peng, T. E. Wilson, C. J. Loweth, M. P. Bruchez and P. G. Schultz, *Nature,* 1996, **382**, 609.

305. C. A. Mirkin, R. L. Letsinger, R. C. Mucic and J. J. Storhoff, *Nature,* 1996, **382**, 607.

306. C. A. Mirkin, *Inorg. Chem.,* 2000, **39**, 2258.

307. W. R. Algar and U. J. Krull, *Langmuir,* 2006, **22**, 11346.

308. D. Zhou, J. D. Piper, C. Abell, D. Klenerman, D.-J. Kang and L. Ying, *Chem. Commun.,* 2005, 4807.

309. R. Gill, I. Willner, I. Shewky and U. Banin, *J. Phys. Chem. C,* 2005, **109**, 23715.

310. Y. Zheng, Z. Yang, Y. Li and J. Y. Ying, *Adv. Mater.,* 2008, **20**, 3410.

311. T. Jin, Y. Yoshioka, F. Fujii, Y. Komai, J. Seki and A. Seiyama, *Chem. Commun.,* 2008, 5764.

312. Y. Wei, J. Yang and J. Y. Ying, *Chem. Commun.,* 2010, **46**, 3179.

313. X. Ai, Q. Xu, M. Jones, Q. Song, S.-Y. Ding, R. J. Ellingson, M. Himmel and G. Rumbles, *Photochem. Photobiol. Sci.,* 2007, **6**, 1027.

314. K. E. Sapsford, T. Pons, I. L. Medintz, S. Higashiya, F. M. Brunel, P. E. Dawson and H. Mattoussi, *J. Phys. Chem. C,* 2007, **111**, 11528.

315. S.-Y. Ding, M. Jones, M. P. Tucker, J. M. Nedeljkovic, J. Wall, M. N. Simon, G. Rumbles and M. E. Himmel, *Nano Lett.,* 2003, **3**, 1581.

316. J. B. Blanco-Canosa, I. L. Medintz, D. Farrell, H. Mattoussi and P. E. Dawson, *J. Am. Chem. Soc.,* 2010, **132**, 10027.

317. I. L. Medintz, M. H. Stewart, S. A. Trammell, K. Susumu, J. B. Delehanty, B. C. Mei, J. S. Melinger, J. B. Blanco-Canosa, P. E. Dawson and H. Mattoussi, *Nat. Mater.,* 2010, **9**, 676.

318. L. E. Euliss, S. G. Grancharov, S. O'Brien, T. J. Deming, G. D. Stucky, C. B. Murray and G. A. Held, *Nano Lett.,* 2003, **3**, 1489.

319. Y. Wu, S. Chakrabortty, R. A. Gropeanu, J. Wilhelmi, Y. Xu, K. S. Er, S. L. Kuan, K. Koynov, Y. Chan and T. Weil, *J. Am. Chem. Soc.,* 2010, **132**, 5012.

320. N. R. Jana, N. Erathodiyil, J. Jiang and J. Y. Ying, *Langmuir,* 2010, **26**, 6503.
321. F. Pinaud, D. King, H.-P. Moore and S. Weiss, *J. Am. Chem. Soc.,* 2004, **126**, 6115.
322. J. M. Tsay, S. Doose, F. Pinaud and S. Weiss, *J. Phys. Chem. B,* 2005, **109**, 1669.

CHAPTER 7

The Use of Single-Source Precursors in Nanoparticle Synthesis

7.1 II–VI-Based Materials

The preparation of nanomaterials described in preceding chapters mainly relied on a binary approach, *i.e.* the cationic and anionic constituents (often termed monomers) were supplied by separate precursors in the form of organometallic or inorganic compounds. These organometallic-based routes to nanoparticles initially drew inspiration from precursors designed for use in chemical vapour deposition (CVD)[1] processes and were often pyrophoric and highly toxic.[2]

Green chemical routes to nanomaterials utilising precursors such as carboxylic acid salts and solutions of elemental chalcogens provide a safer alternative, as discussed in Chapter 1, but cadmium salts, selenium and tellurium still present an exposure hazard. An elegant alternative is the use of a single-source precursor, where the constituent moieties of the target semiconductor are present in an inorganic complex with the M–X bond already in place (M = metal cation, X = anion, such as a chalcogen or pnictide). Decomposition, usually thermally induced, causes the elimination of side groups on the complex leaving the semiconductor phase, either deposited on a substrate or free in solution. A key advantage of this method of synthesis is that the majority of single-source precursors are stable solids at room temperature, which allows rigorous purification and almost indefinite storage. The use of single-source precursors in CVD is well established, with numerous groups worldwide developing precursor chemistry. Precursor

RSC Nanoscience & Nanotechnology No. 33
Semiconductor Quantum Dots: Organometallic and Inorganic Synthesis
By Mark Green
© Mark Green 2014
Published by the Royal Society of Chemistry, www.rsc.org

design covers numerous issues such as volatility, design of simple organic leaving groups, stoichiometry control and programmed orientation of the deposited crystalline phase.[3]

Despite the fact that single-source precursors are often viewed as a later development in the field of nanoparticle preparation, much of the initial work in the synthesis of nanomaterials by organometallics was in fact single-source based. Early work on alternative routes for the preparation of CdTe and HgTe thin films utilised bis(organotelluro)cadmium and bis(organotelluro)mercury compounds as precursors.[4] This methodology was used in the synthesis of a range of nanomaterials using complexes of $M(ER)_2$ (M = Cd, Zn, Hg; E = S, Se, Te; R = organic group) which were thermolysed in 4-ethylpyridine (except $Hg(TeBu)_2$, which decomposed photolytically).[5,6] This showed that single molecular precursors could be thermolysed in Lewis base solvents to give nanomaterials, although few investigations were carried out into the optical properties or the nature of the surface capping agents. The particles produced were ill-defined, yet clearly crystalline and displayed evidence of the size quantisation effects. The relatively poor quality of these materials can in part be attributed to the low boiling point of 4-ethylpyridine (167 °C) and the tendency for pyridines to tar and cross-link upon prolonged heating.[7,8] The use of single-source precursors is not confined to high-temperature routes with coordinating solvents; $[(^tBu)GaSe]_4$ and $[(EtMe_2C)InSe]_4$ have been used to prepare GaSe and InSe nanoparticles respectively,[9] and $(X_2GaP(SiMe_3)_2)_2$ (X = Br, I) and $(Cl_3Ga_2P)_n$ have both been used to prepare GaP nanoparticles[10] although these reactions are not solution-based. Thermolysis of a related compound, $[H_2GaE(SiMe_3)_2]_3$, (E = P, As), in xylenes yielded 5 nm GaP or GaAs nanoparticles.[11]

7.2 Dithio- and Diselenocarbamates

The advent of high-temperature solution routes based on trioctylphosphine oxide (TOPO) allowed single-source precursors to be used in reaction conditions that favoured the growth of high-quality particles. The single-source route to TOPO-capped nanoparticles was pioneered by O'Brien,[12,13] who utilised dithio- and diselenocarbamates on the basis of their successful use in the formation of metal sulfide/selenide thin films. The dimeric complexes can in most cases be prepared from readily available starting material. This route relies on the availability of the sodium (or ammonium) salt of the carbamate ligand, which is only commercially available with symmetrical alkyl groups such as ethyl. A more useful route involves the reaction between the metal salts, CS_2 and the required, often asymmetrical, secondary amine. The range of alkyl groups available on secondary amines allows a wide range of symmetrical and asymmetrical precursors to be prepared. It is worth noting that although carbon disulfide is commercially available, carbon diselenide is not; it is toxic and difficult to prepare.[14] In this case, preparing nanoparticles by the green routes described by Peng is simpler, although single-source precursors have distinct benefits.

Initial reports on the use of dithio- and diselenocarbamates as precursors described the thermolysis of the air-stable, symmetrical compounds $[Cd(S_2CNEt_2)_2]_2$ and $[Cd(Se_2CNEt_2)_2]_2$ in 4-ethylpyridine.[15] Although the dithiocarbamate decomposed cleanly to produce quantum dots (QDs) of CdS with a hexagonal crystal core, the diselenocarbamate formed both nano-particulate CdSe and micrometre-sized selenium particles, both of which exhibited hexagonal crystal cores. Absorption band edges consistent with nanosized CdS and CdSe were observed, although no emission data was presented in this case. The formation of impurities from the simple dis-elenocarbamate presented a significant problem, which was overcome by using a related diselenocarbamate, $[MeCd(Se_2CNEt_2)_2]_2$, prepared by the comproportionation reaction[16] between Me_2Cd and the simple cadmium(diethyldiselenocarbamate).

Simple dissolution of $[MeCd(Se_2CNEt_2)_2]_2$ in trioctylphosphine (TOP) fol-lowed by injection into TOPO at 200–250 °C resulted in clean decomposition yielding CdSe nanoparticles *ca.* 5 nm in diameter which displayed band edge emission comparable with CdSe particles prepared by binary organometallic-based methods.[17] This can be thought of as the first single-source route to high-quality semiconductor nanoparticles. The neopentyl analogue $([(CH_3)_3CH_2Cd(Se_2CNEt_2)_2]_2)$ was also prepared and found to produce nanoparticles of a similar quality. The thermolysis of the thio-analogues $[RCd(S_2CNEt_2)_2]_2$ (R = Me, $(CH_3)_3CH_2$) in TOPO produced nanoparticles of CdS which were again of comparable quality to materials produced by the binary route, as were nanoparticles obtained from the simple dithiocarba-mate $[Cd(S_2CNEt_2)_2]_2$. In all cases, particles were *ca.* 4–5 nm in diameter, exhibiting blue-shifted absorption edges and near band edge emission.[18]

Despite the clean decomposition of $[Me/(CH_3)_3CH_2Cd(Se_2CNEt_2)_2]_2$, the alkylated precursors are air sensitive, unlike the simple dithio/selenocarba-mates which exhibit excellent stability when stored in ambient conditions; ideally, a precursor should be air stable and decompose cleanly to the desired product.

In-depth investigations into the decomposition of symmetrical and asymmetrical dithio/selenocarbamates showed an intimate link between decomposition pathway and the constituent alkyl groups.[19,20] Thermolysis of symmetrical diethyldiselenocarbamates proceeded by the elimination of $EtSe_2Et$, which decomposed further to elemental selenium as determined by gas chromatography-mass spectrometry (GC-MS). Ethane elimination on particle-bound diethyldiselenocarbamate groups was also identified as a potential decomposition pathway that resulted in a selenium-rich product (Figure 7.1a).

Mass spectrometry on the asymmetric diselenocarbamate thermolysis by-products identified a five-membered heterocycle, 5-butyl-3-methyl-2,3-dihy-dro-1,3-selenazole, which was formed by the elimination of a hydrogen atom in the β position to the nitrogen (Figure 7.1b), effectively removing the excess selenium and giving a clean product. No evidence of $EtSe_2Se$ was found in the decomposition by-products of asymmetrical diselenocarbamates.

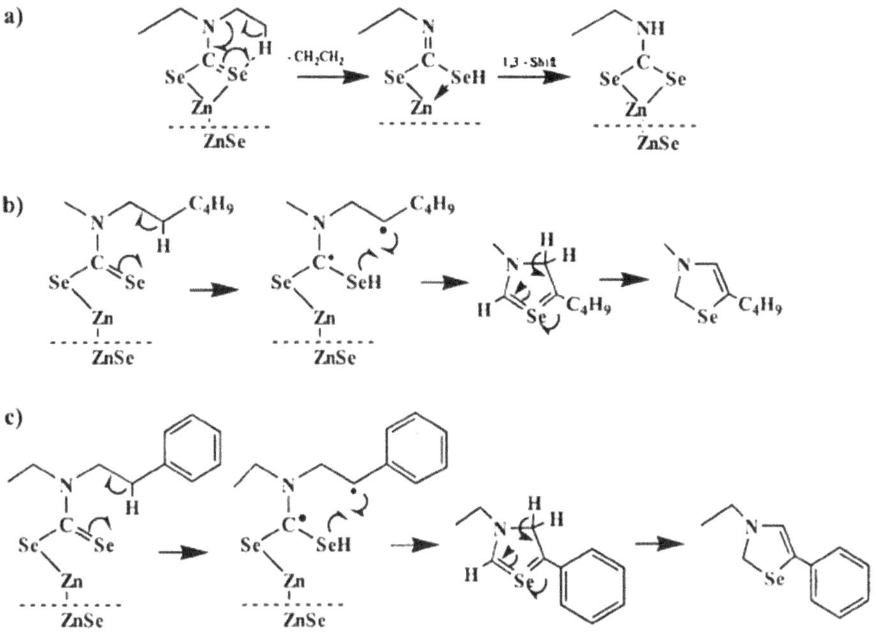

a)

b)

c)

Figure   7.1   (a)   Suggested   ethane   elimination   for   symmetrical diethyldiselenocarbamates. (b) Suggested decomposition mechanism for the asymmetric [Cd(SeCNMe(n-hex))$_2$]$_2$ showing the eliminated selenium containing heterocycle; suggested decomposition mechanism for the asymmetric [Cd(Se$_2$CNMe(CH$_2$CH$_2$Ph))$_2$]$_2$. Reprinted with permission from N. L. Pickett and P. O'Brien, *Chem. Record*, 2001, **1**, 467. Copyright 2001 John Wiley & Sons, Inc. and The Japan Chemical Journal Forum.

Other investigations into the decomposition mechanisms of zinc dithio-carbamates in the presence of long-chain amines, using nuclear magnetic resonance (NMR) and MS, highlighted that nucleophilic attack of the amine on the electron-deficient thiocarbonyl carbon atom induced precursor decomposition.[21]

A further precursor, [Cd(Se$_2$CNMe(n-hex))$_2$]$_2$, was designed with reference to the mechanistic data and prepared as described above.[22] The complex was found to be indefinitely stable when stored in air at room temperature, yet decomposed cleanly at 200 °C in TOPO to give CdSe particles 3–5 nm in diameter with a hexagonal crystalline core. Absorption spectra showed the blue-shifted absorption edge with clear excitonic features and band edge emission with, in some cases, slight evidence of trap emission. The synthesis of CdS and ZnE (E = S, Se) nanoparticles in TOPO using the analogous asymmetrical dithio/selenocarbamate was also successful, giving similar-sized particles.[23] With reference to the decomposition studies, other precursors such as [Cd(Se$_2$CNMe(CH$_2$CH$_2$Ph))$_2$]$_2$ were designed with substituents on the β-carbon (Figure 7.1c) which imparted greater stability to the subsequent decomposition product (3-ethyl-5-phenyl-2,3-dihydro-1,3-selenazole), to enhance elimination.[24,25] Compounds of the formula $[M(S_2CNHR)_2]$ (M = Cd, Zn; R = alkyl chain) have also shown to be effective single-source precursors to CdS and ZnS nanoparticles of varying morphologies.[26]

Lazell and O'Brien reported an interesting synthetic route to CdS nanoparticles[27] using an unusual asymmetric dithiocarbamate, which was inspired by 'self-capping' silver particles.[28] The novel precursor, $[Cd(S_2CNMe(C_{18}H_{37}))_2]_2$, was prepared which possessed a significantly longer side group than normally used. The material was then thermolysed *in vacuo* inside a metal–organic chemical vapour deposition (MOCVD) chamber at 200 °C and the resulting material extracted using pyridine. Fourier transform infrared (FTIR) spectroscopy suggested that one of the decomposition products, $HNMe(C_{18}H_{37})$, passivated the particles. The QDs produced were crystalline, *ca.* 4–5 nm in diameter and displayed the expected shift in band edge. Emission was broad with a slight low-energy tail, but predominantly band edge in origin. This was extended to the preparation of self-capped Bi_2S_3 particles using the precursor $Bi(S_2CN(C_{18}H_{37})(CH_3))_3$ using similar conditions.[29] Bismuth sulfide, Bi_2S_3, has also been prepared by asymmetric dithiocarbamates. Initially investigations used $Bi(S_2CNMe(n\text{-}hex))_3$ in 2-ethoxyethanol yielding nanofibres,[30] but this was extended to a wider range of complexes and solvents.[31] The thermolysis of $Bi(S_2CNMe(n\text{-}hex))_3$ in TOPO led to a mixture of Bi_2S_3 and elemental bismuth, while 4-ethylpyridine and ethylene glycol resulted in the pure nanostructured semiconductor as required, although 2-ethoxyethanol gave the largest product yield. Bi_2S_3 prepared in ethylene glycol gave very distinct rods that appeared to be hundreds of nanometres long, and gave a distinct feature in the absorption spectra at *ca.* 550 nm, unlike material prepared in 4-ethylpyridine or 2-ethoxyethanol which gave featureless spectra, although all were clearly blue-shifted from the bulk bandgap of 1.3 eV (*ca.* 950 nm). Use of the adduct complex $Bi(S_2CNMe(n\text{-}hex))_3(C_{12}H_8N_2)$ resulted in slightly longer fibres of the same morphology. Again, the asymmetric $M(S_2CNMe(n\text{-}hex))_x$ system was used to prepare nanoparticles of PtS (3 nm) and PdS (5 nm) using the compounds $Pt(S_2CNMe(n\text{-}hex))_2$ and $Pd(S_2CNMe(n\text{-}hex))_2$ respectively in TOPO at 250 °C.[32] Both sets of particles gave broad emission profiles at *ca.* 490 nm, the origin of which was unclear, but may be related to the thermolysis of the capping agent. Nanoparticles of PtS had a clear absorption feature at *ca.* 360 nm.

The potential application of ZnE (E = S, Se) particles in optoelectronics has been covered in Chapter 1 and single-source routes based on zinc dithio/selenocarbamates have been developed. The compound $EtZnSe_2CNEt_2$ was used to prepare particles of ZnSe 3–6 nm in diameter with a hexagonal crystalline core, by injecting a TOP solution of the precursor into TOPO at between 200–250 °C.[33] The particles showed a band edge shifted from the bulk value by between 0.15–0.25 eV and exhibited band edge emission, although the emission was significantly broader than reports of ZnSe particles described in Chapter 1. Particles of ZnS prepared in a similar manner using $EtZnS_2CNEt_2$ as a precursor displayed similar optical characteristics with similar shifts in the band edge.[34] The particles appeared to have a smaller size range of 3.0–4.5 nm and possessed a cubic crystalline core.

Dithio- and diselenocarbamates have been used to prepare materials other than II–VI semiconductors. The simple copper dithiocarbamate $Cu(S_2CNEt_2)_2$ has been used to prepare $Cu_{1.8}S$ nanoparticles capped with TOPO.[35] During the synthesis, TOPS was also added to facilitate formation of the sulfide, due to the stability of the carbon–sulfur bond. Failure to add the sulfide resulted in a material rich in copper metal impurities. The materials prepared exhibited band edge of *ca.* 2.35 eV (*ca.* 530 nm) and displayed broad emission. Similarly, $Cu(S_2CNEt_2)_2$ has been used to make Cu_2S particles in the presence of octylamine (OA) and dodecanethiol and used in the preparation of efficient photovoltaic cells.[36]

Using a similar precursor, Revaprasadu *et al.* prepared nanoparticles of Cu_2S in solution using $Cu(S_2CNMe(n\text{-hex}))_2$, *ca.* 4 nm in diameter with a blue shift of 2.09 eV.[37] The particles exhibited broad emission, with a maximum at 477 nm. The use of $Cu(S_2CNEt_2)_2$ was extended to make nanodiscs of Cu_2S on a silicon substrate, by thermolysing the precursor in TOP with the substrate submerged in the reaction solvent.[38] By using a bismuth-covered substrate, nanowires could be grown using the same technique, due to the formation of bismuth catalyst particles resulting in solution–liquid–solid (SLS) growth as described earlier.

The simple diselenocarbamate of copper, $Cu(Se_2CNEt_2)_2$, has been used to prepare CuSe nanoparticles capped with TOPO.[39] The particles, *ca.* 16 nm in diameter, exhibited an increase in the bandgap although the emission profile was not described. The particles exhibited a diffraction pattern consistent with a hexagonal crystalline core.

7.3 Single-Source Routes to Anisotropic Particles

The preparation of anisotropic particles has been extensively examined, and a complicated relationship between monomer concentration, capping agents, reaction temperature and particle morphology has been uncovered (Chapter 1). The thermolysis of the simple cadmium diethyldithiocarbamate in TOPO has shown to give spherical particles, but the use of the same precursor in hexadecylamine (HDA) produced a range of anisotropic particles.[40] This varies from the usual method of producing anisotropic particles as only one surfactant and one precursor/monomer is used. Growth at 300 °C resulted in wurtzite nanorods of CdS, 6 nm wide with an aspect ratio of *ca.* 4. Increasing precursor concentration resulted in the increase in width up to 25 nm, yet still maintained the aspect ratio of 4. By varying the temperature at a fixed concentration different-shaped particles could be prepared, with bipods, tripods and tetrapods forming at lower temperatures. Interestingly, the formation of tetrapods is primarily at 120 °C and dominates at least 80%

of the formed structures. The core of the tetrapod was found to be zinc blende and the arms wurtzite, consistent with other II–VI tetrapod nano-particles. Clearly, the surfactant plays a major role in the formation of anisotropic materials as the low-temperature synthesis of CdS in TOPO using the same precursor resulted in small, spherical particles. Cadmium and lead diethyldithiocarbamates were also used as precursors in the SLS growth of CdS[41] and PbS[35] nanorods, using bismuth nanoparticles as catalysts for anisotropic growth. The seed growth of anisotropic particles is described in Chapter 1. In this case, the introduction of bismuth nanoparticles catalysed anisotropic growth, with nanowire dimension controlled by the reaction temperature, surfactant chemistry and the diameter of the catalyst particles. Interestingly, the wire diameters were larger than the catalyst particle diameters, with the thickest wires obtained from the smallest particles.

Bismuth and antimony dithiophosphates, $M[S_2P(OC_8H_{17})_2]_3$, $M = Bi$, Sb, have been used as precursors for rod formation at relatively low temperatures (160 °C) in OAm. Rods of Bi_2S_3, several hundred nanometres in length with diameters of *ca.* 12 nm, were simply prepared and found to clump into bundles similar to other reports of Bi_2S_3 reported in Chapter 4. Investigations into the growth mechanism uncovered the initial growth (1 minute growth time) of 7 nm diameter particles with aspect ratios of less than 3, which then grew along the (001) axis to give rods with aspect ratios of *ca.* 5 after 8 minutes further growth, which then proceeded to rod lengths of hundreds of nanometres over 1–1.5 hours. The reaction was also found to proceed faster at higher temperatures, producing rods of slightly altered dimensions. Altering the concentration of precursors still resulted in rod formation, but surprisingly decreased the length of the rods, attributed to increased seed growth rather than increasing the length of the rod at the reaction's earlier stages. Slightly larger rods of Sb_2S_3, 1 μm in length and 45 nm in diameter, were grown in similar conditions.[42] Similarly, zinc selenide nanorods, *ca.* 24 nm in length and 6 nm in diameter displaying emission at *ca.* 400 nm, have been prepared from the thermolysis of $[Zn(^iPr_2PSe_2)_2]$ in TOP and HDA.[43]

A similar route was developed to the preparation of MnS particles using $Mn(S_2CNEt_2)_2$ in HDA.[44] The structures prepared and conditions required are shown in Figure 7.2. In this case, time of synthesis was also a key variable, with prolonged heating resulting in the growth of larger cubes or the conversion of rods into spheres. The particles exhibited either a wurtzite (γ-MnS, hexagonal) or zinc blende (β-MnS, cubic) structure at synthesis temperatures below 200 °C, which allowed growth of branched structures, or a rock salt (α-MnS) structure at temperatures above 200 °C. Investigations into the optics of the rods and spheres showed evidence of quantum confinement, with band edges between 3.3 eV (375 nm) and 3.5 eV (354 nm), slightly wider than the bulk value of 3.2 eV (*ca.* 390 nm). Notably, the Stokes shift for the wires was approximately twice as large as that observed for the spheres. Emission was band edge and showed no evidence of trapping states. By mixing cadmium and manganese diethyldithiocarbamates prior to ther-molysis, dilute magnetic semiconducting nanorods of $Cd_{1-x}Mn_xS$ could be

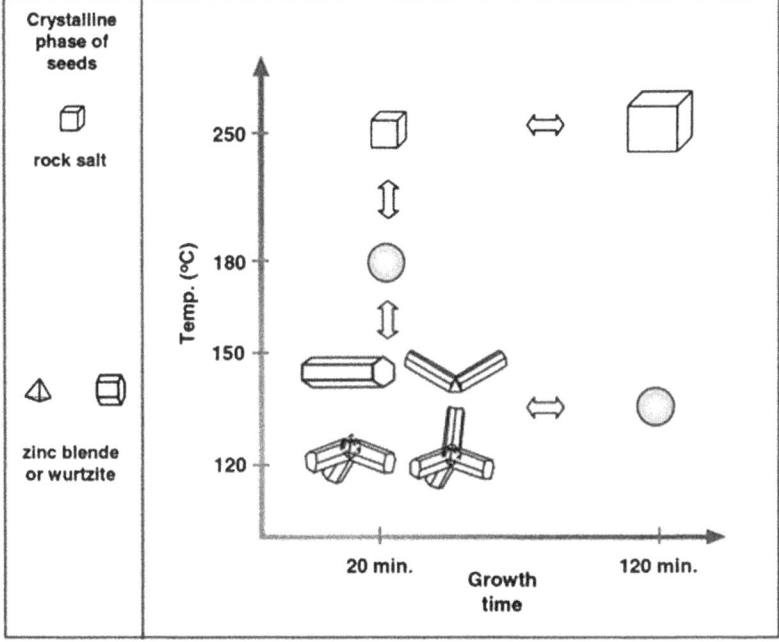

Figure 7.2 Shape change map of MnS particles for varying growth temperatures and reaction times. Reprinted with permission from Y.-W. Jun, Y.-Y. Jung and J. Cheon, *J. Am. Chem. Soc.*, 2002, **124**, 615. Copyright 2002 American Chemical Society.

prepared, *ca.* 7 nm wide with aspects ratios of *ca.* 4. Up to 12% manganese could be doped in the rods, and surface exchange experiments suggested that the dopant ions were situated throughout the body of the particle and not just on the surface. Ag_2S has also been prepared from a slurry of silver diethyldithiocarbamate in octadecene (ODE) with OA and OAm at 200 °C. The resulting particles were *ca.* 10 nm in diameter, monodispersed, and exhibited near infrared absorption and an emission maximum at *ca.* 1060 nm.[45]

A notable study on a range of single-source precursors suggested that most primarily gave spherical particles, whilst cadmium thiosemicarbazide and cadmium selenosemicarbazide had lower activation barriers for decomposition relative to other single-source precursors, and released a nitrogen-containing side product that coordinates to the particle surface and aided anisotropic growth.[46,47]

7.4 Single-Source Routes to Core/Shell Particles

The use of asymmetric precursors has been applied to the formation of CdSe/CdS core shell particles in a simple one-pot reaction.[48] Cores of CdSe, 4.5 nm in diameter, were grown in TOPO at 250 °C using [Cd(SeCNMe(n-hex))$_2$]$_2$. A shell of CdS 0.4 nm thick was deposited using [Cd(SCNMe(n-hex))$_2$]$_2$ which was simply added to the reaction vessel after 30 minutes of core particle growth. Although quantum yield measurements were not reported, the exciton leakage into the CdS shell was clearly visible in the optical spectrum. This work was advanced to cover the preparation of CdSe/ZnS and CdSe/ZnSe core/shell particles using asymmetrical single-source precursors, [M(ECNMe(n-hex))$_2$]$_2$ (M = Cd, Zn; E = S, Se).[49] Again, the main evidence of the formation of core/shell materials was the optical spectrum, which showed leakage of the excitons into the shell; the shift was only slight for CdSe/ZnS, yet significantly more pronounced for CdSe/ZnSe. All emission was near band edge and reportedly strong, with no evidence of deep trap emission. High-resolution transmission electron microscopy (HRTEM) studies showed crystalline particles with clear grain boundaries, highlighting that the growth of CdSe/ZnS particles is not always a simple clean epitaxial process. Interestingly, in this report, the authors attempted to grow CdSe/CdS composites by thermolysing two precursors simultaneously. The material obtained exhibited the expected optical and physical properties, but the X-ray diffraction (XRD) pattern was shifted to lower 2θ values, indicative of alloy particles, as described in Chapter 2. The use of the symmetrical zinc diethyldithiocarbamate has also been reported as a method of depositing a ZnS shell on CdSe core particles prepared by the green route,[50] and has become more popular as a reproducible method of shell deposition,[51] even for other semiconducting phases such as CuInS$_2$.[52] The use of zinc diethyldithiocarbamate as a shelling precursor has also been applied to the thermal cycling technique of depositing ZnS shells.[53] In this case, CdS core particles were prepared and purified by standard green techniques, then the single-source precursor added (in a OAm solution, used to reduce the decomposition temperature) at 120 °C, then the temperature raised to 160–200 °C for shell growth. This was then repeated to grow several layers of the shell. The quantum yield of the resulting CdS/ZnS particles rose gradually upon layer addition, peaking at four monolayers of ZnS coverage (50% quantum yield), with no evidence of trap emission, unlike CdS/ZnS grown by the successive ion layer adsorption and reaction (SILAR) route.

Single-source precursors have also been used to grow QD-quantum well structures, where a wide-bandgap core is grown, followed by a relatively narrow-bandgap shell, which is capped by another wide-bandgap material, confining the charge carriers to the middle shell.[54]

7.5 Other Single-Source Precursors

Although compounds with the general formula M(ER)$_2$ were used in the formation of nanomaterials in seminal work described earlier, the materials obtained were substandard relative to the high-quality TOPO-capped particles described in Chapter 1. This can in part be attributed to the low temperature of growth. The thermolysis of related compounds in TOPO has proved to be an effective route to high-quality nanoparticles. The air-stable adduct compound $[Zn(SePh)_2][(CH_3)_2NCH_2CH_2N(CH_3)_2]$ has been used in the preparation of TOPO-capped ZnSe nanoparticles by injection of a TOP solution of the precursor into the solvent at temperatures between 320 and 385 °C. The particles formed were 2.7–4.9 nm in diameter, with emission between 2.75 eV (*ca.* 450 nm) and 3.20 eV (*ca.* 390 nm) dependent on the temperature of synthesis. Excitonic shoulders were clearly observed in the absorption spectra of particles prepared at lower temperatures, and all emission was band edge with only slight trap emission observed.[55,56] Thermolysis of the tellurium analogue, $[Zn(TePh)_2][(CH_3)_2NCH_2CH_2N(CH_3)_2]$ in dodecylamine (DDA) and TOP at the lower temperature of 180–240 °C resulted in the formation of spherical particles of ZnTe with a cubic crystalline core. The substitution of DDA for a mixture of trioctyl amine (TOA) and dimethylhexylamine resulted in the formation of spherical particles, which quickly grew into rods and bulk ZnTe. Higher synthesis temperatures resulted in particles with a band edge of 3.57 eV (*ca.* 350 nm), whereas particles prepared at the lower temperature exhibited a bandgap of 2.79 eV (*ca.* 450 nm), a shift from the bulk value of 2.26 eV (*ca.* 550 nm). These results appear counter-intuitive, as they indicate that the higher-temperature synthesis yielded smaller particles. The emission was, again, close to band edge with no evidence of trap emission.[57]

Metal xanthate complexes, structurally similar to dithiocarbamates, have found use in the preparation of metal sulfide nanoparticles. In initial work, the precursor cadmium diethylxanthate was easily prepared by the reaction between cadmium chloride and potassium ethylxanthate. The resulting precursor was dissolved in TOP and injected into TOPO at 180 °C. Particle formation was observed after 5 minutes, with growth of an hour resulting in only a small red shift in the absorption spectra, an indication of slow particle growth. The routes yielded spherical CdS particles of *ca.* 4 nm diameter with a hexagonal crystalline core, which exhibited band edge luminescence.[58] An investigation into varying reaction concentrations, times and temperatures using HDA as a solvent instead of TOPO resulted in the controllable synthesis of anisotropic particles.[59] Using high monomer (precursor) concentrations, long reaction times and temperatures of not more than 200 °C, multiarm particles formed with rods being the preferred morphology at synthesis temperatures above 200 °C. Increasing the monomer concentration also increased the aspect ratio of the resulting particles, and prolonged reaction times resulted in the bent and high-aspect-ratio particles forming spherical particles.

Similar results were obtained in work reported by Pradhan *et al.*, who prepared particles[60] and rods of CdS from ethylxanthate precursors by altering the monomer concentration.[61–63] In this work, particles of MS (M = Pb, Zn, Hg, Ni, Mn, Cu) were also prepared by alkylxanthate precursors, as were core/shell particles of CdS/ZnS and ZnS/CdS. Notably, lead and mercury ethylxanthates decomposed at room temperature, making preparation of high-quality particles difficult (although in other studies, lead hexadecylxanthate has been used at 65 °C to prepare ultrathin PbS wires[64] and ultrasmall PbS dots[65]). Ethyl-, hexyl- and hexadecylxanthates were used to prepare CdS and ZnS particles in HDA, whereas tertiary amines were used to synthesise CuS, NiS and MnS due to the displacement of the xanthate from the metal by the strongly binding primary amine. A one-pot reaction was used to prepare ZnS/CdS particles, relying on the differing decomposition temperatures of the compounds. By slowly heating a mixture of the two complexes in HDA, the zinc-containing precursor decomposed first (70 °C), followed by the cadmium-containing precursor at 120–170 °C. It has been suggested that these precursors may give better-quality particles than the dithiocarbamate analogues because of their lower decomposition temperature. In other work, the thermolysis of the zinc and cadmium ethylxanthates in a range of solvents resulted in the formation of a $Zn_xCd_{x-1}S$ alloy particle system, the composition (and hence the optical properties) of which was controllable by varying the precursor ratio. Decreasing the zinc content also resulted in a change of shape, from sphere, to rod, to multipod.[66] Zinc ethylxanthate has also been used as a precursor for ZnS shells (deposited on preformed CdSe particles).[67] In this case, the precursor was dissolved in tributyl phosphate (TBP), added to the CdSe particles at 60 °C and sonicated for *ca.* 10 minutes, whereupon the emission quantum yielded reached a maximum of *ca.* 50%.

The solvents/capping agents used had a key role in lowering the decomposition temperature of the precursors, which was explained with reference to the Chugaev reaction, which has similarities to the decomposition of the precursor (Figure 7.3). The introduction of the amine may have several effects; activating the $O–C–S_2–Cd$ group by shifting charge from the C–S bond to the Cd–S bond (Figure 7.3A); acting as a protonated intermediate aiding the required proton transfer (Figure 7.3B) and coordinating directly to the cadmium, weakening one Cd–S bond, activating it as a leaving group and hence forming one Cd–S bond (Figure 7.3C).

Zinc ethylxanthate has also been used in the synthesis of ZnS rods and dots where the stabilising solvent was used to control particle morphology and crystalline phase.[68] Using the precursor in HDA and OA between 150 and 200 °C, wurtzite (hexagonal crystalline core) rods were formed with tuneable diameters and aspect ratios (20–200 nm in length, 2.5–4.5 nm in width) which self-assembled into 2D lattices. Using HDA and TOP at 200 °C resulted in spherical particles with a zinc blende core. Using TOP alone resulted in spherical particles with a wurtzite core. The growth of dots occurred even with large monomer concentrations, indicating that anisotropic growth still required the presence of ligands that bound to a specific crystal facet. Zinc ethylxanthate has also been used to deposit a ZnS shell on $CuInS_2$ particles, increasing the emission quantum yield from 8% to 60%, although $Zn(CO_2(CH_2)_{16}CH_3)_2$ was also present during shell deposition.[69]

Wurtzite rods of CdS up to 24 nm long and 5 nm wide have also been grown using cadmium thiosemicarbazide $[Cd(NH_2CSNHNH_2)_2Cl_2]$ as a precursor in TOPO.[70] In this case, a precursor concentration consistent

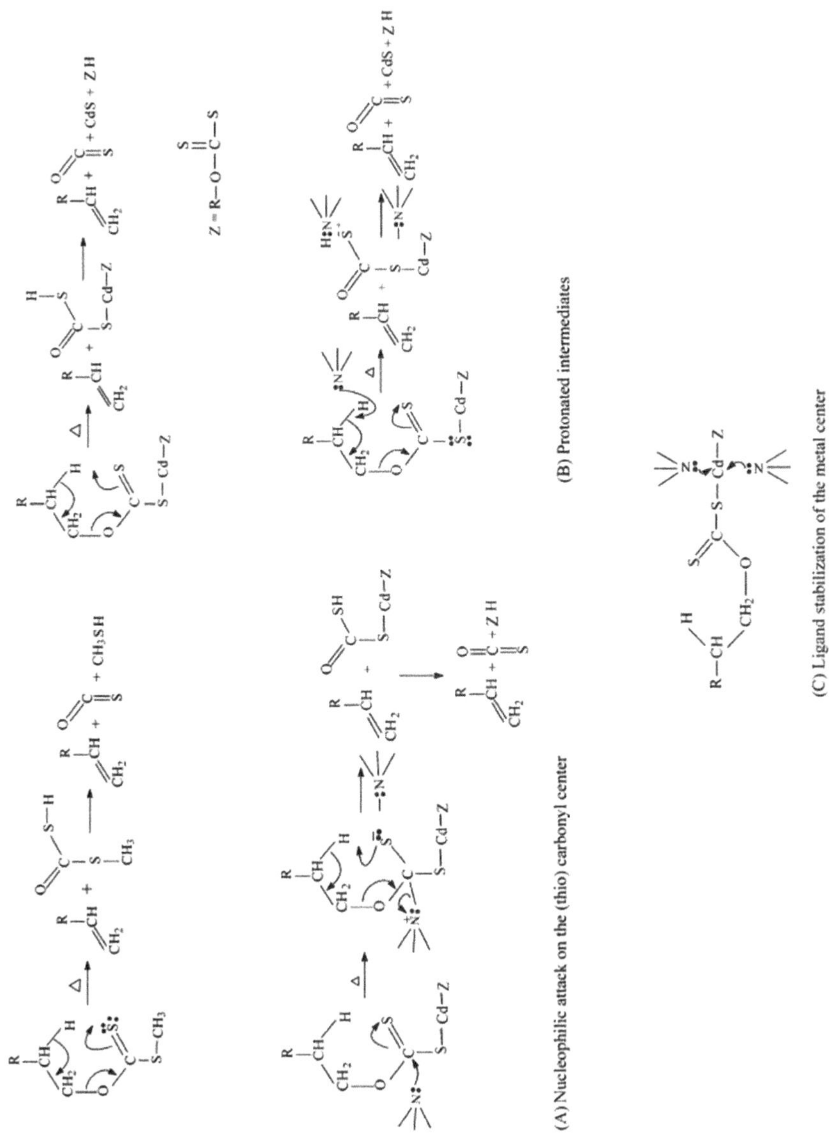

Figure 7.3 (Top left) The Chugaev reaction. (Top right) Related decomposition of cadmium ethylxanthate: (A) nucleophilic attack on thiocarbonyl centre by triethylamine; (B) protonated intermediates; (C) triethylamine stabilisation of the metal centre. Reprinted with permission from N. Pradhan, B. Katz and S. Efrima, *J. Phys. Chem. B*, 2003, **107**, 13843. Copyright 2003 American Chemical Society.

with dot growth was used, and it is possible that an amine decomposition product of the precursor may have coordinated to a crystal facet and induced the anisotropic growth. Using similar precursors, differing structures were also grown; spherical particles of CdS *ca.* 4 nm in diameter with a wurtzite core were grown using cadmium *N,N'*-bis(thiocarbamoyl)hydrazine (Cd(SCNHNH$_2$)$_2$Cl$_2$) as a precursor under similar growth conditions[71] while thermolysis of a cadmium complex of dithiobiurea Cd(NH$_2$CSNHNHCSNH$_2$)Cl$_2$, again in TOPO, resulted in irregular particles when grown at 150 °C and large aggregates of *ca.* 50 nm when grown at 240 °C.[72] This wide variation in particle morphology when using subtly different precursors under the same growth condition highlights the role precursors (and maybe decomposition products) play in defining nanostructure composition.

An interesting reaction pathway is the heterocumulene metathesis route to Cd(ESiMe$_3$)$_2$ (E = S, Se) which can be used as an *in situ* single-source precursor to CdSe or CdS.[73] The reaction between Cd[N(SiMe$_3$)$_2$]$_2$ and ECNR (where E = S, R = *t*Bu; E = Se, R = cyclohexyl) proceeds to give Cd(ESiMe$_3$)$_2$. Attempts to isolate adducts of the two compounds using TOPO or dodecanethiol as a Lewis base in a toluene solution resulted in the room-temperature preparation of TOPO-capped CdSe or thiol-capped CdS. Although few details regarding the optical properties were given, the route is notable for proceeding at a low temperature. Also, the precursor was relatively simple to prepare, and the selenium source, cyclohexylisoselenocyanate, was easily obtainable from the reaction between cyclohexylisonitrile and selenium metal. This is in fact a key point to note, as in some cases the preparation of a single-source precursor (such as diselenocarbamates) is often more difficult or dangerous than the preparation of the nanoparticles by simple binary methods. Likewise, a homoleptic cadmium selenolate, [Cd(SeCH$_2$CH$_2$NMe$_2$)$_2$], has been shown to be an effective precursor for TOPO-capped CdSe with a cubic crystal core,[74] but required the use of bis(2-dimethylaminoethyl)diselenide as a precursor. Although not difficult to prepare, it was still substantially harder than the simple binary routes to CdSe.

In the case of single-source precursors for cadmium selenide, the use of cadmium diselenocarbamates is limited by the synthesis of carbon diselenide, a toxic and evil-smelling compound that is difficult to prepare. A precursor system has been developed that avoids the use of such selenium precursors. Cadmium imino-bis(diisopropylphosphine selenide), Cd[N(SeP^{i}Pr$_2$)$_2$]$_2$, was prepared by deprotonating NH(SeP^{i}Pr$_2$)$_2$ in the presence of cadmium chloride, yielding an air-stable complex.[75]

Dissolution of Cd[N(SeP^{i}Pr$_2$)$_2$]$_2$ in TOP, followed by injection into TOPO at 250 °C and 30 minutes growth, resulted in CdSe nanoparticles *ca.* 5 nm in diameter with a wurtzite structure. Similarly, diselenophosphinato-complexes[76] of cadmium, [R$_2$PSe$_2$]$_2$Cd (R = Ph, ^{i}Pr, ^{t}Bu), have been used as precursors for CdSe QDs,[77] and the cobalt analogues have been used in the synthesis of either CoSe$_2$ (when thermolysed in TOPO or HDA) or Co$_2$P (when thermolysed in TOPO and TOP). Extended thermolysis in TOPO/TOP also resulted in CoP. The differing reaction products were attributed to the presence of TOP, which might therefore be classed as a reagent as well as a capping agent. The related compound $[^tBu_2PS_2]_2Co$ was found to give Co_9S_8 when reacted under similar conditions.[78] This reaction has been extended to the preparation of NiSe or Ni_2P or $Ni_{12}P_5$ using $[Ni(Se_2PR_2)_2]$, (R = ^iPr, ^tBu, Ph).[79] These precursors are interesting due to the ability to prepare differing materials using differing reaction conditions. The cobalt precursors and the nickel variants of this specific family of precursors have also exhibited mixed chalcogen/phosphide products when used as CVD precursors, again dictated by reaction conditions. In this case, TOP was absent so the same mechanism cannot be responsible for the phosphide phase.[80–82] Related to these precursors are dimorpholinodithioacetylaceto-nate compounds, which form a similar system and have been used as precursors for CdS nanorods.[83]

A useful and convenient family of single-source precursors has been described by Cumberland *et al.*[84] The cluster-based precursors, $(X)_4[M_{10}Se_4(SPh)_{16}]$ (X = Li^+, $(CH_3)_3NH^+$; M = Cd, Zn) are simply prepared by an established route[85] and utilised elemental selenium during synthesis. Similar materials have been investigated as precursors to the larger cluster $Cd_{32}S_{14}(SPh)_{36}$, a 15 Å diameter crystal which is midway between a cluster and a small QD.[86] Once prepared, the cadmium selenide precursor $(Li)_4[Cd_{10}Se_4(SPh)_{16}]$ was mixed with HDA and heated to 240 °C, inducing precursor decomposition and passivation of the resulting particles with the amine. Once the desired size of particle was obtained (as determined by absorption spectroscopy), the CdSe nanoparticles were cooled to 20 °C and left overnight to anneal and reduce the size distribution, then flocculated by addition of methanol. A shell of ZnS could be deposited using Me_2Zn and $S(SiMe_3)_2$ as described in Chapter 5.

Interestingly, blue-emitting CdSe nanorods up to 12 nm in length with a diameter of 2.5 nm which self-assembled have been prepared using the same method. Typically, the concentration of the precursor solution may often be the driving force behind anisotropic growth, but in this case the surfactant system and more importantly the reaction temperature was the defining factor inducing anisotropic growth. The authors also stated a slightly different precursor—$(Li)_2[Cd_{10}Se_4(SPh)_{16}]$.[87] The same precursors, when used in dimethylsulfoxide (DMSO), have shown larger clusters (Cd_{32} and Cd_{54}) in the stepwise growth of CdS.[88]

A similar method was used to prepare ZnSe nanoparticles using the precursor $((CH_3)_3NH)_4[Zn_{10}Se_4(SPh)_{16}]$ in HDA at 220–280 °C. The ZnSe particles were 2–5 nm in diameter, again exhibited band edge emission and displayed a cubic lattice. This route is notable for the ease of preparation of the precursors and particles. A detailed discussion regarding the growth mechanism was reported, and a mechanism incorporating fragmentation of the cluster was proposed, as opposed to the intact clusters acting as the nuclei, as previously suggested for the growth of GaS by MOCVD using single-source precursors.[89,90]

Molecular cluster compounds are particularly interesting as they might be considered the smallest structural units of a bulk solid, yet are not a classical solid-state material and as such are analysed using techniques usually associated with pure inorganic chemistry, such as single-crystal XRD. As mentioned above, they can be effective precursors, yet clusters such as $[(N,N'-(CH_3)_2NCH_2CH_2N(CH_3)_2)_5Zn_5Cd_{11}E_{13}(EPh)_6(tetrahydrofuran)_n]$ (E = Se, $n = 2$; E = Te, $n = 1$) exhibit optical properties consistent with quantum-confined CdSe (when E = Se).[91,92] In this case, the cluster can be envisaged as a $Cd_{11}Se_{13}$ core with a $(N,N'-(CH_3)_2NCH_2CH_2N(CH_3)_2)ZnSe_2$ shell. This should not be too surprising, as the clusters are prepared using silylated chalcogenides, metal salts and phosphines—similar precursors to II–VI nanomaterials, with the reaction proceeding at much lower temperatures (room temperature or lower). The first excitonic transition in $[(N,N'-(CH_3)_2NCH_2CH_2N(CH_3)_2)_5Zn_5Cd_{11}Se_{13}(SePh)_6(tetrahydrofuran)_2]$ is found at 379 nm, with broad 'band edge' emission visible at room temperature. Thermolysing related precursors such as $[Zn_xCd_{10-x}E_4(EPh)_{12}(R)_y]$ (E = Se, Te; R = propyl, $y = 4$; R = $(CH_3)_2NCH_2CH_2N(CH_3)_2$, $y = 5$) in HDA resulted in the formation of alloyed particles of $Zn_xCd_{1-x}E$, up to 5 nm in diameter. The emission spectra showed some evidence of trap emission when E = Te, yet appeared band edge when E = Se. Thermolysis of $[Zn_5Cd_{11}Se_4(-SePh)_6((CH_3)_2NCH_2CH_2N(CH_3)_2)_5]$ in the long-chain amine resulted in a material consistent with a CdSe/ZnSe core/shell structure (or a highly graded alloy structure).[93] Other materials used as single-source precursor include the polysulfide *N*-methylimidazole complexes $[M(N$-methylimidazole$)_6]S_8$ (M = Fe, Ni) and $ZnS_6(N$-methylimidazole$)_2$, which have been used in the synthesis of spherical or irregular NiS_2 particles, Fe_3S_4 particles and nanorods of ZnS respectively when decomposed in OAm at 300 °C.[94]

The single-source route to HgE (E = chalcogen) nanoparticles has also been examined; $Hg(TeCH_2CH_2NMe_2)_2$ has been thermolysed in a toluene solution of HDA at 80 °C yielding HgTe particles with a cubic crystalline core, *ca.* 6 nm in diameter. Introducing $Mn(OAc)_2 \cdot 4H_2O$ into the reaction resulted in the formation of ferromagnetic $Hg_{0.973}Mn_{0.027}Te$ particles.[95] Likewise, HgSe particles have been prepared by the thermolysis of $[Hg(Se-COR)_2(Me_2NCH_2CH_2NMe_2)]$ (R = Ph, C_6H_4-*p*-CH_3) in HDA although few properties were provided.[96]

7.6 Single-Source Precursors to Other Semiconducting Systems

There are semiconducting systems, other than cadmium and zinc chalcogenides, where the use of single-source precursors provided more obvious benefits. Lead chalcogenide QDs are of immense interest due to their emission in the infrared region, but early studies (before the use of metal carboxylate salts) were hindered by the lack of a convenient metal precursor: whereas group II metal alkyls were routinely used, lead alkyls were avoided because of their extreme toxicity. The earliest routes to TOPO-capped lead chalcogenide nanoparticles were single-source based, where monomeric lead(II) diseleno- or dithiocarbamates were thermolysed to give the required particles.[97,98] Again, in a similar manner to the cadmium dithiocarbamate system, the alkyl constituents dictated the final product. The use of symmetrical lead diselenocarbamates such as $Pb(Se_2CNEt_2)_2$ resulted in the generation of micrometre-sized selenium particles as well as PbSe nanoparticles, whereas the use of $Pb(Se_2CNMe(n\text{-hex}))_2$ resulted solely in the formation of PbSe nanoparticles with as cubic crystal core as determined by X-ray powder diffraction. The thermolysis of the symmetrical thiocarbamate system, $Pb(S_2CNEt_2)_2$ resulted in the formation of PbS particles with either a hexagonal or cubic morphology, and later studies also revealed quasi-spherical particles when prepared at low temperatures (60 °C), and cubic at higher temperatures (80 °C).[99] The PbE (E = S, Se) particles produced were crystalline, relatively large and polydispersed, displaying various shapes from spherical to cubic structures, despite the relatively low temperature of synthesis (*ca.* 150 °C for PbS, 180 °C for PbSe). The band edges of the resulting particles did not display any sharp features, but were significantly shifted from their bulk values. In these cases no emission was reported, but what is important, in this case and at the time of publication, is that the single-source precursors provided a synthetic pathway to these otherwise unavailable technologically important materials. Later publications also reported the thermolysis of $Pb(S_2CNEt_2)_2$ in diphenylether using thiols and amines as capping agents, giving PbS exhibiting a wide range of shapes.[100] Other precursors used for the synthesis of PbS include $Pb(SCOC_6H_5)_2$, which was dissolved in long-chain amines which catalysed the decomposition of the precursor, giving a range of materials, although no optical properties were reported.[101]

The dithio/diselenocarbamate precursors have been extended to other semiconducting systems. Indium chalcogenide nanoparticles may find use in solar cell applications as the copper indium gallium selenide (CIGS)-based materials are radiation-hard and are the focus of much research. The simple, symmetrical diethyl dithio- and diselenocarbamates were prepared using the standard techniques, and thermolysed in either TOPO or 4-ethylpyridine. Although TOPO might at first appear the more suitable surfactant system due to the higher temperature, isolation of the nanoparticles proved difficult and better-defined particles were obtained from 4-ethylpyridine.[102]

Dithiocarbamates are also applicable to other more unusual systems. Europium sulfide (EuS) is a ferromagnetic semiconductor, with a bandgap of 1.7 eV (*ca.* 730 nm)[103] and a small exciton diameter of *ca.* 0.5 nm.[104] A range of lanthanide dithiocarbamates (including Nd, Sm, Gd, Ho and Er dithiocarbamates) were prepared as potential precursors for rare-earth sulfides, with EuS being the main material of interest.[105] The precursor $Eu(S_2CNEt_2)_3$:co-ligand was used in the synthesis of EuS nanoparticle with a cubic crystalline core, exhibiting various morphologies between 5 and 50 nm, with the larger particles exhibiting a cubic shape.[106] Due to the large coordination sphere of lanthanides, co-ligands such as 1,10-phenanthroline and 2,2′-bipyridine, commonly used in the preparation of rare-earth complexes were employed during the preparation of the precursors, and these ligands were found to play an important role in nanoparticle synthesis, coordinating to the particle surface after precursor decomposition. These precursors were thermolysed in either TOP, OAm or an ODE solution of OAm and OA, producing a range of materials with morphologies ranging from cubes to spherical particles, depending upon the system used. Similarly, xanthates with similar co-ligands or a potassium ethylxanthate were used, although nanoparticles prepared by this route were found to be of a lower quality than those prepared by dithiocarbamates. The optical characteristic of the nanoparticles were dominated by the electronic structure of the rare-earth ion, not the overall band structure of the solid-state material. Notably the nanoparticles decomposed in polar solvents such as tetrahydrofuran, yielding a lumines-cent trivalent europium complex. The particles, in general, displayed optical properties similar to bulk materials.[107]

In a similar synthesis, EuS was prepared from a slurry of $Eu(S_2CNEt_2)_3$:1,10-phenanthroline in OAm at 200 °C, giving a material that was easily isolated and redispersed in non-polar solvents. The resulting nanoparticles were compared to EuS nanoparticles prepared by a furnace route using the same precursor and found to be significantly smaller (5.5 nm compared to at least 50 nm, and up to 120 nm), and exhibited cluster glass-like behaviour and quasi-ferrimagnetic behaviour upon surface oxidation.[108]

Nanoparticulate EuS has also been prepared by the photolytic decompo-sition of $Na[Eu(S_2CNEt_2)_4] \cdot 3.5H_2O$, by irradiating the ligand to metal charge transfer band of the precursor in solution, with no capping agents. The particles, *ca.* 9 nm in diameter, exhibited weak luminescence although this was the first time the emission spectrum of nanoparticulate EuS was reported and was attributed to an f–d transition with a blue-shifted band edge of 1.9 eV.[109]

Several groups have reported similar routes to EuS, exploring the magnetic properties,[110,111] polymer encapsulation,[112] superlattice formation[113] and the different products obtained when the precursors were thermolysed in air, observing that Eu_2O_2S was obtained instead.[114] Hasegawa *et al.* also reported one of the few preparations of EuSe using a diselenophosphinate, and it was found that the tetraphenylphosphine cation, a decomposition product of the precursor, coordinated to the particle surface, affecting the particle morphology and electronic properties.[115]

The use of single-source precursors to binary compounds has been extended; Castro *et al.* reported a notable synthesis, where all three elements of a ternary system were provided by a single-source precursor.[116] The compound, $(PPh_3)_2CuIn(SEt)_4$, was dissolved in dioctylphthalate and heated with hexanethiol, and decomposed above 200 °C to give $CuInS_2$ between 2 and 4 nm in diameter. The reaction did not proceed when using TOPO, carboxylic acid or long-chain amines as coordinating ligands, but TOPO can be used to replace the thiol after the reaction has been completed and the nanoparticles isolated. Absorption and emission spectroscopies showed a shift from the usual absorption edge (1.55 eV to *ca.* 2.0 eV, 800 nm to 620 nm) with a maximum quantum yield of 4.4%, and XRD of the particles showed agreement with the patterns from the bulk material. This is an excellent example of where single-source precursors can greatly simplify the preparation of solid-state materials (one might argue that preparing the precursor is itself relatively difficult and overly time consuming, but once it is prepared further steps are greatly simplified). The same compound, combined with another related single-source precursor, $(PPh_3)_2CuGa(SEt)_4$, when heated by microwave in benzylacetate with 1,2-ethanediol yielded $CuIn_xGa_{1-x}S_2$.[117] The ratio between indium and gallium was found to be controllable by varying the ratio of precursors, and the presence of the diol was found to be essential, suggested to link the two precursors together. The optical band edge was controlled by varying the metal content, with band edges between *ca.* 550 nm ($CuGaS_2$) and *ca.* 800 nm ($CuInS_2$) obtained. Similarly, $[(Ph_3)CuIn(SC\{O\}Ph)_4]$ was used as a single-source precursor for $CuInS_2$, but this time TOPO and TOP were used as surfactants alongside dodecanethiol. The ratio of surfactants was altered to control the crystallinity of the material, with the diffraction patterns displaying the dominant wurtzite phase along with evidence of zinc blende particles in some reactions. It was also shown that $CuInS_2$ could be prepared from the reaction of two separate single-source precursors, $[Cu(SC\{O\}Ph)]$ and $[In(bipy)(SC\{O\}Ph_3]$ (where bipy = 2,2′-bipyridine) under similar experimental conditions.[118]

A similar precursor, $[(PPh_3)_2Cu(\mu\text{-}SePh)_2In(SePh)_2]$, has been used in the catalytic SLS growth preparation of high-quality stoichiometric $CuInSe_2$ wires, 33 nm in diameter and several micrometres long.[119] In this reaction, the precursor was dissolved in TOP, then injected in a mixture of OA and TOP at 300 °C immediately after the injection of Au/Bi core/shell catalysts particles. After under 5 minutes growth, the wires were isolated by rapid cooling and the addition of a non-solvent. The wires, clearly the ternary phase as determined by X-ray powder diffraction, had an absorption edge in the infrared region. Another related ternary compound, $CuInS_2$, has been prepared by the thermolysis of two single-source precursors, $Cu(S_2CNEt_2)_2$ and $In(S_2CNEt_2)_3$, in various capping agents which dictated the crystalline phase of the nanoparticles.[120] Bandgap-sensitive elements, gallium and tellurium, have also been introduced into the $CuInS_2$ system by preparing a range of quaternary materials by single-source precursor.[121] $Cu_{1.0}Ga_xIn_{2-x}S_{3.5}$ was prepared by the thermolysis of diethyldithiocarbamates of copper, indium and gallium in toluene with OA and OAm, for 90 minutes at 180 °C in an autoclave. Likewise, $Cu_{1.0}In_xTl_{2-x}S_{3.5}$ was prepared using single-source precursors in OA and ODE, followed by the addition of OAm at 200 °C for 1 minute, although no autoclave was necessary. OAm was found to be essential, and no reactions occurred without it. The inclusion of gallium and tellurium did not affect the cubic crystalline structure. Electron microscopy of $Cu_{1.0}Ga_{1.0}In_{1.0}S_{3.5}$ revealed approximately spherical particles *ca.* 6 nm in diameter, whereas particles of

$Cu_{1.0}In_{1.0}Tl_{1.0}S_{3.5}$ were *ca.* 16 nm in diameter with relatively narrow size distributions. The absorption spectra of $Cu_{1.0}Ga_xIn_{2-x}S_{3.5}$ was also shown to red-shift with increasing indium content in agreement with Vegard's law, from a band edge at *ca.* 450 nm to *ca.* 900 nm.

The use of simple dithiocarbamates has been used to make $ZnS–AgInS_2$ solid-solution nanoparticles. Thermolysis of $(AgIn)_xZn_{2(1-x)}(S_2CNEt_2)_4$ in OAm resulted in particles *ca.* 4 nm in diameter, which exhibited an XRD pattern consistent with a solid solution. By varying the precursor composition, the emission could be tuned across the entire visible region of the spectrum.[122] There are other single-source routes to related ternary semiconducting systems such as $AgInSe_2$.[123] Thermolysis of $[(PPh_3)_2AgIn(SeC\{O\}Ph_4)]$ in a mixture of OAm and dodecanethiol caused the formation of a hitherto unknown orthorhombic phase of $AgInSe_2$ rods, *ca.* 14.5×50 nm in size, isostructural with $AgInS_2$. The use of both surfactants was essential, as removal of the amine resulted in formation of the bulk phase, while removal of the thiol resulted in the formation of impure, irregular-shaped particles. Similarly, $[(Ph_3P)_2AgIn(SCOPh)_4]$ has been used as a single-source precursor to $AgInS_2$, when thermolysed in OA and dodecanethiol at between 125–220 °C.[124,125] Thermolysis of a related precursor $[(PPh_3)_3Ag_2(SeC\{O\}Ph)_2]$ in a mixture of TOP and HDA at temperatures between 95 and 180 °C gave Ag_2Se nanoparticles of varying morphology depending on reaction temperature and nuclei concentration. It was suggested that at lower temperatures the phosphine stabilised the precursor and inhibited the decomposition process, slowing the nucleation process and controlling the morphology observed. The amine was responsible for the activation of the precursor, providing a β-elimination pathway, giving the pure Ag_2Se phase and separating the nucleation and growth steps.[126] Similarly, thermolysis of Ag(S-COPh) in TOP at 160 °C resulted in the formation of both silver nanoparticles, 80 nm in diameter, and Ag_2S nanoparticles, *ca.* 25 nm in diameter. The formation of silver particles was reportedly due to the absence of alkylamine activators.[127,128] The bismuth analogue of the precursor, Bi(SCOPh), has been thermolysed in TOPO and dodecanethiol at 150 °C for 1 hour, yielding rods *ca.* 300 nm × 21 nm, although longer rods which agglomerated could be obtained by replacing TOPO with OAm.[129]

Although group-VI-containing QDs are the most common, others such as group III–V materials (*e.g.* InP, InAs) are also important and in many way, these systems may well benefit more from a single-source approach than the II–VI materials. Phosphide and arsenide precursors are usually toxic and difficult to handle, but the preparation of sulfide, selenide and telluride nanoparticles is much simpler and can use the elements themselves in solution as precursors.

Cadmium phosphide, Cd_3P_2, is a narrow-bandgap semiconductor (0.55 eV, *ca.* 2250 nm) with a large excitonic diameter of 36 nm, providing the potential to tune the emission across a wide spectral range. The diorganophosphide $[MeCdP^tBu_2]_3$, prepared by the direct reaction of Me_2Cd and HP^tBu_2 in a range of solvents, was thermolysed in either TOPO or 4-ethylpyridine to give nanoparticles *ca.* 4–5 nm in diameter with emission tuneable between *ca.* 400 nm and 600 nm depending on the temperature of the reaction.[130,131] A cadmium : phosphorus ratio of 1.6 : 1 was obtained at lower temperature; at higher temperatures, the ratio was as high as 7 : 1. Reactions with a related diorganophosphide, $[MeCd(\mu\text{-}PPh_2)]_3(HPPh_2)_2$, with no β-hydrogen atoms, resulted in predominantly elemental cadmium being formed, highlighting the importance of the organic groups to the decomposition mechanism. Of note is the lack of XRD data, as the material gave no useable reflections, unlike the analogous III–V QDs. It is worth noting that cadmium phosphide has been investigated by a number of groups; Weller has described an essentially colloidal route to Cd_3P_2 using a variety of metal salts and phosphine gas in triethylamine and methyl methacrylate, followed by room-temperature vacuum polymerisation in a sealed quartz cuvette.[132] The resultant QDs were characterised primarily by measurements of lattice fringes with no XRD data reported. Buhro has developed organometallic routes to nanoparticulate Cd_3P_2 using a variety of routes, although the material was not reported in depth.[133] One notable route involves the methanolysis of a silylated single-source precursor, $Cd[P(SiPh_3)_2]_2$.[134] The resulting nanocrystals were capped by Ph_3Si at the phosphide sites and by OMe at the metal sites. Replacement of the phenyl groups on the precursors for methyl groups resulted in bulk material being formed.

The use of diorganophosphides was extended to the preparation of InP and GaP.[135] $In(P^tBu_2)_3$ was thermolysed in 4-ethylpyridine which gave capped nanoparticles *ca.* 7 nm in diameter with a blue-shifted band edge at *ca.* 2.7 eV (*ca.* 460 nm) and broad, near band edge luminescence.[136] GaP nanoparticles *ca.* 7 nm in diameter could be also be prepared using $Ga(P^tBu_2)_3$ in 4-ethylpyridine, the resulting nanoparticles showing similar optical characteristics to the analogous InP particles. Similar experiments where the precursors were thermolysed in TOPO were unsuccessful. The particles had a large amount of metal-based impurities, either elemental metal or the metal oxide, which was found on prolonged heating. MS investigations into the decomposition products suggested reductive elimination, β-hydrogen elimination and strain-induced bond rupture as the relevant decomposition mechanisms. This supported the use of organic side chains such as tertiary butyl groups, in which all hydrogen atoms are in the β position. Despite this, unpublished work has demonstrated the preparation of InP particles by thermolysis of $In(PPh_2)_3$.[137] Different-shaped nanoparticles of GaP could also be formed when $Ga(P^tBu_2)_3$ was thermolysed in a mixture of TOA and HDA. When the ratio of TOA : HDA was 1 : 1, a mixture of rods (*ca.* 42 × 8 nm) and dots were obtained. Increasing the ratio to 1 : 1.25, wurtzite-structured rods (45 × 8 nm) were uniquely formed. By using just TOA alone as a capping agent, spherical particles of *ca.* 8 nm diameter GaP with a zinc blende structure were obtained. The difference in structures was attributed to the steric properties of the capping agents, *i.e.* the sterically bulky TOA drove the particle growth towards the staggered zinc blende GaP confirmation, whereas the less sterically demanding HDA reduced rotational barriers to wurtzite growth.[138] HDA was also used as a capping agent and reaction solvent in the preparation of GaAs QDs by the thermolysis of $[^tBu_2AsGaMe_2]_2$. The resulting particles were *ca.* 3.2 nm in diameter, with a blue-shifted band edge and broad emission. Powder XRD confirmed small amounts of Ga_2O_3 and In_2O_3, and heating the material in air resulted in the formation of $GaAsO_4$.[139] In related work, the precursor $[^tBu_2AsInEt_2]_2$ was thermolysed in hot HDA to give InAs particles *ca.* 9 nm in diameter with similar optical properties to those of GaAs described previously.[140] Notably, in this study the XRD pattern is extremely clear relative to other III–V materials prepared by single-source precursors described above and can easily be compared to the bulk diffraction pattern, possibly due to the large crystal domains in the bigger particles.

Group III–nitride materials made by solvothermal routes have been discussed in only a few reports, because of the lack of obvious precursors and the metastability of certain materials, such as InN which decomposes at 400 °C. Despite all efforts, nitride nanomaterials remain difficult to prepare by solution routes. Several examples exist, based on azide-type structures; in a route described by Dingman *et al.*, one of the first reports of single-source solution routes to nitrides used the precursors $[R_2InN_3]_n$ (where R = alkyl group), a ladder-type polymeric structure, which was found to be insoluble in most hydrocarbons, yet soluble in Lewis base solvents. Thermolysis of the isopropyl precursor in diisopropylbenzene resulted in amorphous material, but inclusion of H_2NNMe_2 resulted in InN nanofibres, grown from an indium nanodroplet (produced by the reduction of the precursor by the H_2NNMe_2). The material, characterised by XRD and electron microscopy, appeared to grow *via* the SLS mechanism described in Chapter 1.[141] Gallium azides have also been explored as potential precursors to GaN; $[Et_2Ga(N_3)]_3$, $(N_3)_2Ga$ $[(CH_2)_3NMe_2]$ and $(Et_3N)Ga(N_3)_3$ have been used in refluxing triglyme, yielding particles *ca.* 200 nm in diameter, significantly larger than other similarly prepared nanoparticles. Despite the large particle size, no XRD patterns were obtained, indicating the poor crystallinity of the materials. Broad emission towards the blue end of the optical spectrum was observed despite the amorphous character.[142]

InN with a cubic crystalline core has also been prepared from the thermolysis of $InN_3(CH_2CH_2CH_2NMe_2)_2$ in TOPO, yielding particles *ca.* 2–10 nm in diameter. The band edge was determined to be *ca.* 570 nm, with emission observed at 690 nm, although emission from the capping agent at 550 nm was also observed. This is notable, as a definite capping agent was utilised, unlike the previous example.[143] Nanoparticles of GaN, InN and AlN have also been prepared by thermolysis of $M(H_2NCONH_2)_3Cl_3$ (M = Al, In, Ga) in TOA, although few details were provided.[144] In a detailed study, InN particles with a hexagonal crystalline core were produced by the reaction between $InBr_3$ in toluene with NaN_3 yielding an azide intermediate which was gently thermolysed to yield the desired product. As well as a high-pressure-based route, a bench-top ambient-pressure synthesis using TOA as a capping agent was described, which was possible due to the slight solubility of $InBr_3$ in toluene (previously, other halides were found to be insoluble and unsuitable for solution-based reactions with azides). The ambient-pressure route produced small particles (*ca.* 3 nm) with poor crystallinity relative to the high-pressure route, possibly due to the high-pressure route's constant exposure to high-pressure nitrogen as opposed to the constantly vented low-pressure synthesis. The particles appeared larger and more aggregated when compared to the II–VI analogues, or even other III–V nanomaterials. Notably, the use of the amine reportedly resulted in the decomposition of the nitride structures, resulting in indium metal.[145]

The single-source route does have its limitations. The stoichiometry of the particles prepared by just one precursor cannot be easily manipulated, unlike the binary approach where altering the ratio of precursors can markedly affect the optical properties.[146] The development of the SILAR route described in Chapter 5 is an excellent example of where a specific nanoparticle structure can be prepared by the addition of binary precursors. The preparation of single-source precursors is, in some cases, more difficult than the preparation of nanoparticles themselves from the simple binary routes, although some new phases of materials may only be accessible through single-source complexes. The chemistry is undoubtedly different for single-source and binary approaches and this may impact on the resulting materials, and one has to choose the optimum approach as befits the situation and the nanoparticles desired.

References

1. M. Steigerwald, *Polyhedron*, 1994, **13**, 1245.
2. The seminal work on organometallic-based routes to TOPO-capped nanoparticles was carried out by C. B. Murray, who reported initial results into the synthesis of HgE (E = S, Se, Te) materials using mercury alkyls (C. B. Murray, PhD thesis, MIT, 1995). The neurotoxic nature of mercury alkyls is well documented (*e.g.* T. W. Clarkson, *Annu. Rev. Public Health*, 1983, **4**, 375).
3. For example, see: P. O'Brien and R. Nomura, *J. Mater. Chem.,* 1995, **5**, 1761.
4. M. L. Steigerwald and C. R. Sprinkle, *J. Am. Chem. Soc.,* 1987, **109**, 7200.
5. J. G. Brennan, T. Siegrist, P. J. Carroll, S. M. Stuczynski, P. Reynders, L. E. Brus and M. L. Steigerwald, *Chem. Mater.,* 1990, **2**, 403.
6. J. G. Brennan, T. Siegrist, P. J. Carroll, S. M. Stuczynski, L. E. Brus and M. L. Steigerwald, *J. Am. Chem. Soc.,* 1989, **111**, 4141.
7. H. Noglik and W. J. Pietro, *Chem. Mater.,* 1994, **6**, 1593.
8. N. Félidj, G. Lévi, J. Pantigny and J. Aubard, *New J. Chem.,* 1998, **22**, 725.
9. S. L. Stoll, E. G. Gillan and A. R. Barron, *Chem. Vap. Deposition,* 1996, **2**, 182.
10. R. L. Wells, M. F. Self, A. T. McPhail, S. R. Aubuchon, R. C. Woudenberg and J. P. Jasinska, *Organometallics,* 1993, **12**, 2832.
11. J. F. Janik, R. L. Wells, V. G. Young, A. L. Rhinegold and L. A. Guzei, *J. Am. Chem. Soc.,* 1998, **120**, 532.

12. D. Crouch, S. Norager, P. O'Brien, J.-H. Park and N. Pickett, *Philos. Trans. R. Soc. London, Ser. A,* 2003, **361**, 297.
13. T. Trindade, P. O'Brien and N. L. Pickett, *Chem. Mater.,* 2001, **13**, 3843.
14. L. Henriksen and E. S. S. Kristiansen, *Int. J. Sulfur Chem., Part A,* 1972, **2**, 133.
15. T. Trindade and P. O'Brien, *J. Mater. Chem.,* 1996, **6**, 343.
16. M. B. Hursthouse, M. A. Malik, M. Motevalli and P. O'Brien, *Polyhedron,* 1992, **11**, 45.
17. T. Trindade and P. O'Brien, *Adv. Mater.,* 1996, **8**, 161.
18. T. Trindade, P. O'Brien and X. M. Zhang, *Chem. Mater.,* 1997, **9**, 523.
19. M. Chunggaze, J. McAleese, P. O'Brien and D. J. Otway, *Chem. Commun.,* 1998, 833.
20. M. Chunggaze, M. A. Malik and P. O'Brien, *J. Mater. Chem.,* 1999, **9**, 2433.
21. Y. K. Jung, J. I. Kim and J.-K. Lee, *J. Am. Chem. Soc.,* 2010, **132**, 178.
22. B. Ludolph, M. A. Malik, P. O'Brien and N. Revaprasadu, *Chem. Commun.,* 1998, 1849.
23. M. A. Malik, N. Revaprasadu and P. O'Brien, *Chem. Mater.,* 2001, **13**, 913.
24. N. L. Pickett and P. O'Brien, *Chem. Rec.,* 2001, **1**, 467.
25. D. J. Binks, D. P. West, S. Norager and P. O'Brien, *J. Chem. Phys.,* 2002, **117**, 7335.
26. A. A. Memon, M. Afzaal, M. A. Malik, C. Q. Nguyen, P. O'Brien and J. Raftery, *Dalton Trans.,* 2006, 4499.
27. M. Lazell and P. O'Brien, *Chem. Commun.,* 1999, 2041.
28. K. Abe, T. Handa, Y. Yoshida, N. Tanigaki, H. Takiguchi, H. Nagasawa, M. Nakamoto, T. Yamaguchi and K. Yase, *Thin Solid Films,* 1998, **329**, 524.
29. M. Lazell, S. J. Norager, P. O'Brien and N. Revaprasadu, *Mater. Sci. Eng., C,* 2001, **16**, 129.
30. O. C. Monteiro and T. Trindade, *J. Mater. Sci. Lett.,* 2000, **19**, 859.
31. O. C. Monteiro, H. I. S. Nogueira, T. Trindade and M. Motevalli, *Chem. Mater.,* 2001, **13**, 2103.
32. M. A. Malik, P. O'Brien and N. Revaprasadu, *J. Mater. Chem.,* 2002, **12**, 92.
33. N. Revaprasadu, M. A. Malik, P. O'Brien, M. M. Zulu and G. Wakefield, *J. Mater. Chem.,* 1998, **8**, 1885.
34. N. Revaprasadu, M. A. Malik, P. O'Brien and G. Wakefield, *J. Mater. Res.,* 1999, **14**, 3237.
35. V. Lou, A. C. S. Samia, J. Cowen, K. Banger, X. Chen, H. Lee and C. Burda, *Phys. Chem. Chem. Phys.,* 2003, **5**, 1091.
36. Y. Wu, C. Wadia, W. Ma, B. Sadtler and A. P. Alivisatos, *Nano Lett.,* 2008, **8**, 2551.
37. N. Revaprasadu, M. A. Malik and P. O'Brien, *S. Afr. J. Chem.,* 2004, **57**, 40.
38. I. J.-L. Plante, T. W. Zeid, P. Yang and T. Mokari, *J. Mater. Chem.,* 2010, **20**, 6612.

39. M. A. Malik, P. O'Brien, N. Revaprasadu and G. Wakefield, *Mater. Res. Soc. Symp. Proc.,* 1999, **536**, 371.
40. Y.-W. Jun, S.-M. Lee, N.-J. Kang and J. Cheon, *J. Am. Chem. Soc.,* 2001, **123**, 5150.
41. J. Sun and W. E. Buhro, *Angew. Chem., Int. Ed.,* 2008, **47**, 3215.
42. W. Lou, M. Chen, X. Wang and W. Liu, *Chem. Mater.,* 2007, **19**, 872.
43. C. Q. Nguyen, M. Afzaal, M. A. Malik, M. Helliwell, J. Raftery and P. O'Brien, *J. Organomet. Chem.,* 2007, **692**, 2669.
44. Y.-W. Jun, Y.-Y. Jung and J. Cheon, *J. Am. Chem. Soc.,* 2002, **124**, 615.
45. Y. Du, B. Xu, T. Fu, M. Cai, F. Li, Y. Zhang and Q. Wang, *J. Am. Chem. Soc.,* 2010, **132**, 1470.
46. P. Sreekumari Nair and G. D. Scholes, *J. Mater. Chem.,* 2006, **16**, 467.
47. T. Mandal, V. Stavila, I. Rusakova, S. Ghosh and K. H. Whitmire, *Chem. Mater.,* 2009, **21**, 5617.
48. N. Revaprasadu, M. A. Malik, P. O'Brien and G. Wakefield, *Chem. Commun.,* 1999, 1573.
49. M. A. Malik, P. O'Brien and N. Revaprasadu, *Chem. Mater.,* 2002, **14**, 2004.
50. A. Komoto, S. Maenosono and Y. Yamaguchi, *Langmuir,* 2004, **20**, 8916.
51. J. R. Dethlefsen and A. Dossing, *Nano Lett.,* 2011, **11**, 1964.
52. K. Nose, Y. Soma, T. Omata and S. Otsuka-Yao-Matsuo, *Chem. Mater.,* 2009, **21**, 2607.
53. D. Chen, F. Zhao, H. Qi, M. Rutherford and X. Peng, *Chem. Mater.,* 2010, **22**, 1437.
54. Y. Li, J. Feng, S. Daniels, N. L. Pickett and P. O'Brien, *J. Nanosci. Nanotechnol.,* 2007, **7**, 2301.
55. Y.-W. Jun, J.-E. Koo and J. Cheon, *Chem. Commun.,* 2000, 1243.
56. S. Schlecht, S. Tan, M. Yosef, R. Dersch, J. H. G. Wendorff, Z. Jia and A. Schaper, *Chem. Mater.,* 2005, **17**, 809.
57. Y.-W. Jun, C.-S. Choi and J. Cheon, *Chem. Commun.,* 2001, 101.
58. P. Sreekumari Nair, T. Radhakrishnan, N. Revaprasadu, G. Kolawole and P. O'Brien, *J. Mater. Chem.,* 2002, **12**, 2722.
59. Y. Li, X. Li, C. Yang and Y. Li, *J. Mater. Chem.,* 2003, **13**, 2641.
60. N. Pradhan and S. Efrima, *J. Am. Chem. Soc.,* 2003, **125**, 2050.
61. N. Pradhan, B. Katz and S. Efrima, *J. Phys. Chem. B,* 2003, **107**, 13843.
62. S. Efrima and N. Pradhan, *C. R. Chim.,* 2003, **6**, 1035.
63. M. Protière and P. Reiss, *Nanoscale Res. Lett.,* 2006, **1**, 62.
64. S. Acharya, U. K. Gautam, T. Sasaki, Y. Bando, Y. Golan and K. Ariga, *J. Am. Chem. Soc.,* 2008, **130**, 4594.
65. S. Acharya, D. D. Sarma, Y. Golan, S. Sengupta and K. Ariga, *J. Am. Chem. Soc.,* 2009, **131**, 11282.
66. Y. Li, M. Ye, C. Yang, X. Li and Y. Li, *Adv. Funct. Mater.,* 2005, **15**, 433.
67. M. J. Murcia, D. L. Shaw, H. Woodruff, C. A. Naumann, B. A. Young and E. C. Long, *Chem. Mater.,* 2006, **18**, 2219.
68. Y. Li, X. Li, C. Yang and Y. Li, *J. Phys. Chem. B,* 2004, **108**, 16002.

69. L. Li, T. Jean Daou, I. Texier, T. T. K. Chi, N. Q. Liem and P. Reiss, *Chem. Mater.,* 2009, **21**, 2422.

70. P. Sreekumari Nair, T. Radhakrishnan, N. Revaprasadu, G. A. Kolawole and P. O'Brien, *Chem. Commun.,* 2002, 564.

71. P. Sreekumari Nair, N. Revaprasadu, T. Radhskrishnan and G. A. Kolawole, *J. Mater. Chem.,* 2001, **11**, 1555.

72. P. Sreekumari Nair, T. Radhskrishnan, N. Revaprasadu, G. A. Kolawole and P. O'Brien, *Polyhedron,* 2003, **22**, 3129.

73. J. R. Babcock, R. W. Zehner and L. R. Sita, *Chem. Mater.,* 1998, **10**, 2027.

74. G. Kedarnath, S. Dey, V. K. Jain, G. K. Dey and B. Varghese, *Polyhedron,* 2006, **25**, 2383.

75. D. J. Crouch, P. O'Brien, M. A. Malik, P. J. Skabara and S. P. Wright, *Chem. Commun.,* 2003, 1454.

76. C. Q. Nguyen, A. Adeogun, M. Afzaal, M. A. Malik and P. O'Brien, *Chem. Commun.,* 2006, 2182.

77. S. K. Stubbs, D. J. Binks, F. Aslam, C. Q. Nguyen, A. Malik, P. O'Brien, C. C. Byeon, D.-K. Ko and J. Lee, Optical characterisation of CdSe nanocrystal quantum dots grown from new single-source precursors, *Proc. Conf. Lasers and Electro-optics,* CLEO/Pacific Rim, 2007.

78. W. Maneeprakorn, M. A. Malik and P. O'Brien, *J. Mater. Chem.,* 2010, **20**, 2329.

79. W. Maneeprakorn, C. Q. Nguyen, M. A. Malik, P. O'Brien and J. Raftery, *Dalton Trans.,* 2009, 2103.

80. A. Panneerselvam, G. Periyasamy, K. Ramasamy, M. Afzaal, M. A. Malik, P. O'Brien, N. A. Burton, J. Waters and B. E. Van Dongen, *Dalton Trans.,* 2010, **39**, 6080.

81. A. Panneerselvam, C. Q. Nguyen, J. Waters, M. A. Malik, P. O'Brien, J. Raftery and M. Helliwell, *Dalton Trans.,* 2008, 4499.

82. A. Panneerselvam, M. A. Malik, M. Afzaal, P. O'Brien and M. Helliwell, *J. Am. Chem. Soc.,* 2008, **130**, 2420.

83. K. Ramasamy, M. A. Malik, P. O'Brien and J. Raftery, *Dalton Trans.,* 2009, 2196.

84. S. L. Cumberland, K. M. Hanif, A. Javier, G. A. Khitrov, G. F. Strouse, S. M. Woessner and C. S. Yun, *Chem. Mater.,* 2002, **14**, 1576.

85. I. G. Dance, A. Choy and M. L. Scudder, *J. Am. Chem. Soc.,* 1984, **106**, 6285.

86. N. Herron, J. C. Calbrese, W. E. Farneth and Y. Wang, *Science,* 1993, **259**, 1426.

87. S. G. Thoma, A. Sanchez, P. P. Provencio, B. L. Abrams and J. P. Wilcoxon, *J. Am. Chem. Soc.,* 2005, **127**, 7611.

88. M. Bendova, M. Puchberger, S. Pabisch, H. Peterlik and U. Schubert, *Eur. J. Inorg. Chem.,* 2010, 2266.

89. A. N. MacInnes, M. B. Power and A. R. Barron, *Chem. Mater.,* 1993, **5**, 1344.

90. A. N. MacInnes, M. B. Power and A. R. Barron, *Chem. Mater.,* 1992, **4**, 11.

91. V. N. Soloviev, A. Eichhöfer, D. Fenske and U. Banin, *J. Am. Chem. Soc.,* 2001, **123**, 2354.

92. M. W. DeGroot, N. J. Taylor and J. F. Corrigan, *J. Mater. Chem.,* 2004, **14**, 654.

93. M. W. DeGroot, H. Rösner and J. F. Corrigan, *Chem.–Eur. J.,* 2006, **12**, 1547.

94. J. H. L. Beal, P. G. Etchegoin and R. D. Tilley, *J. Phys. Chem. C,* 2010, **114**, 3817.

95. G. Kedarnath, S. Dey, V. K. Jain, G. K. Dey and R. M. Kadam, *J. Nanosci. Nanotechnol.,* 2008, **8**, 4500.

96. G. Kedarnath, L. B. Kumnhare, V. K. Jain, P. P. Phadnis and M. Nethaji, *Dalton Trans.,* 2006, 2714.

97. T. Trindade, P. O'Brien, X.-M. Zhang and M. Motevalli, *J. Mater. Chem.,* 1997, **7**, 1011.

98. T. Trindade, O. C. Monteiro, P. O'Brien and M. Motevalli, *Polyhedron,* 1999, **18**, 1171.

99. J. Akhtar, M. A. Malik, P. O'Brien and M. Helliwell, *J. Mater. Chem.,* 2010, **20**, 6116.

100. S.-M. Lee, Y.-W. Jun, S.-N. Cho and J. Cheon, *J. Am. Chem. Soc.,* 2002, **124**, 11244.

101. Z. Zhang, S. H. Lee, J. J. Vittal and W. S. Chin, *J. Phys. Chem. B,* 2006, **110**, 6649.

102. N. Revaprasadu, M. A. Malik, J. Carstens and P. O'Brien, *J. Mater. Chem.,* 1999, **9**, 2885.

103. D. E. Eastman, F. Holtzberg and S. Methfessel, *Phys. Rev. Lett.,* 1969, **23**, 226.

104. W. Chen, X. Zhang and Y. Huang, *Appl. Phys. Lett.,* 2000, **76**, 2328.

105. M. D. Regulacio, N. Tomson and S. L. Stoll, *Chem. Mater.,* 2005, **17**, 3114.

106. T. Mirkovic, M. A. Hines, P. S. Nair and G. D. Scholes, *Chem. Mater.,* 2005, **17**, 3451.

107. V. M. Huxter, T. Mirkovic, P. Sreekumari and G. D. Scholes, *Adv. Mater.,* 2008, **20**, 2439.

108. F. Zhao, H.-L. Sun, S. Gao and G. Su, *J. Mater. Chem.,* 2005, **15**, 4209.

109. Y. Hasegawa, M. Afzaal, P. O'Brien, Y. Wada and S. Yanagida, *Chem. Commun.,* 2005, 242.

110. M. D. Regulacio, K. Bussmann, B. Lewis and S. L. Stoll, *J. Am. Chem. Soc.,* 2006, **128**, 11173.

111. M. L. Redígolo, D. S. Koktysh, S. J. Rosenthal, J. H. Dickerson, Z. Gai, L. Gao and J. Shen, *Appl. Phys. Lett.,* 2006, **89**, 222501.

112. A. S. Pereira, P. Rauwel, M. S. Reis, N. J. O. Silva, A. Barros-Timmons and T. Trindade, *J. Mater. Chem.,* 2008, **18**, 4572.

113. A. Tanaka, H. Kamikubo, Y. Doi, Y. Hinatsu, M. Kataoka, T. Kawai and Y. Hasegawa, *Chem. Mater.,* 2010, **22**, 1776.

114. F. Zhao and S. Gao, *J. Mater. Chem.,* 2008, **18**, 949.

115. Y. Hasegawa, T.-A. Adachi, A. Tanaka, M. Afzaal, P. O'Brien, T. Doi, Y. Hinatsu, K. Fujita, K. Tanaka and T. Kawai, *J. Am. Chem. Soc.,* 2008, **130**, 5710.

116. S. L. Castro, S. G. Bailey, R. P. Raffaelle, K. K. Banger and A. F. Hepp, *J. Phys. Chem. B,* 2004, **108**, 12429.

117. C. Sun, J. S. Gardner, G. Long, C. Bajracharya, A. Thurber, A. Punnoose, R. G. Rodriguez and J. J. Pak, *Chem. Mater.,* 2010, **22**, 2699.

118. S. K. Batabyal, L. Tian, N. Venkatram, W. Ji and J. J. Vittal, *J. Phys. Chem. C,* 2009, **113**, 15037.

119. A. J. Wooten, D. J. Werder, D. J. Williams, J. L. Casson and J. A. Hollingsworth, *J. Am. Chem. Soc.,* 2009, **131**, 16177.

120. D. Pan, L. An, Z. Sun, W. Hou, Y. Yang, Z. Yang and Y. Lu, *J. Am. Chem. Soc.,* 2008, **130**, 5620.

121. D. Pan, X. Wang, Z. H. Zhou, W. Chen, C. Xu and Y. Lu, *Chem. Mater.,* 2009, **21**, 2489.

122. T. Torimoto, T. Adachi, K.-I. Okazaki, M. Sakuraoka, T. Shibayama, B. Ohtani, A. Kudo and S. Kuwabata, *J. Am. Chem. Soc.,* 2007, **129**, 12388.

123. M. T. Ng, C. B. Boothroyd and J. J. Vittal, *J. Am. Chem. Soc.,* 2006, **128**, 7118.

124. L. Tian, H. I. Elim, W. Ji and J. J. Vittal, *Chem. Commun.,* 2006, 4276.

125. L. Tian, M. T. Ng, N. Venkatram, W. Ji and J. J. Vittal, *Cryst. Growth Des.,* 2010, **10**, 1237.

126. M. T. Ng, C. Boothroyd and J. J. Vittal, *Chem. Commun.,* 2005, 3820.

127. Q. Tang, S. M. Yoon, H. J. Yang, Y. Lee, H. J. Song, H. R. Byon and H. C. Choi, *Langmuir,* 2006, **22**, 2802.

128. W. P. Lim, Z. Zhang, H. Y. Low and W. S. Chin, *Angew. Chem., Int. Ed.,* 2004, **43**, 5685.

129. L. Tian, H. Y. Tan and J. J. Vittal, *Cryst. Growth Des.,* 2008, **8**, 734.

130. M. Green and P. O'Brien, *Adv. Mater.,* 1998, **10**, 527.

131. M. Green and P. O'Brien, *J. Mater. Chem.,* 1999, **9**, 243.

132. A. Kornowski, R. Eichberger, M. Giersig, H. Weller and E. Eychmuller, *J. Phys. Chem.,* 1996, **100**, 12467.

133. W. E. Buhro, *Polyhedron,* 1994, **13**, 1131.

134. S. C. Goel, M. Y. Chang and W. E. Buhro, *J. Am. Chem. Soc.,* 1990, **112**, 5636.

135. M. Green and P. O'Brien, *J. Mater. Chem.,* 2004, **14**, 629.

136. M. Green and P. O'Brien, *Chem. Commun.,* 1998, 2459.

137. S. Felber and F. Willig, unpublished results.

138. Y.-H. Kim, Y.-W. Jun, B.-H. Jun, S.-M. Lee and J. Cheon, *J. Am. Chem. Soc.,* 2002, **124**, 13656.

139. M. A. Malik, M. Afzaal, P. O'Brien, U. Bangert and B. Hamilton, *Mater. Sci. Technol.,* 2004, **20**, 959.

140. M. A. Malik, P. O'Brien and M. Heliwell, *J. Mater. Chem.,* 2005, **5**, 1463.

141. S. D. Dingman, N. P. Rath, P. D. Markowitz, P. C. Gibbons and W. E. Buhro, *Angew. Chem., Int. Ed.,* 2000, **39**, 1470.

142. A. Manz, A. Birkner, M. Kolbe and R. A. Fischer, *Adv. Mater.*, 2000, **12**, 569.
143. P. S. Schofield, W. Zhou, P. Wood, I. D. W. Samuel and D. J. Cole-Hamilton, *J. Mater. Chem.*, 2004, **14**, 3124.
144. K. Sardar, M. Dan, B. Schwenzer and C. N. R. Rao, *J. Mater. Chem.*, 2005, **15**, 2175.
145. J. Choi and E. G. Gillan, *J. Mater. Chem.*, 2006, **16**, 3774.
146. C. D. M Donegá, S. G. Hickey, S. F. Wuister, D. Vanmaekelbergh and A. Meijerink, *J. Phys. Chem. B,* 2003, **107**, 489.

Subject Index

References to figures are given in *italic* type. Textual references to endnotes are indicated with the page number followed by n and the endnote number.